Cross-Cultural Issues in Bioethics

The Example of Human Cloning

At the Interface

Volume 27

A volume in the *Making Sense Of:* project
'Health, Illness and Disease'

Probing the Bounderies

Cross-Cultural Issues in Bioethics

The Example of Human Cloning

Edited by

Heiner Roetz

Amsterdam - New York, NY 2006

The paper on which this book is printed meets the requirements of "ISO 9706:1994, Information and documentation - Paper for documents - Requirements for permanence".

ISBN: 90-420-1609-4

Printed in the Netherlands

Contents

Welcome to a *Making Sense Of:* Project

Welcome to the *Making Sense Of:* series of projects. These are innovative, cutting-edge inter-disciplinary and multi-disciplinary research forums designed to bring multiple insights and perspectives to bear on the meaning of what it is to be persons.

The main aim of the projects is to examine and explore the impact on our understandings of persons in light of significant experiences and conditions which arise in the course of living in the world. The projects will also seek to begin to assess the implications for our understandings of what it is to be human in relation to other persons, human community, and advances in all areas of technology.

The projects are built around an annual conference which seeks to draw a global audience of people from differing academic disciplines, professions, vocations and organisations to engage in cutting edge dialogue and conversation. The projects are supported via an active publication series, eForum and discussion groups.

Cross-Cultural Issues in Bioethics, The Example of Human Cloning is a volume which focuses on an important area of contemporary reflection and living, and brings to bear inter-disciplinary insights and perspectives on the vexed question of human cloning.

Dr Robert Fisher
Inter-Disciplinary.Net
www.inter-disciplinary.net

Introduction

In December 2003, the research group "Kulturübergreifende Bioethik" (Culture-transcending Bioethics), sponsored by the German Research Foundation (DFG), held an international symposium on "Cross-Cultural Issues in Bioethics: The Example of Human Cloning" at Ruhr University Bochum. The symposium brought together leading bioethics scholars from Asia, Africa and the West to address an urgent bioethical question within the context of different cultural, religious and regional settings and against the background of globalising biotechnology. These scholars were invited to attend the symposium not only in their capacity as reputed bioethicists of the different countries and organisations but also because they are actively involved in political decision making on both the national and international level.

The symposium was thus meant to explore on a cross-cultural level the problems and opportunities of global bioethics and of global regulations in light of the rapid world-wide advances in biotechnology. The present volume contains a number of articles based on the papers presented at the conference and revised in the light of recent developments and additional essays by members of the research group "Culture-transcending Bioethics." Together they give an overview, an analysis and an evaluation of present positions and tendencies in the international discussions on human cloning.

The research group "Culture-transcending Bioethics," which comprises a number of individual projects working together in a combination of regional, cultural, religious and systematic approaches (see www.rub.de/kbe), proceeds from the assumption that a global agreement in bioethics cannot be achieved without a critical understanding of different cultures and a common global discourse on equal terms. The predominance of the Western influence does not reflect the multitude of voices.

"Culture-transcending bioethics" includes empirical research and comparison as methodical steps, but in the final analysis it explores the conditions and opportunities for a trans-cultural, international understanding and consensus, and, on this basis, for a global regulation of fundamental ethical issues - inasmuch as there is agreement about what these fundamental issues are in the first place. This is an important topic in itself, as can be shown by the example of human cloning. Still there are a number of obvious reasons for such a global perspective in bioethics.

The first of these is that due to global migration, countries have become increasingly multicultural. In Germany, for instance, the relationship between doctors and patients is no longer based almost exclusively on a common system of values as it was in the past, but it has to take into account, for example, the beliefs of the growing Islamic community.

The second reason is that modern biotechnology is itself a global phenomenon in terms of research, production and trade. It is quickly mastered outside the traditional industrial countries and is being promoted in many parts of the world since it promises great social and economic benefit. Asia in particular has become a booming region where large biotechnological capacities have been built up, where challenging and ambitious experiments have been carried out, and where embryonic stem cells are being produced for research and for export to countries where their production is - at least for the time being - forbidden. Biotechnological research is moveable and can be exported in order to escape so-called "overregulation" or a critical public sphere. Restrictions in one country may quickly lead to the relocation of the research to more permissive regions. A relatively low degree of public information, less strict legal standards and less rigid enforcement of such standards might be exploited for such endeavours. In any case, local partners offer their capacities and enter into joint ventures with the foreign investors. Whose standards and whose values count in these co-operations? Would a solution be to simply opt for the lowest common denominator? Or do we need broader orientations?

Furthermore, genetic resources, including human genetic resources, in all parts of the world attract pharmacological interests, for example, in places where due to the non-existence of a public health system diseases can be studied in their "natural" state, without the "distortion" of medical treatment. This practice raises serious questions of informed consent and global justice. Obviously, without accepted common standards, there will be no way to prevent rape research, gene piracy and the plundering of biological resources in so-called developing countries.

Next to migration and economic globalisation, the third and most important reason for a global perspective in bioethics is that modern biotechnology with regard to human medicine, at least in some aspects - e.g., cloning, germline engineering, and genetic enhancement - might well in the long run affect not only the destiny of individuals, but that of human biology itself and ultimately the evolutionary future of humankind. From a principled point of view, positive decisions on these issues cannot, therefore, be taken by individuals, corporate groups, scientific communities or even countries alone. Ideally, every human being, including in theory future generations, should have the right to consent or not to consent. This is the most important reason for transcending regional discussions.

It is because of this global nature of the problem that we should not content ourselves with isolated decision finding in specific national contexts. At least in the case of techniques which affect the whole of humankind, in the case of biotechnological joint ventures, and in the case of international trade with human biological "material," common global ethical and legal standards are a must - even if the opportunities inherent in the new technologies should far outweigh the risks. The principle of informed consent in this perspective applies not only to decisions on individual therapy, but also to decisions on the overall direction of the biomedical sciences.

This is also true of human cloning. Human cloning, both so-called therapeutic or scientific and reproductive cloning, is a culmination point of bioethical questions, and it is a topic of global relevance in that it directly concerns fundamental problems of humanity. To quote Thomas S. Hoffmann in his contribution to this volume, "The ability to clone humans touches much more basically upon issues not only of individual human self-understanding, but also upon questions concerning the constitution of modern societies, it touches upon fundamental issues concerning ethics and law - it addresses dimensions which greatly exceed the need for a mere dispute among professional ethics experts."

The search for global standards has been ongoing for some time now, thanks to the initiative of international organisations like the World Health Organisation and the UN, where cloning has been on the agenda for several years - albeit without a final binding agreement. Still, what is taking place on these levels is first of all negotiation between negotiators. This is of course an indispensable step in international decision finding, but it cannot substitute a thoroughgoing process to establish mutual understanding based on cultural hermeneutical knowledge and, what is more, an open ethical discourse aiming at consent.

However, there is scepticism with regard to such an endeavour, and its meaningfulness has been called into question. Sceptics argue, in short, that negotiation is all that we can achieve on the international level, that it moreover has to be quick given the speed at which the natural sciences are progressing, and that the outcome will at best be a low-level common denominator. And since the more parties are involved, the lower this denominator will be, the final result would even be the dissolution of existing national standards. In one of the reactions to a recent symposium of the German National Ethics Council on "The treatment of pre-natal life in other cultures," it was even suggested that the hidden agenda behind attempts to broaden the occidental perspective was actually to get rid of inconvenient regulations. The strategy would be to marginalize the value system rooted in Christianity in a pluralism of world views. It is true that knowledge of other cultures and opinions can foster relativism, with the corresponding "tolerant" or "liberal" practical conclusions. The problem is complicated by the fact that cultural differences are often presented in terms of different concepts of the human being which might, in turn, directly affect the bioethical decisions in question. There has probably been no other field in the history of science in which ethics and questions of culture are so closely interwoven as in modern biomedical technology. Are we stuck in this dilemma, which is aggravated by the fact that questions of culture are loaded with sensitive questions of identity?

My answer would be as follows: First of all, as stated above, not only for practical, but also for principled, ethical reasons there is no alternative to globalising our approach to biomedical ethics. Secondly, it is only at first glance that intercultural discourse hopelessly complicates the issue; upon closer inspection it does not add to the confusion that we already find in our national discourses. Despite the existence of established different normative outlooks, above all those rooted in religion, the reactions to the new biotechnological developments in all countries and cultural and religious settings are in fact very much of the same kind from a structural point of view: there are pros and cons, hopes and fears, engagement and disinterest, utilitarian and deontological arguments, traditionalist and modernist responses, re-interpretations and adaptations of traditions, and lobbying of all kinds. And there is no argument whatsoever from the non-Western discussions that in its consequences would outweigh the "technological imperative," the notorious "can implies ought" theorem which was invented in the West. Therefore, there is no reason to fear that standards will be dissolved by taking other discourses seriously. Furthermore, scepticism about the fruitfulness of intercultural dialogue is obviously fuelled by a monolithic and essentialist understanding or misunderstanding of "culture."

Admittedly, it is a frequent phenomenon of so-called intercultural dialogue that we take roles - the role of "the" Westerner, "the" Buddhist, "the" Chinese, "the" Christian, "the" Moslem and so on. This kind of collective identity formation is a fact that cannot easily be dismissed. However, it is only half the story. For we do not simply bear cultural identities; rather, we respond to social and political surroundings which are not necessarily culture-specific, and we are of course individuals who speak for themselves and whose biographies cannot be reduced to all of those external factors. Moreover, on closer inspection "the" Christian or "the" Confucian view, to take these as examples, simply does not exist. There are many conflicting interpretations as to what Christianity and Confucianism mean, not least in the context of bioethical questions. And these interpretations are themselves challenged and possibly transformed by new experiences, in particular by experiences like those we are undergoing today in the wake of biotechnological progress - something that was not on the agenda of any of the traditions that are referred and appealed to. Ethics is not only challenged by culture, but it is also in the very nature of ethics to bring about cultural change.

Despite these principal considerations concerning the possibility of a global consensus, the starting point of any discourse on human cloning is a broad spectrum of opinions which is documented in this volume.

The overview starts with Japan. Japan is one of the global players in biotechnology, and in order to boost the country's stagnant economy, the government is keen on developing this promising sector. In this context, a law for the regulation of human cloning has been passed, leaving the concrete measures to guidelines issued by the government. *Masahiro Morioka* ("The ethics of human cloning and the sprouts of human life") gives an account and evaluation of these procedures and discusses the corresponding debates. *Robert Horres, Hans Dieter Ölschleger*, and *Christian Steineck* ("Cloning in Japan: Public opinion, expert counselling, and bioethical reasoning") describe Japanese public opinion on human cloning and the process of political decision making that led to the new law. They then focus on the recently released administrative report on "Fundamental Principles Concerning the Handling of Human Embryos." The report, which touches upon bioethical issues like the moral status of embryos and human dignity, contains the current view of the Japanese government on this matter and will most probably determine future guidelines, with a tendency to allow therapeutic uses of cloning for high medical goals. In the final part of their paper, Horres, Ölschleger and Steineck compare the position of the report with the broader range of opinions in the general Japanese ethical debate. The spectrum resembles that to be found in other industrialised countries, with a conspicuous absence of "fundamentalist" stances.

As one of the most ambitious biotechnological superpowers of the future, China occupies a position of growing importance, also in international decision making. *Qiu Renzong* ("Cloning Issues in China"), for many years one of the leading figures in the Asian bioethics community, gives a comprehensive and detailed overview of the Chinese legislation and the different bioethical positions on human cloning during recent years. He pleads for a regulation of cloning in accord with Confucianism and the yin-yang cosmology and comes to the conclusion that reproductive cloning should be banned as incompatible with both. Cloning for developing therapies, however, should be allowed, since due to the "relational concept of personhood" which Qiu Renzong attributes to

Confucianism, the embryo that would be destroyed in this research is not regarded as a full person and would only enjoy "due" respect.

Ole Döring ("Culture and Bioethics in the Debate on the Ethics of Human Cloning in China"), after elaborating on methodological and hermeneutical issues of a global study on bioethics and the notoriously difficult relationship between culture and ethics, gives a further outline of Chinese cloning research and the normative standpoints of researchers and bioethicists. There is an overall "emphasis on consequentialist or utilitarian concerns" within the context of what Döring calls "pragmatic conservatism." Criticism of cloning is based above all on kinship concerns, i.e., possible threats to the family. Moreover, there is a "primacy of the born human" as standing in full communal relationships which in ethical terms constitute the human being. This leads to a permissive attitude towards consumptive embryo research. Yet there are also diverging opinions in China concerning the status of the embryo. The true practical implications of the "relational approach" have yet to be determined.

China, with special regard to Hong Kong, is also the topic of *Gerhold Becker* ("Chinese ethics and human cloning: A view from Hong Kong"). In Hong Kong, the "Human Reproductive Technology Bill" of 2000 prohibits reproductive cloning, while a door is left open for "therapeutic" use of the technique. The bill mirrors the plurality of values in Hong Kong, but it is also based on traditional Chinese family values. The legislation process has stimulated an ethical discourse in which Confucian voices, again stressing the "relational" concept of personhood, can be detected. However, Becker finds doubts that something like "Confucian bioethics" can be construed, let alone that it would be suitable to the pluralistic conditions of Hong Kong. He also throws light on "Asian" criticism of "Western" bioethics, pointing out its ambiguities and its "myopic," static view of culture.

South Korea has been in the headlines since the *Science* report of February 2004 about experiments that proved the possibility of human cloning through somatic cell nuclear transfer. This "breakthrough" demonstrated the geographic shift that is taking place today in biological engineering, and it has raised speculations about possible cultural factors that create a favourable environment for biotechnology. However, as the articles in this volume by *Chin Kyo-hun* ("Current debates on human cloning in Korea") and *Marion Eggert* and *Phillan Joung* ("The cloning debate in South Korea") show from different perspectives, there is a public bioethical debate in Korea of almost unparalleled intensity that is very far removed from a pro-cloning "cultural" consensus. In fact, South Korea is a society in which different cultural influences from indigenous and Asian traditions as well as from the West merge with the sociopolitical impacts of modernity and the rapid economic development. The country can thus be considered a key example of the constellation in which the challenges of biotechnology have to be tackled in the world at large. Because of their exemplary nature, the Korean debates should be followed closely.

Woo Suk Hwang, the leader of the South Korean cloning team, legitimizes his research by his Buddhist faith. His arguments are discussed by *Jens Schlieter* ("Some observations on Buddhist thoughts on human cloning") who does not find them consistent with Buddhist teachings. The Korean cloning experiments involve killing which poses a greater challenge to Buddhist ethics than does the idea of bringing into existence a human being for reproductive purposes. As Schlieter points out, on the basis of the constitutive elements of Buddhism, it is reproductive cloning for childless couples which meets with acceptance among Buddhist ethicists. Interestingly enough, in the UN debates on a possible convention against human cloning, Thailand, a country with over ninety percent Theravada-Buddhist population and rapidly growing biomedical centres, joined the camp which voted against reproductive and in favour of "therapeutic" cloning. The contradiction might be partly explained by the fact that Buddhism has exerted rather limited influence on politics and legislation.

The Thai discourse, which is the main source of Schlieter's presentation, is also the topic of *Pinit Ratanakul* ("Human cloning: Thai Buddhist pespectives"). Ratanakul, one of the prominent participants in this discourse, outlines the basic premises of Buddhist ethics and comes to the conclusion that cloning should not be seen as an act against nature but as a part of the world process at large, the very possibility of cloning speaking for itself. To reject this process would amount to "clinging," which again would bring suffering. At the same time, however, there is also a harmful *kammic* implication of killing embryos for scientific purposes, there are slippery slope effects, and there are possible negative consequences for the social order, above all for kinship relations. According to Ratanakul, these implications seem in the final analysis to outweigh the "benefits cloning will bring to humanity."

Like the "Far East," the South-East Asian nations, too, are developing biotechnological capacities. Apart from Thailand and Singapore, one of these countries is Malaysia, a melting pot of cultures and religions, with Islam as its dominant tradition. In 2002, a nation-wide public conference organised by the Ministry of Foreign Affairs and the Institute of Strategic and International Studies of Malaysia was held in Kuala Lumpur in order to discuss questions of human cloning and to formulate Malaysia's position in the negotiations of the United Nations. Centring around an analysis of this conference, *Siti Nurani Mohd Nor* ("The ethics of human cloning with reference to the Malaysian bioethical discourse") draws a complex picture of the Malaysian discussions, with special focus on Islam in its Malaysian form. Rather than setting up a fixed system of orthodox rules, Islam has historically developed criteria and devices for flexible responses to novel challenges that directs the reading of the primary sources and is used in bioethical decision finding. However, there is a common denominator - the protection of family lineage. Together with other considerations like interference with God's role as supreme creator, it is the lineage argument above all which leads to the rejection of reproductive cloning technology.

According to *Heinz Werner Wessler's* assessment ("The charm of biotechnology: Human cloning and Hindu bioethics in perspective"), India rather than China might become "the world's cloning superpower in the near future." There is a biotechnology-friendly mood in India, and although there is no federal funding, the country's private industry is a player on the global embryonic stem cell market. This contradicts the traditional Hindu view of human life which ascribes a moral status to the unborn from the time of conception and cannot be tolerant of cloning as long as it implies the death of the embryo. However, India is "modernizing" its traditions, and there is even a traditional basis for hailing cloning as a method of procreation. Rather than being strict, Hindu voices begin to differ, and there is an "atmosphere of permissiveness in bioethics" in India.

Islam, the Malaysian form of which was already addressed by Siti Nurani Mohd Nor, is also the topic of *Abdulaziz Sachedina* ("Cultural and religious in Islamic biomedicine: The case of human cloning"). As Sachedina points out, Islam has been pluralistic in the interpretation of its sources, and today there are different judicial opinions about human cloning. In any case, since religion is primarily seen as concerning human relationships, the deliberations focus on questions of the possible hereditary status of a cloned person and on family concerns. Lineage rather than the isolated individual dominates Muslim sensitivity. This means that human cloning would be acceptable if it is an "aid to fertility strictly within the bounds of marriage." Possible harm to social norms is a more serious concern than tampering with nature. There is a science-friendly atmosphere in Islam, and Shiite legists even endorse cloning technology as being "within the natural sphere to bring about conception."

That most Muslim scholars are concerned more about the social effects of cloning than about the act itself is also stressed by *Thomas Eich* ("The debate about human cloning among Muslim religious scholars since 1997"). The act is seen as a discovery of new techniques within the created world and not as an alteration of creation. It is therefore legitimate, provided that it is scientifically safe and that other means of reproductive medicine have failed. Arguments against cloning focus on the disturbance of genealogic relations. Hitherto, the argument of human dignity has has not ranked high in the discussions of Muslim scholars. However, there are a growing number of voices that see human dignity infringed upon by "therapeutic" cloning inasmuch as it implies the destruction of embryos.

Michael Barilan ("The debate on human cloning: some contributions from the Jewish tradition") gives an overview of the discourse in Israel, again an important factor on the stem cell market, and positions on cloning in current Judaism. There are unique views in the Jewish tradition not only with regard to the notion of dignity, nature and the development human life, but also on the status of ethical principles that do not easily accord with common "Western" ideas. And there is emphasis in Israel on Jewish self-determination rather than on conformity with "Christian" ethics that failed to stop the Nazis. As Barilan explains, there is a general tendency in the Rabbinic discourse to allow what is not forbidden by the Torah in a literal and casuistic reading, and nothing else can restrict the liberty of individuals. In principle, there is no ban on cloning either for scientific or reproductive purposes that could be justified on this ground, provided there is no confusion over identity and paternity, hence family relationships. This approach does not lend itself to generalization on a global level, since it is based on the contract between God and the Jewish people, which contains very few rules that pertain to all mankind, such as the prohibition of murder. Yet Barilan thinks that precisely this division between a "particularized system of conduct" on the one hand and few fundamental precepts on the other might embody a model for getting along with each other in a multicultural world without subscribing to mere relativism.

Christofer Frey ("Bioethics in the Perspective of Universalization") suggests that there is more universalistic potential in the Jewish tradition, namely the Torah, than might be realized in the present Israeli discussions. Awareness of the "universal appeal" of one's own moral belief is essential for universalization in bioethics. This universalization cannot be achieved by a deductive "top-down" approach from abstract rules. Ethical reason has be developed out of "life," which links it to history and traditions. According to Frey, this does not lead to "empiricist ethics" with relativistic conceptions of the "good." There are "human essentials" transcending culture although culturally implemented, and they would be the starting point for developing universal norms.

Nigel M. de S. Cameron evaluates the "American debate on human cloning." In the U.S., many states are passing cloning regulations, from the very liberal "Stem Cell Research Bill" of New Jersey to bans on cloning in other states, with a restrictive policy on the federal level. The accompanying discussions show remarkable coalitions, with advocates as well as critics of cloning on both sides of the abortion debate. Cameron rejects the widespread distinction between "reproductive" and "therapeutic" cloning as mere rhetoric in order to make embryonic consumptive research acceptable. He votes for a comprehensive ban on cloning, since a partial ban in the name of "prudence" would only pave the way for a further commodification of human nature. What is at stake is the "primacy of human dignity in the development of biotechnology."

Thomas S. Hoffmann ("Primordial Ownership versus Dispossession of the Body. A contribution to the problem of cloning from the perspective of classical European Philosophy of Law") analyses the legitimacy of claims to the implementation of human cloning by the yardstick of classical philosophical ethics, drawing above all on the rational law theories of Kant and Fichte. He argues that there is neither a right to reproduction nor a right to therapy nor a right to free research that would be so far-reaching as to outweigh the principle of a person's primordial ownership of the body. To put this ownership at society's disposal would be to undermine the foundation of the modern rule of law - the individual "concrete bodiliness" which cannot be administered by, but sets limits to, other claims. Hoffmann points out that this concrete bodiliness, which begins with the "mute" and "not yet culturally interpreted" early phase of human development, might also serve as an independent reference point for the interculturality of ethics.

Ilhan Ilkilic ("Human Cloning as a Challenge to Traditional Health Care Cultures") and *Hans-Martin Sass* ("Let probands and patients decide about moral risks in stem cell research and medical treatment") address a specific aspect of dealing with bioethical problems in pluralistic and multicultural societies: the role of the layperson as against the medical expert, the bioethicist and the "biopolitician." As Ilkilic argues, based on an "effective information policy" concerning bio-scientific knowledge, the "health literacy" of the individual and his or her personal value system, which might reflect cultural influences, should play an important role in finding solutions to bioethical problems. Sass stresses that "different moral intuitions and traditions" rule out uniform answers and recommends a "principle of subsidiary" rather than decision by another authority. If there is an "undisputed moral good" in any culture, Sass argues, it is the reduction of pain and suffering. This value outweighs possible interests of technically produced "embryonic constructs." Sass therefore pleads for allowing somatic cell nuclear transfer for therapeutic purposes while putting a ban on reproductive cloning, at least as long as the technique is not safe. The dignity involved is primarily on the side of the individual conscience.

Florian Braune, Nikola Biller-Andorno and *Claudia Wiesemann* ("Human reproductive cloning: A test case for individual rights?") discuss how a possible ban on cloning relates to individual rights like the right to reproduction. They argue that the way in which individual and societal demands are balanced might reflect different cultural concepts of morality, and that cloning is a test case in this respect. They discuss cultural variability by means of the example of informed consent in the Chinese context, suggesting that differences in the understanding of this concept might also underlie approaches to human cloning.

In an epilogue to this volume, *Nicola Biller-Andorno* ("Cross-cultural discourse in bioethics: It's a small world after all?") advocates a "self-critical inquiry into the possibilities and limits of a cross-cultural bioethics" in view of temptations to quickly draw up global declarations, reproaches of "moral imperialism," mutual misunderstandings and Western/Non-Western stereotypes. According to Biller-Andorno, the methodology of how to achieve a global bioethics is far from resolved or uncontroversial. In order to prevent the discourse from being hijacked by all kinds of interests, procedural rules are necessary. It is only through tolerance, careful listening, goodwill and transparency that the development of parallel moral worlds can be precluded.

What conclusion can be drawn from the articles in this volume in regard to the global constellation of bioethics? It appears that "culture," although figuring prominently in the various debates, never functions as an unequivocal factor in terms of a one-to-one relationship between a would-be cultural tenet and a bioethical position. What culture means is a matter of debate and, moreover, subject to re-evaluation and change, in particular under the impact of the modern technological challenges. Even adherents to one and the same ethical tradition, like Confucianism, Buddhism or Islam, can come to diverging opinions on such bioethical issues as human cloning. However, beyond all ambiguity, there is also a common concern which is conspicuous in most of the bioethical and legal decision finding that we have studied. Rather than substantialistic assessments of the human being, this pertains to family and the family-based social structure. And there is another surprising finding which might have a direct bearing on the discussions about the prohibition or legalisation of human cloning: The sharp separation between therapeutic and reproductive cloning which has been called into question because of the identical nature of the basic procedures might also not be tenable for another reason. Despite concern about the erosion of family values, there is greater acceptance of reproductive cloning worldwide than might be expected, provided the technique is perfected. This leads one to question whether there is an insurmountable conceptual and psychological barrier deeply rooted in the established moral convictions of humankind that would prevent scientific research on cloning from proceeding along the "slippery slope" towards the "brave new world."

Heiner Roetz, Bochum, Germany

The Ethics of Human Cloning and the Sprout of Human Life

Masahiro Morioka

Abstract: In 1998, the Council for Science and Technology established the Bioethics Committee and asked its members to examine the ethical and legal aspects of human cloning. The Committee concluded in 1999 that human cloning should be prohibited, and, based on its report, the government presented a bill for the regulation of human cloning in 2000. After a debate in the Diet, the original bill was slightly modified and issued on December 6, 2000. In this paper, I take a closer look at this process and discuss some of the ethical problems that were debated. Also, I make a brief analysis of the concept "the sprout of human life." Not only people who object to human cloning, but also many of those who seek to promote research on human cloning admit that a human embryo is the sprout of human life and, hence, it should be highly respected. I also discuss the function of the language of utilitarianism, the language of scepticism, and religious language in the discussion of human cloning in Japan.

Key Words: Human cloning, Japanese bioethics, ES cells.

1. Introduction

The birth of Dolly, the first mammal cloned from a somatic sell, attracted wide public attention in Japan, and the words "cloned human being" became a popular notion. However, the "public debate" on the ethics of human cloning was considerably less heated than that relating to brain death and organ transplantation. Scientists and commentators repeatedly stated that while the cloning of a sheep was acceptable, human cloning should be prohibited. A well-known female scientist said that she could not imagine a scientist who would try to clone a human being.

In 1998, the Council for Science and Technology established the Bioethics Committee and asked its members to examine the ethical and legal aspects of human cloning. The Committee concluded, in 1999, that human cloning should be prohibited, and, based on the report, the government presented a bill for the regulation of human cloning in 2000. After a debate in the Diet, the original bill was slightly modified and issued on December 6, 2000.

In the following chapters, I take a closer look at this process and discuss some of the ethical problems that were debated. Also, I make a brief analysis of the concept "the sprout of human life."

2. The Report of the Sub-committee on Cloning

In 1998, a sub-committee on cloning was established within the Bioethics Committee. The sub-committee published its report *Fundamental Thoughts on the Production of a Human Being by Cloning Technology*[1] on November 17, 1999. This was the first report to deal with human cloning and its regulation.

The report highlighted two problems, namely, the "violation of human dignity" and the "safety problem." With regard to the "violation of human dignity," it makes two points: 1) Human cloning techniques may open the floodgates for the creation of people with a particular ability in order to attain a particular goal ("breeding of human beings") and for regarding people as a means or a tool with which to attain a particular goal ("human beings as means or tools"). 2) While a cloned individual has a separate personhood from the donor of a somatic cell, he or she is constantly forced to be aware of his/her relationship to that donor. This is a violation of human rights, both for the cloned individual and for the donor. This problem, together with the "breeding of human beings" and "human beings as means or tools," leads to a violation of respect for an individual's free will and existence. It is totally against constitutional principles ("violation of respect for an individual"). 3) Human cloning is asexual reproduction. It deviates altogether from our basic understanding of human reproduction, and it is expected to cause confusion of the familial order, such as, e.g., the parent-child relationship.

With regard to the "safety problem," the report concluded that present cloning techniques cannot guarantee the safe production of a human clone individual.

In the light of these problems, the report concluded that the production of a human clone individual must be legally prohibited. Concerning research on human somatic clone embryos, the report stated that this should be permissible within certain limitations if a justifiable ground is to be found, because it may bring great benefit to humans in the field of medicine. But at the same time, the report stressed that a human somatic clone embryo has significance as the "sprout of human life" (hito no seimei no hôga), like a human embryo, and should therefore be handled with the utmost care.

Based on this report, the Bioethics Committee of the Council for Science and Technology announced that the production of a clone individual, together with chimeric/hybrid human individuals, must be legally penalised, and that research on human somatic clone embryos should be regulated in some way (December 21, 1999). This announcement signalised the government's decision to legally regulate the production of a clone human individual and other chimeric/hybrid human individuals but not "therapeutic cloning" and other research. In other words, the government had abandoned the idea of establishing a comprehensive law dealing with assisted reproductive technology and research on human germline cells.

3. The Establishment of the Law

The government presented the bill *The Law Concerning Regulation Relating to Human Cloning Techniques and Other Similar Techniques*[2] to the Diet on April 4, 2000. It was discussed in the Commission for Science and Technology but failed to pass during that session. On November 7, 2000, two different bills were presented to the Diet, one by the government, which was similar to the first one, and the other by the Democratic Party of Japan, which was fundamentally different on some points. The most important parts of the government's bill are presented below.

> *The Law Concerning Regulation Relating to Human Cloning Techniques and Other Similar Techniques* (Original Version, November 7, 2000)
>
> *Article 1* (Purpose of the law) The cloning techniques and other similar techniques ..... could have a severe influence on preservation of *human dignity, safety* for human life and body, and maintenance of *social order.* Based upon these understandings, the purpose of this law is to prevent and restrain *creation of a human clone individual* and an amphimictic individual, and to regulate artificial creation of individuals similar to such individuals set forth herein, by means *of prohibiting transfer of embryos* produced by the cloning techniques or the Specific Fusion/Aggregation Techniques into a human or an animal uterus, by means of regulating production, assignment and import of such embryos, and by means of taking

other necessary measures to secure appropriate handling of such embryos.
Article 3 (Prohibited Acts) No person shall transfer *a human somatic clone embryo, a human-animal amphimictic embryo, a human-animal hybrid embryo or a human-animal chimeric embryo* into a uterus of a human or an animal.

Article 4 ..... the Minister of Education, Culture, Sports, Science and Technology shall prescribe *guidelines* in relation to handling of *Specified Embryos* (*See *Table 1* p. 7).

Supplementary Provisions
Article 2 (Study and Examination) The Government shall, within five years of enforcement of this Law, take necessary measures in accordance with the results of its study and examination of the system of handling Specified Embryos with consideration to the circumstances in which this Law is enforced or to any change of the situation surrounding the cloning techniques and other similar techniques.

* (Translation by the government, except Supplementary Provisions Article 2. Emphases added by Morioka)

The main characteristic of this bill was that it prohibited only the "transfer" of four types of Specified Embryos, including a human somatic clone embryo, into the uterus of a human or an animal. The reason for this prohibition was that the transfer of these embryos leads to the production of an individual with the same genetic structure as another specific individual (in the case of a human somatic clone embryo) or an embryo belonging to a subspecies of humans (in the case of the other three embryos). The bill put the consideration of the "production" of these embryos into future guidelines. It is also worth noting that the bill imposed the penalty of "imprisonment" for violation.

The bill presented by the Democratic Party of Japan, *The Bill Concerning Regulation Relating to the Production and the Use of a Human Clone Embryo and other Embryos*,[3] showed some important differences, especially in the following articles. *Article 1-1* declares that the bill deals with the regulation of a "human embryo" and an "embryo that has characteristics specific to humans." This stipulation covers larger areas than the government's bill, which only deals with Specified Embryos. Article 1-1 refers to "preservation of human dignity" and "safety for human life and body," but does not mention "maintenance of social order." Article 1-3 (1-3), Article 2-1 (1) and Article 3-1 (1) are as follows.

Article 1-3 (Basic Ideas)

1) A human embryo is a *"sprout of human life"* (hito no seimei no hôga). No person shall produce or use it without permission.
2) When handling a human embryo, a person must *handle it honestly and carefully* so as not to violate *human dignity*.
3) The production and use of an embryo that has characteristics specific to humans must not lead to the production of an individual.

Article 2-1 (Prohibited Acts on a Human Embryo)

1) *No person shall produce a human embryo* outside a human uterus. However, the production for the purpose of *assisted reproductive medicine or medical research on assisted reproductive medicine* (hereinafter referred to as "research on assisted reproductive medicine") can be an exception.

2), 3), 4), 5) omitted.

Article 3-1 (Prohibited Acts on an Embryo that has Characteristics Specific to Humans)

1) No person shall transfer an embryo that has characteristics specific to humans into the uterus of a human or an animal.

2), 3) omitted.

(*Translation and Emphases by Morioka)

The Democratic Party's bill had some important characteristics. First, it places "basic ideas" in Article 1, and in this provision it is stressed that a human embryo is a "sprout of human life." The bill sought to put a special value on a human embryo, which was not found in the government's bill. (The words "sprout of human life" first appeared in the report of the sub-committee on cloning in 1999, mentioned in section 2 of this paper, and reappeared in the government report on human embryo research in March 2000.) Second, the bill prohibits the "production" of a human embryo except for the purpose of assisted reproductive medicine or medical research on assisted reproductive medicine. This means that the production of a human clone embryo for assisted reproductive medicine can be allowed, but the production of cloned ES cells or cloned organs for transplantation is prohibited. Anyway, the basic idea was that a human embryo is a valuable and precious sprout of human life, hence it should be exploited as little as possible. Third, the bill prohibits the transfer of an embryo that has characteristics specific to humans into the uterus of a human or an animal. This means that the transfer of the animal embryo in which human genes or cells are inserted is prohibited and, consequently, that xenotransplantation without the rejection of an organ transplant becomes impossible.

After a series of debates in the Diet, the government's bill was slightly revised and passed the Diet on November 30, 2000. The Democratic Party of Japan finally agreed to the government's revised bill. The phrase "within five years" from Supplementary Provisions Article 2 was revised to "within three years," and the new words "the sprout of human life" (hito no seimei no hôga), which the Democratic Party had stressed in their bill, were inserted.

Accordingly, the revised article was as follows:

> *Supplementary Provisions*
> *Article2* (Study and Examination) The Government shall, within three years of enforcement of this Law, take necessary measures in accordance with the results of its study and examination of the provisions under this law, on the basis of the results of the study and examination by the Council for Science and Technology Policy, Cabinet Office concerning the method of handling of a human fertilised embryo as the sprout of human life with consideration to the circumstances in which this Law is enforced or to any change of the situation surrounding the cloning techniques and other similar techniques.

(*The translation of the words "hito no seimei no hôga" by the government was "the beginning of a human life," but I believe this translation loses subtle nuances that are present in the literal translation "the sprout of human life." In the above translation I have used the latter.)

Article 4 of this law stipulated that the Minister of Education, Culture, Sports, Science and Technology shall prescribe guidelines in relation to the handling of Specified Embryos. In response to this article, the Ministry began to establish guidelines concerning specific embryos. After a heated debate in a committee, the Ministry announced *The Guidelines for the Handling of a Specific Embryo (Mainichi Shimbun, 2002, Yomiuri Shimbun 2002)*[4] on December 5, 2001. Important parts of the guidelines are as follows.

The Guidelines for Handling of a Specific Embryo (December 5, 2001)
Article 1 Production of a Specified Embryo shall be allowed only when the following requirements are satisfied:
1) Scientific knowledge, which cannot be acquired from research with only animal embryos or other research without Specific Embryos, is acquired from production of such a Specified Embryo
2) omitted.

Article 2
1) Regardless of the provision in Article 1 above, only an animal-human chimeric embryo shall be allowed to be produced among nine categories of Specified Embryos, and the purpose of its production shall be limited to the research concerning production of human cell-derived organs translatable to a human being.
2) A Producer shall not use any human fertilised embryos or human unfertilised eggs in order to produce an animal-human chimeric embryo.

> *Article 9* Specified Embryos, except for ones prescribed in Article 3 of 'the Law Concerning Regulation Relating Human Cloning Techniques and Other Similar Techniques (Law No. 146, 2000)' (hereinafter referred to as "the law"), shall not be transferred into the uterus of a human or animal for the present.

(*Translation by the Ministry of Education, Culture, Sports, Science and Technology)

List of Specified Embryos	*Transfer* prohibited by the *law*	*Transfer* prohibited by the *guidelines*	*Research* prohibited by the law	*Research* prohibited by the *guidelines*
Human somatic clone embryo	prohibited			prohibited
Human-animal amphimictic embryo	prohibited			prohibited
Human-animal chimeric embryo	prohibited			prohibited
Human-animal hybrid embryo	prohibited			prohibited
Human split embryo		prohibited		prohibited
Human embryonic clone embryo		prohibited		prohibited
Human-human chimeric embryo		prohibited		prohibited
Animal-human hybrid embryo		prohibited		prohibited
Animal-human chimeric embryo		prohibited		*approved*

Table 1

The controversy over the legal regulation of human cloning was settled by the establishment of the law and the guidelines. However, the content of the regulation is very complicated and hard to understand, even for specialists. Below is a table showing the variation of regulations.

The only approved way of handling Specified Embryos is, at present, research on animal-human chimeric embryos, that is to say, research on an embryo produced by unification as a result of 1) inserting human somatic cells into an animal embryo, 2) inserting embryonic cells of a human fertilised embryo into an animal embryo, or 3) inserting embryonic cells of other Specified Embryos into an animal embryo. This means that the insertion of human ES cells into an animal embryo (in order to create transplantable organs) is considered to be approved in Japan.

The most striking feature of the Japanese regulation is that it is governed by a two-layered system, consisting of the law and the guidelines. One of the implications is that the guidelines can be "swiftly" altered when the circumstances surrounding human cloning technologies greatly change. For example, if a company in a foreign country begins to make tremendous profits from data acquired from research using human somatic clone embryos in a laboratory, the Ministry of Education, Culture, Sports, Science and Technology may revise the guidelines and lift the ban on research on human somatic clone embryos without a long drawn-out debate in the Diet. The extraction of ES cells from a human somatic clone embryo is currently prohibited, but if many countries begin to do research on therapeutic cloning of this kind, the Ministry may revise the guidelines "swiftly" and allow researchers to study ES cells acquired from a human somatic clone embryo. It is also possible for the Ministry to lift the ban on the transfer of a human-human chimeric embryo into a uterus, leading to the creation of a human individual made of two (or more than two) different human embryos.

4. Debate on Human Cloning

There was not much "public" discussion of human cloning after the establishment of the law. The public response was indifferent, reflecting general disinterest in the legal regulation of human cloning. People believed that the government would support their conviction that the creation of a cloned human individual should be prohibited.

The Prime Minister's Office conducted an opinion survey on cloning in 1998, the year in which the Sub-committee on cloning was established within the Council for Science and Technology. Respondents were scholars, journalists, physicians, researchers and so on (N = 2,114). The result was considered to reflect the general Japaneseattitude toward human cloning. 92.3% had an interest in cloning, and more than 93.5% thought that the creation of a cloned individual was questionable in terms of bioethics. The reasons were:

- Human cloning should not be allowed in terms of *human dignity*, because humans should be conceived by the involvement of both sexes. 67.7%
- The cloned individual will be regarded as a *means* for attaining a predefined goal, not as a free individual. 43.6%

- It should not be allowed to *intentionally determine* the characteristics of a human being in advance. 29.8%
- The creation of an individual endowed with *specific excellent characteristics* might be preferred in the future society. 26.1%
- A cloned individual may be exposed to social *discrimination.* 14.9%
- It is not guaranteed that the cloned individual can grow up *in safety*. 10%

(* Emphases added by Morioka)

The words "human dignity," "means," "intentional determination," "preference of excellent characteristics" were used as reasons to prohibit the creation of a cloned individual. These ideas were reflected in the report of the Sub-committee on cloning.

More than 90% of the respondents were against the creation of a cloned individual, an attitude shared by the government. This is the main reason why a heated public debate on human cloning has not occurred up hitherto. Of course, there were a series of discussions in the Diet, but a compromise was soon reached between the government and the Democratic Party of Japan; the discussion has never grown into a public debate.

However, after the establishment of the law and the guidelines, the topic has been fiercely debated between scientists who wish to promote "regenerative medicine" and specialists who want to put the brakes on the rapid advance of scientific technology. The central point of the debate is whether to remove the ban on "therapeutic human cloning" to acquire ES cells from a human clone embryo.

Junji Kayukawa, a journalist specialising in human reproductive technology, urges us to pay attention to the fact that "regenerative medicine" was included as one of the major topics of the Japanese government's "Millennium Project," announced in 1999, together with other advanced technologies in the field of information science, medicine, and environmental science. This Millennium Project was launched to facilitate technological innovation for a new industry, and the government spent 250 billion yen on research in these technologies in the fiscal year 2000.[5] This implies that research on regenerative medicine is strongly supported by the Japanese government, medical researchers, and the industry sector (San, Kan, Gaku, in Japanese).

The first meeting of the Japanese Society for Regenerative Medicine was held in 2002. The news media reported that members of the

Society objected strongly to the ban on "therapeutic human cloning." In the second meeting held in 2003, Makoto Oohama, chairperson of the board of directors, Japan Spinal Cord Foundation, stressed that research on therapeutic human cloning should be allowed because it may lead to the regeneration of an injured spinal cord.

On the other hand, journalists and researchers who are sceptical about therapeutic human cloning and human ES cell research have published papers and books criticising the argument that aimed to promote these technologies.

Kumiko Ogoshi, research associate at Nara Medical University, calls the current law "Human Cloning Techniques Promotion Law" because it will result in encouraging research on therapeutic human cloning and ES cells, which may violate "human dignity" and "human rights."[6] She thinks that the most problematic point in this law is that it was established without sufficient discussion about the value of human life, and without hearing the voices of women, disabled people, and the general public. She laments that if the government had heard their voices, such an "inhumane" law would never have passed the Diet. The government, she stresses, should have discussed the problems arising from research on human female eggs, especially the problem of extracting eggs from a female body. She also says that the two-layered system consisting of the law and the guidelines was a "shrewd" way of regulating, because the government can mitigate the ban whenever it wishes, without revising the law itself.[7]

Junji Kayukawa pointed out in his book that there are at least three fundamental ethical problems surrounding research on "therapeutic human cloning."[8] The first of these is that it may support, or sometimes promote, "eugenic ideas" that we all harbour deep down. By this, Kayukawa means our inclination to think that some people (e.g., healthy, talented, smart. etc.) are superior or preferable to others (e.g., disabled, mediocre, rude, etc.). He quoted the words of an American couple who wanted to have a cloned baby. In an interview, the wife said she did not wish to adopt a child whose parent might be a killer, and that her own parents had a strong gene, but if her baby was to be born disabled, she would abort it. Kayukawa detects "eugenic ideas" in her words. He also detects them in the opinion that human cloning should not be allowed because a cloned baby is going to have a severe "disability." Kayukawa's conclusion is that "eugenic ideas" shape our attitudes toward human cloning, or even therapeutic human cloning, and hence, these techniques are problematic in terms of ethics.

The second problem is that there has not been enough discussion about how we obtain human eggs for therapeutic human cloning. The extraction of eggs puts extreme physical and psychological pressure on the female donor. And while therapeutic human cloning imposes a severe burden on females, the leaders in regenerative medicine appear to be unaware of this kind of gender imbalance. For example, a research questionnaire by a self-help group for infertile women shows that fertility drugs produce various side-effects in more than half of the drug users. In this sense, therapeutic human cloning is considered to be a heavily gender-biased medicine. As Kayukawa and Ogoshi pointed out, this has not been sufficiently discussed.

The third problem is that research on therapeutic human cloning (and research on human embryos in general) is inevitably going to regard a woman's body as a mere "resource" to be exploited for scientific technology, and a woman's body is going to be treated as "material" to produce a profit, even if money is paid to her as donor. Kayukawa presents two different opinions: one is from a researcher who said "an ES cell is a mere cell," and the other is from an infertile woman who said "if we donate our surplus eggs for research, our eggs will become a mere 'instrument' for people." Kayukawa urges us to discuss the gap between these two opinions, or in other words, the gap between these two worldviews concerning human life.

In 2001, the Council for Science and Technology Policy, Cabinet Office, was established in the government, and the Expert Research Commission on Bioethics was established in the Council. Its mission was to comprehensively discuss the ethical issues concerning human cloning, ES cells, and other reproductive technologies. In November 2003, a member of the Commission, Susumu Shimazono, professor of religious studies at the University of Tokyo, published a paper in a popular magazine in which he severely criticised the discussion in the Commission.[9]

Shimazono first pointed out that "*A Draft for Fundamental Thoughts on the Handling of Human Embryos*,[10] circulated in the Commission on August 23, 2003, sought to compare two values, namely, "the value of a human embryo on which human dignity is reflected" and "the value created by scientific technology." Two proposals, for and against promotion, were formulated in the draft. In the case of both ES cells and therapeutic human cloning, the proposal says that the value created by scientific technology clearly surpasses that of a human embryo.

Shimazono insisted that the Commission had never discussed whether or not research on a human embryo and ES cells violates "human dignity," and it had never discussed what "the sprout of human life" is and how it is different from "human life." He argued that the artificial creation of an "animal-human chimeric embryo" might violate "human dignity," but they had never discussed the ethical aspect of this handling. The draft uses the words "the sprout of human life" and "human dignity" many times, but the Commission had never considered the ethical and philosophical meaning of these terms in any depth. He laments the fact that the country, which seriously discussed the issue of brain death and organ transplantation, has not discussed this topic earnestly. He suspects that the consideration of economic aspects might have influenced the discussion in the Commission.

On December 26, 2003, the Commission published *Fundamental Principles on the Handling of Human Embryos, an Interim Report.* This report concluded that the production and use of cloned human embryos should not be completely prohibited, but the Commission members failed to reach consensus about whether a moratorium should be placed on research until further scientific knowledge is acquired. For more information about the Interim Report, see the chapter Cloning in Japan by Robert Horres, Hans Dieter Ölschleger, and Christian Steineck in this book.

5. Discussion

One of the most interesting terms in the Japanese discussion on human cloning is "the sprout of human life" which appears in the Japanese law and many other materials. Not only people who object to human cloning, but also many of those who seek to promote research on human cloning admit that a human embryo is the sprout of human life and, hence, it should be highly respected.

The government translated the term as "the beginning of human life," but this translation loses an important nuance. When they hear the words "the sprout of human life," many Japanese feel some kind of vigorous energy moving inside the embryo. It might be biological energy, or it might be spiritual. This energy does not mean the mere "future possibility" of becoming a person. It is something that actually exists inside the embryo.

It is also interesting that the locus of human dignity is expressed as "sprout," because this word means the bud of a "plant," not an animal. However, Shizuka Shirakawa, a prominent linguist, insists that the Chinese character meaning "sprout" contains that of "fang," and this means the sprout of a plant has a wild, animal-like energy.[11] I presume that the energy in the sprout of human life is probably something that is shared by plants, animals and humans. Hence, many Japanese feel that it should be respected as much as possible. This concept is reminiscent of Masao Maruyama's well-known words, "tsugi tsugi ni nariyuku ikihohi" (flowing energy that transforms and develops itself one after another) to be found in the ancient layer of Japanese consciousness of history.[12] Maruyama came upon this concept in *Kojiki*. In this sense, ancient Japanese writings and contemporary bioethics literature might share similar ideas on life and death.

Now let us turn our attention to "language" or "discourse." People who wish to maintain the ban on therapeutic human cloning are journalists, feminists, and researchers critical of the "progress" of scientific technology. Their "language" is based on the "language of scepticism": scepticism about the propaganda that the progress of science and medicine brings us "health and happiness." They do not believe this kind of optimism. And they try to keep away from "religion" as much as possible, because in Japan "religious language" has not worked as an instrument of criticism. But precisely because of this, their arguments have not been as persuasive as they had anticipated.

By contrast, the "language of utilitarianism" used by the advocates of advanced medicine seems very powerful. Supporters of regenerative medicine emphasise the benefit of research to the general public, particularly patients with intractable diseases. Not only researchers but also patients themselves talk about their expectations from medical progress. Their language is simple, direct, and forceful. We see an echo of this utilitarianism in the Commission's Interim Report.

It is striking that we encounter no important comments or opinions on this topic in the religious sector. In its Interim Report, the Commission reported that they could find no important opinions in Japanese Buddhism, Shinto, or Japanese Christianity. My own impression is similar. To my knowledge, they have published no reports on human cloning or other related topics. I can offer no explanation for their silence on human cloning research.

Interestingly, both supporters and opponents use the words "human rights" and "human dignity." They do not debate these concepts because they accept their importance. Instead, the debate is between the "language of utilitarianism" and the "language of scepticism." And the "language of religion" remains silent. Even disabled people seem to be torn between support and opposition. We should be aware of the fact that many Japanese disabled people have been critical of the "progress" of medical technology and of "eugenic ideas" (see my paper "Disability Movement and Inner Eugenic Thought.")[13] At the same time, however, there are disabled people's groups that look forward to the development of new technology (e.g., Japan Spinal Cord Foundation). This is the rough sketch of the Japanese discourse on research on human cloning.

My personal view is that a stronger argument is needed for protecting the value of the human embryo, including a cloned human embryo, especially in Japan where the "language of religion" has little clout in the discussion. Instead of religious language, we need "philosophical language" to affirm the value of a human embryo or "the sprout of human life."

What is it we wish to protect when we use the word "the sprout of human life"? The answer would be "a vigourous energy to develop and transform itself" that we once were, that we came from, and that we still have at the basis of our existence. This is what we have to protect, even if its destruction would be beneficial to the progress of medicine. Why then should we protect it? The answer would be that its destruction means the destruction of something very important which we actually "share" at the basis of our lives; hence, its destruction might lead to the destruction of ourselves. The ultimate danger of research on human embryo is that in the long run it might erode something very important inside us in the name of social welfare and the progress of medicine. We need "philosophical language" to explain the core meaning of the words "something very important" in a way that can be easily understood by the general public. In this sense, we need a new "philosophy of life," or "life studies,"[14] which will give us the wisdom to protect "something very important" from our own selfish desire to live a long and healthy life.

Notes

1. *Kurôn Gijutsu ni yoru Hito Kotai no Sansei Tou ni kansuru Kihonteki Kangaekata*, published by Kagaku Gijutsu Kaigi Seimei Rinri Iinkai, Kurôn Shô Iinkai.
2. *Hito ni Kansuru Kurôn Gujutsu Tou ni Kansuru Hôritsu.*
3. *Hito Hai Tou no Sakusei Oyobi Riyô ni Kansuru Hôritsu An.*
4. *Hito ES Saibô no Jûritsu Oyobi Shiyô Kansuru Shishin.*
5. Junji Kayukawa, 2001, 133; *Mainichi Shimbun* (April 18, 2002); *Yomiuri Shimbun* (July 29, 2002).
6. Kumiko Ogoshi, 2001a, 4.
7. Kumiko Ogoshi, 2001b, 40-43.

8. Kayukawa, 2001.
9. Shimazono, 2003, 134-43.
10. *Hito Hai no Toriatsukai ni Kansuru Kihonteki Kangaekata no Soan.*
11. Shirakawa, 2003.
12. Maruyama, 1992, 334.
13. Morioka, 2002, 94-97.
14. About "life studies" and "philosophy of life," see my website International Network for Life Studies: http://www.lifestudies.org.

References

Kayukawa, Junji. *Kurôn Ningen.* Kôbunsha Shinsho, 2001.

Mainichi Shimbun, April 18, 2002.

Maruyama, Masao. "Rekishi Ishiki no Kosô." In *Chûsei to Hangyaku*, Masao Maruyama (ed.). Chikuma Shobô, 1992.

Morioka, Masahiro. "Disability Movement and Inner Eugenic Thought: A Philosophical Aspect of Independent Living and Bioethics." *Eubios Journal of Asian and International Bioethics* 12 (May 2002): 94-97. (see also http://www.lifestudies.org/disability01.html)

Ogoshi, Kumiko. "Hajimeni." In *Hito Kurôn Gijutsu wa Yurusareru ka*, Ogoshi Kumiko, Koichi Nishimura, Ryoko Suzuki, Hideko Fukumoto, Renko Kitagawa, Junji Kayukawa (eds.). Ryoku Fû Shuppan, 2001a).

Ogoshi, Kumiko. "Hito Kurôn Kisei Hô Dokkai." In *Hito Kurôn Gijutsu wa Yurusareru ka*, Ogoshi Koichi, Nishimura, Ryoko Suzuki, Hideko Fukumoto, Renko Kitagawa, Junji Kayukawa (eds.). Ryoku Fû Shuppan, 2001b).

Shimazono, Susumu. "How to discuss the ethics of advanced life sciences?" *Sekai*, December, 2003.

Shirakawa, Shizuka. *Jiyû Jikai.* Heibon Sha, 2003.

Yomiuri Shimbun, July 29, 2002.

Cloning in Japan: Public Opinion, Expert Counselling, and Bioethical Reasoning

Robert Horres, Hans Dieter Ölschleger, and Christian Steineck

Abstract: In Japan, biomedical research is seen as a key factor for the future development of the economy as well as social welfare. The Japanese government, seeking to secure Japan's position as one of the leading nations in science and technology, provides generous funding. At the same time, it seeks to ensure that such research will be in line with the fundamental values of Japanese – and international – society, and not prove socially or politically disruptive. The issue of human cloning is a show case in this respect. An analysis of public opinion, government decision making and bioethical reasoning reveals different attitudes in these spheres of discourse. Public opinion is generally favourable and focuses on the medical benefits promised by reproductive medicine. The government's attitude is additionally shaped by concerns for the international viability of such research. Expert opinions vary greatly, with the strongest opposition to cloning and the destruction of human embryos coming from bioethicists with a background in the humanities. However, their arguments have so far been virtually ignored by the public, the government, and scientists working in the field. Remarkably, „therapeutic cloning" draws more criticism than „reproductive cloning," although the latter is penalised by law.

Key Words: Japan, Cloning, Research on human embryos, Bioethics, Acceptance of new biotechnologies, Public opinion, Legislation, Ethical reasoning.

Japan is one of the most highly modernised countries in the world. It is an important industrial and economic power and, in fact, the second largest economy on the planet. However, in recent years Japan has undergone a prolonged economic crisis, and the Japanese government is striving hard to find ways that may spark a new wave of economic growth. Generous funding for research into new technologies, such as nanotechnology and biotechnology, is one such government initiative, formulated in the so-called "Millenium Plan" that was announced in 1999. Clearly, the Japanese government aspires to ensure Japan's position as one of the leading global players in all areas of advanced science and technology. This is seen as a golden path to securing both the welfare and affluence of the nation and its political and economic clout in the world.

However, there is also an awareness that some fields of advanced biomedical research are highly controversial. This is especially true of human cloning, be it for purposes of research on new medical therapies or for reproduction. Thus, government initiatives to encourage research in the therapeutic uses of reproductive medicine from the outset have been accompanied by strict regulation, in order to keep this research, and its applications, on a path that would be ethically acceptable both to the Japanese people and to the international public. A *Law Concerning Regulation Relating to Human Cloning and Similar Techniques* (*Hito ni kansuru kurôn gijutsu nado no kisei ni kansuru hôritsu*; quoted below as *Law on Human Cloning*) was passed by the Diet in 2000, which is the subject of M. Morioka's article in this volume. The law penalises the transfer of cloned human embryos to human or animal uteruses. Regulation concerning the creation of cloned human embryos etc. is left to ministerial guidelines. The current guidelines prohibit the production of human embryos for research as well as many other similar techniques. However, these guidelines are now being revised. The new guidelines are to permit the production of cloned embryos, and research on fertilized and cloned embryos for specific purposes. But research will be strictly regulated and the production of embryos with the aim of producing human embryonic stem cells will remain prohibited.

In this article, we present an analysis of the situation in Japan concerning human cloning from three different points of view. Firstly, we examine public opinion on the subject in an international comparison and relate it to the general attitudes of the Japanese towards science and technology. Secondly, we describe how, through various governmental committees and expert hearings, the Japanese administration developed its position, up to the formulation of the currend law and the "old," still valid guidelines.

Thirdly, we explore the ethical arguments currently discussed in Japan with respect to human cloning and embryo research,: focusing on the arguments behind the new government principles on human embryo research that will, in all probability, guide future regulations in this field. In the final part of the article, we draw some conclusions from the data gathered, especially highlighting a degree of acceptance that differs according to the social and occupational positioning of the individual and the (missing) role of Japanese tradition in shaping the attitude toward human cloning.

1. Attitudes Toward Human Cloning: The Case of Japan in International Comparison[1]

In the year 2000, a nationwide survey in Japan asked: "What comes to mind when you think about modern biotechnology in a broad sense […]?" The example most commonly mentioned was Dolly the cloned sheep, which, of course, had been in the headlines of the Japanese press three years earlier.[2] Cloning has, to say the least, become part of the Japanese awareness of biotechnology. And biotechnology, exemplified for many respondents by cloning, fares surprisingly well in its acceptance compared to other fields of modern technology (see Table 1).

Table 1. Perceived benefits of science and technology for society in Japan, 2000 (in %)

	Computers and IT	Biotechnology	Genetic engineering	Telecommunications	Space exploration
Will improve	81.6	66.3	58.6	77.2	63.7
No effect	3.8	4.5	2.4	10.5	19.3
Make worse	8.5	15.1	23.6	7.8	5.8
Don't know	6.1	14.1	15.4	4.4	11.2

Source: Adapted from Ng *et al.*, 2000, Table 4.

This high degree of trust in biotechnology is consistent with the familiarity of the Japanese public with the term 'biotechnology.' According to several studies conducted in the 1990s, this familiarity has always been extremely high, peaking at 97% as early as 1991 and since then hovering above the 90% mark.[3]

Concerning the acceptance of cloning, we get quite a different picture from the results of an opinion poll on basic bioethical problems, the bioethical problems of cloning and the cloning of animals, which was conducted among experts from several fields by the Public Relations Office of the Prime Minister's Office in 1998.[4] Here, too, cloning was a problem with a high level of visibility: 45.6% of the sample saw it as a bioethical issue of major concern, this percentage being topped only by genetic treatment, organ transplant, and brain death (see Table 2; multiple answers were possible).

The opinion poll surveyed bioethical problems of high interest. As can be seen from Table 2, the problem of cloning ranged among the fields of high interest, but it was ranked highest especially by the groups of researchers and bureaucrats.[5]

Table 2. Bioethical problems of high interest (in %)

Bioethical field	% of respondents
Gene therapy	50.4
Organ transplantation	47.7
Brain death	47.1
Cloning	45.6
Euthanasia	44.7
Patient information in case of cancer, etc.	22.7
Extracorporal insemination	15.0
Surrogate motherhood	10.1
Abortion	3.1
Other	1.7
None	1.3

Source: Sôrifu Kôhôshitsu, 1998a, p. 4.

More than 93% of the respondents held the attitude that from a bioethical point of view cloning is unacceptable. As the most important reasons for this unacceptability, the respondents chose:

1. "The application of the technology to human beings would impair human dignity, because the production of human beings should involve both sexes." (67.7%)
2. "Human cloning treats human beings simply as a means of achieving predetermined purposes instead of creating free individuals." (43.6%)
3. "It should never be permitted to predetermine the character of a human being." (29.8%)
4. "There is a possibility that a society might be created where priority is given to producing human beings with superior characteristics." (26.1%)[6]

But despite the overwhelmingly critical attitude towards cloning in general, one-quarter (23.5 %) of the interviewees would accept reproductive cloning under specific regulations, if a couple was unable to have children by other methods.[7] Concerning the implementation of a regulatory system on cloning, 71.2 % of the interviewees preferred a legal base, 16.7 % a system of administrative guidelines.[8]

A fast implementation of regulations was expected by 66,9 % of the interviewees, but only 59,4 % advocated unlimited implementation: 13.0 % would like to see a revision within 2 or 3 years and 23.9 % a revision within 5 years.[9]

Table 3. Benefits of science and technology for society in Japan, perceived by scientists 2000 (in %)

	Computers and IT	Biotechnology	Genetic engineering	Telecommunications	Space exploration
Will improve	86.9	77	72.1	80.9	50
No effect	1.1	3.6	3	8.7	23.9
Make worse	4.6	7.7	12.6	6.6	5.8
Don't know	0.0	0.0	0.0	0.0	0.0

Source: Adapted from Ng *et al.*, 2000, Table 4.

This critical stance toward biotechnology was not found when only scientists were questioned in the year 2000. On the contrary, as can be seen in the comparison of the data of Tables 1 and 3, with the sole exception of space exploration the respondents with a scientific background perceived the influence of the new technologies - including biotechnology - in more positive terms than the Japanese public. In an international comparison, public acceptance of biotechnology, including cloning, is not extraordinarily high in Japan. We find equally high levels of acceptance of human cell cloning (as well as of genetic testing, but a strong rejection of genetically manufactured food) in all European countries covered by the Eurobarometer poll of the year 2002.[10] Located on a scale from -1 (strongest opposition) to +1 (strongest support), the attitudes in 5 countries range between +0.5 and +1, the remainder show values between 0.0 and +0.5. And in contrast to the bioethical discussions in Europe, which emphasise ethical doubts concerning the cloning of human cells, and against the strong political current of rejecting cloning, the public does not only see the usefulness of cloning for medical purposes, but also holds that cloning is morally acceptable and should be encouraged. The underlying reason for this acceptance lies in the promises this new technology holds when applied in medicine. This usefulness overrides not only all ethical objections, but also possible risks which are acknowledged by the public. Further analysis classified the majority of respondents as so-called risk-tolerant supporters, people who "perceive risk but then discount it," do not think it morally unacceptable, and therefore encourage the new technology.[11]

To summarise this short overview of the results of opinion polls on public attitudes toward biotechnology, it may suffice to say that neither Japan nor the member states of the European Union are inhabited by technophobics who unanimously condemn biotechnology. Concerning the use of human cloning, only the sample of Japanese experts - the majority of them not *directly* involved in the research and/or application of the new biotechnologies - expressed their reservations against this new technology and based their opinion on bioethical considerations.

Thus, Macer's remarks about the New Zealanders,[12] namely that "the cloning debate may have filled the media but not influenced the hearts of New Zealanders in their perceptions of the use [of] genetic technology" can be said to be true of other societies as well - including the Japanese society.

2. Factors Behind Acceptance or Rejection of Biotechnology

But although the public acceptance of biotechnology is relatively high, several qualifications should be made to show that the reality of public opinion is far more complex than implied by this simple statement. These may also suggest an answer to the question, what are the factors behind this high degree of acceptance which is in stark contrast to the bioethical discussion among experts (politicians, philosophers, administrators, and religious officials), who to a large extent reject or challenge cloning technologies.

One possible factor might be the risks and benefits connected with the new technology. But public opinion about technological innovations, be they biotechnological or not, is not only influenced by pragmatic reasoning on the risks and benefits - on the contrary, the knowledge of these risks and benefits is often limited to the scientists involved and, probably, the administrative and political elite of a society. The overwhelming majority of the population is, plainly speaking, not able to give a correct account of scientific progress in all its dimensions.[13] And, as has been shown in the case of German society,[14] public opinion toward biotechnology is not founded on personal experience - the new technology is still not able to make good its promises and therefore the ordinary citizen has not yet had the opportunity to benefit from some new medical application (or the misfortune to fall a victim to one of its risks). Thus, attitudes toward new technologies are nourished from other sources, in the first place from general attitudes toward science and technology prevalent in a population.

A. General Attitudes Toward Science and Technology in Japan

At least as far as stereotypes are considered, and this is true of both the autostereotype and the heterostereotype, the Japanese are very fond of technology, or, as one magazine report says: "Japanese people have a strange love for technology."[15] Be that as it may - it is very easy to see a population of technology freaks in a society which is leading in the production, consumption, and exportation of high-tech articles - there are several points concerning the attitudes of the Japanese public toward science and technology worth mentioning which might help to explain the high degree of acceptance of biotechnology.

Table 4/1. Attitudes toward science and technology in general in Japan, 1987 and 1990 (in %)

	More advantages	Advantages and disadvantages are about the same	More disadvantages	Do not know
1987	54.3	28.7	8.3	8.7
1990	52.6	30.5	7.3	9.6

Source: Adapted from NISTEP, 1995, Fig. 7-2-18.[16]

Note: Question: "Are there more advantages or more disadvantages in the progress of science and technology?"

Table 4/2: Attitudes toward science and technology in general in Japan, 1998 (in %)

	Many advantages	Some advantages	About the same	Some disadvantages	Many disadvantages	Do not know
Total	22.1	35.6	26.8	6.6	4.1	4.8
Male	27.3	38.8	21.4	5.8	4.0	2.7
Female	17.8	32.8	31.4	7.3	4.1	6.6

Source: Adapted from NISTEP 2001, Fig. 8-1-3.

First of all, confidence and trust in science and scientists is high and seem to be unchanged since at least 1990.[17]

Furthermore, this acceptance of science and technology is differentiated according to different applications. This fact has been known for many decades.[18] Applications which promise to be of future use to the individual are accepted to a higher degree than those which are of no use or even of destructive use (e.g., weapons). Especially medical applications, whose value is seen in the healing of diseases that were previously considered incurable, are strongly endorsed. Neither ethical considerations that speak against these applications, nor the fact that the application is still technologically immature alter this high degree of acceptance.

A final point concerning the Japanese attitude seems worth mentioning: Science and technology is considered an aspect of Japanese society of which a large section of the public feels proud. Since the burst of Japan's bubble economy in 1989/90 science and technology have even surpassed economic performance as a source of public pride.[19]

Table 5. Attitudes toward science and technology, Japan 2001 (in %)

	S&T has made life more comfortable	Future generations will have better opportunities	Scientists engage in research to improve people's lives	Scientific applications make work interesting
Strongly agree	6	6	4	4
Agree	67	60	56	50
Disagree	13	16	20	25

Strongly disagree	1	1	2	2
Don't know	13	18	18	20

Source: Adapted from Kagaku Gijutsu Seisaku Kenkyūjo, 2002, Table 10.

B. Interest in and Knowledge of Science and Technology in Japan

This positive attitude toward science and technology does not translate into a high degree of interest in scientific and technological issues. With the sole exception of environmental pollution - which, of course, is at the centre of public interest following several environmental disasters which have killed literally thousands of Japanese and maimed even more - science and technology rank low in the interest of Japanese individuals. 'Use of new technological inventions,' 'atomic energy,' 'scientific discoveries,' and 'space exploration' rank far below 'economic and business issues,' and the sense of being well informed of these issues is even more underdeveloped. A comparison with data from the United States and European Union member states shows that in 2001, Japan ranked lowest of all nations surveyed as regards the population's interest in scientific discoveries.[20] Considering this low degree of interest, it is not in the least surprising that in an international comparison knowledge of technological innovations is low in Japan. Again, in a sample of 13 nations (European Union and the United States of America) the Japanese public ranks third last.[21] A direct comparison with New Zealand shows that there the background knowledge and understanding of biotechnology is considerably higher; to cite just one example: 74% of the New Zealanders questioned had heard of 'genetic testing' as against 36% of the Japanese sample.[22]

Figure 1. Interest in information about science and technology in Japan, 1981 to 1998 (in %)

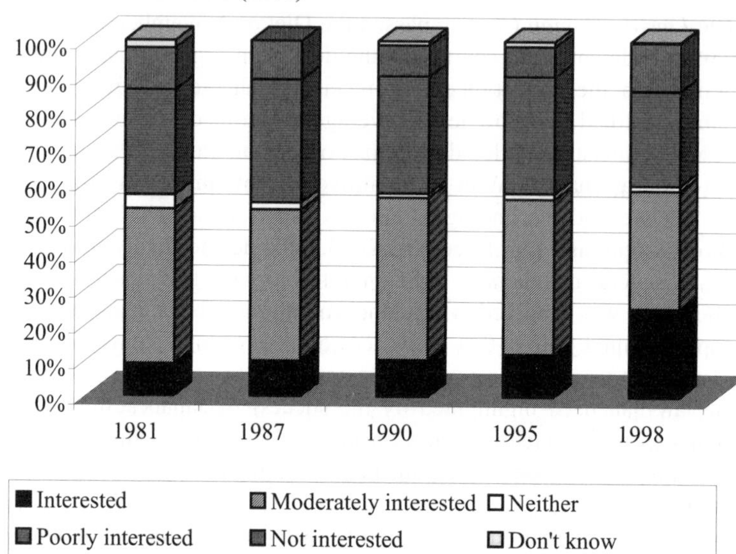

Source: Adapted from Sôrifu Kôhôshitsu, different years.

The high acceptance of biotechnology in Japanese society results from a positive overall view of science and technology especially as they are expected to make the life of the individual easier and healthier. Without proper background knowledge of biotechnology and its applications and only a low degree of interest in information concerning these new technologies, the ("statistically average") Japanese individual evaluates these innovations by their promises of improving life and not by carefully weighing either risks and benefits or ethical considerations and arguments for or against them. The question to consider next is the form and content of the process of political opinion shaping, and if - and in what way - public opinion in its different aspects is reflected in this process.

3. The Counselling Process Leading to the *Law on Human Cloning*

The *Law on Human Cloning* passed the Diet on November 30, 2000, with few amendments.[23] In contrast to the debate on organ transplantation and brain death, the legislation did not receive much attention from the general public and had only low-level coverage in the media. Even among Japanese bioethicists, detailed discussions on human cloning started only after the law was enacted. With the controversial interim report of the Council of Science and Technology Policy's Expert Research Commission on Bioethics published in December 2003,[24] the debate finally started to heat up, as several members of the Commission publicly voiced their dissent. The issue was resolved within the commission by a majority vote, an unprecedented move that drew critical comments from the Japanese press.[25] In contrast, the counselling process prior to the establishment of the law had been highlighted by the success of Japanese researchers in cloning a cow, which was born on July 5, 1998. This underlined the need for a fast decision on the establishment of a regulatory system for cloning in Japan.

The regulation of cloning in Japan - as in other democratic societies - constitutes a show case for the problem of experts and expertise in politics. Since clone-related issues are complex, the government and the general public need the counsel of expert commissions. The content of their reports and studies may serve as an indicator of the state of information of political decision makers.

In the following, we will analyse some important studies and governmental commission reports concerning the regulation of human cloning in Japan, which were published before legislation was passed. This will help us to evaluate the situation prior to the enactment of the *Law on Human Cloning.*

On University Research on Cloning (Report): On April 18, 1997, the Science Commission of the Ministry for Education, Science and Culture established a Life Science Subcommittee within the Special Research Area Promotion Committee (Gakujutsu Shingikai Tokutei Kenkyû Ryôiki Suishin Bunkakai Baiosaiensu Bukai).

The final report of the Life Science Subcommittee examined the situation of research on cloning in Japan and other countries and made suggestions for the establishment of a regulatory system. The suggested draft regulations envisage the prohibition of research on the production of human clones.[26]

Interim Report on Fundamental Principles Concerning Cloning Technology (1998): On June, 15, the Cloning Subcommittee of the Bioethics Committee (Council for Science and Technology) published an interim report *Kurôn gijutsu ni kansuru kihonteki kangaekata ni tsuite* [Fundamental principles concerning cloning technology].[27] In this report, the subcommittee examined attitudes toward cloning in Japan and formulated basic principles for Japan's policy on the cloning issue.

The interim report states that the production of a human person or individual through cloning means a violation of human dignity and concludes that it is "not useful" to commend this practice, since at present the possibilities for research with animal cells were deemed to be sufficient (Section II. 2.1). On the other hand, the report suggests that applications of cloning techniques might be useful if they do not involve the production of a human individual or a human organ. This kind of research does not offend human dignity. Consequently, the authors of the report see no reason to prohibit this kind of research (Section II. 2.2). The report takes the same stance on the cloning of animals. Research on ES cells is considered useful, if the above stated limitations are observed (Section II. 2.3). The production of cloned organs is deemed technically and ethically problematic and consequently evaluated as not useful (Section II. 2.4).

The interim report also makes suggestions concerning the need to regulate the application of cloning techniques: All activities that involve the application of cloning techniques to produce a human individual shall be subject to official government regulations, guidelines, or laws. The need for legal regulation arises from the seriousness of possible abuse, which may violate human dignity and safety. The report states that regulation by guidelines only (without a law) does not effectively rule out abuse by those who do not belong to communities of physicians and researchers.[28]

Legal Regulations on Advanced Science and Technology - Regulations on Life Science (1999): In May 1999, the National Institute of Science and Technology Policy published a study on Japanese and international legal and regulatory systems concerning life science and technology.[29] In this study, named *Legal Regulations on Advanced Science and Technology - Regulations on Life Science*, the efficiency of the regulations and regulatory processes in Japan as well as in the Unites States, the UK, Germany, and France is analysed.

The study concludes that a rational system for the regulation of human cloning must be established within the wider framework of reproductive medicine and technology.[30] It pursues an investigation of the legal conditions for cloning and life sciences in Japan, covering the range from the constitutional law to the civil code (e.g., family act) and the penal code.[31] A third major contribution of this study is the analysis of proceedings for reaching a consensus on cloning. This consensus is thought to be necessary both for the legal and administrative regulation of cloning.[32]

Guidelines for Derivation and Utilization of Human Embryonic Stem Cells (June 14, 2000): In December 1998, the Bioethics Committee of the Council for Science and Technology established a special subcommittee on the Evaluation of Research on Human ES Cells (human embryonic stem cells). From February 1999 till March 2000 the subcommittee developed guidelines for research on human embryonic stem cells in Japan.

These guidelines relied heavily on the statement of the American National Bioethics Advisory Commission *Ethical Issues in Human Stem Cell Research*, published in September 1999. However, they departed from their model in recommending financial support for the establishment of human ES cell lines from so-called "excess embryos" (*yojô hai*), resulting from in-vitro fertilisation.

Concerning research on human ES cells, the guidelines introduce the following principles:

> Human beings shall not be created for research objectives. Only excess embryos from IVF treatment shall be used, if there is written consent from both parents.
> The research activities on human ES cells shall be carried out in a transparent manner (publication of application and research results) and under the control of an ethics committee.

There is a ban on research that aims at a) generating a human individual from ES cells, or b) transferring human ES cells to a human embryo (pre- or post-implantation). The generation of individuals from animal embryos through the transplantation of human ES cells is likewise prohibited.

The papers introduced above discuss a wide range of cloning-related problems and constituted a sufficient basis for a political decision on cloning. They draw an adequate picture of official Japanese attitudes towards cloning and its proposed regulation. Not surprisingly, the suggestions for a Japanese policy on cloning, which can be found in these papers, are divergent in some minor points. The rejection of the draft cloning law by the Democratic Party of Japan (DPJ) and the ensuing revision of the draft law[33] lies within the range of these differences and is at least an indicator of the high level of discussion within the political parties.[34] Equally clear is the fact that the Japanese government was well aware of both the international situation and the possible problems within Japan with regard to the regulation of the cloning issue.

The *Law on Human Cloning* and the subsequently published ministerial guidelines that it prescribed were the result of the process of political opinion formation described above. However, with the exception of the ban on human reproduction through cloning, the regulation was designed to be provisional and subject to reviews according to new scientific and political developments. Thus, the Council of Science and Technology Policy's Expert Research Commission on Bioethics started to discuss the issues of human cloning and embryo research anew in 2001. The first official result of its endeavours was an interim report that was published in December 2003. Further discussion led to the formulation of new "fundamental principles" (*kihonteki kangaekata*) concerning the treatment of human embryos by the Council for Science and Technology Policy in July, 2004. In the following section, we will discuss the content of these documents in relation to bioethical discussions in Japan.

4. The Ethical Reasoning Behind the New Principles for Human Embryo Research

A. Content of The New *Fundamental Principles*

On July 23, 2004 the Council for Science and Technology Policy (*Sôgô kagaku gijutsu kaigi*) published a plan for new "Fundamental Principles Regarding the Treatment of Human Embryos" (*Hitohai no toriatsukai ni kansuru kihonteki kangaekata,* quoted herafter as *Fundamental Principles*). As has been anticipated by critical observers like M. Morioka,[35] these mark a departure from the currently valid regulations. The new principles will allow the production and use (including destruction) of human embryos both through artificial fertilization and cloning techniques, for certain defined research purposes. The overall reasoning behind this move is that human embryos are recognized as "sprouts of human / personal life" (*hito no seimei no hôga*).[36] As such, they are said to require special legal protection, although not enjoying the status of a human person.[37] Manipulation of human embryos is prohibited in principle, because it may adversely affect human dignity.[38] However, if "fundamental human rights" such as the "pursuit of happiness relating to health and welfare" demand so, "harmful treatment" can be allowed "as an exception".[39] Restrictive conditions are 1) scientific reasonability, 2) protection of human personal life and safety, 3) the "social validity" (*shakaiteki datôsei*) of the merits, or expected merits, of such treatment.[40]

Under these considerations, production and use of fertilized human embryos will be permitted for fundamental research in assisted human reproduction. A moratorium is declared on such treatment for research on severe congenital diseases.[41] Establishment of human embryonic stem cell lines is not accepted as a reason for the production of a human embryo by fertilization; however, "excess embryos" (*yojôhai*) from assisted reproduction, as well as cloned embryos may be used for fundamental research on severe diseases, such as Parkinson or diabetes.[42] That is to say, the new principles do allow the production of cloned embryos for research into treatment for severe diseases and afflictions.[43] Should alternative methods, such as those using adult stem cells, prove successful, this exceptional permit is to be withdrawn.[44] A national system of supervision and control for all research involving human embryos is required, and special attention is given to the protection of women in the harvesting of ova.[45]

The ethical reasoning given in the *Fundamental Principles* does not move beyond very general statements. Human dignity is recognized as the fundamental value of Japanese society. It is stated that this value must inform the treatment of human embryos, which have a special status as "sprouts of human / personal life." This status entitles them to some degree of respect and protection, although not giving them the full protection enjoyed by fetusses after the 22nd week of pregnancy and human beings after birth. Existing regulations are cited as a foundation for this assessment.[46] According to the *Fundamental Principles*, the "respect" commanded by human embryos, and grounded in the notion of human dignity, may be outweighed by consideration for the "fundamental human rights" (*kihonteki jinken*) of those with severe afflictions, that is, their aspiration for new therapies as a condition for "living with dignity" (*songen aru seizon*).[47] Social acceptance, termed as "social validity" (*shakaiteki datôsei*) appears as an important foundation for such claims, but it is introduced in this function without further explanation.

In order to understand the reasoning behind the *Fundamental Principles,* more clarity can be gained from going back to the *Interim Report* of the Expert Research Commission on Bioethics (full title: *Hito-*

hai no toriatsukai ni kansuru kihonteki kangaekata: chûkan hôkoku, quoted below as *Interim Report*), which contains a broader scope of opinions and reasoning. The *Fundamental Principles* are based on a final report by the same Expert Commission. This final report was drafted by the new chairman of the commission, T. Yakushiji. Yakushiji expressly wanted the final report to come up with a clear stance and was thus bent to exclude alternative opinions. His draft was approved by a vote within the commission, a move that has been unanimously termed "out of the ordinary" (*irei*) by major Japanese newspapers.[48] In contrast, the *Interim report* allows us to grasp the scope of arguments that were considered within the commission, and to follow to some extent the way the majority position was formed.

B. Definition of Problems in the *Interim Report*

The main purpose of the *Interim Report* is to devise a fundamental "way of thinking" in respect to human embryos that will guide the future policy of the Japanese administration on research in this field. The issue of human reproduction through cloning does not come within the report's scope, as it had already been settled in the *Law on Human Cloning,* issued in 2000. While the *Law* penalises the transfer of cloned human embryos to a human or animal uterus (*Law on Human Cloning,* Art. 3), the regulation of human cloning for medical research is left to ministerial guidelines (*Law on Human Cloning,* Art. 4). Cloning of human beings for reproduction is therefore considered to be out of question.

The *Interim Report* discusses production and use (including destruction) of cloned human embryos as parts of medical research involving human embryos (both fertilized and cloned) as well as human-animal chimeric or amphimictic embryos. It states that "manipulative interference" (*sôsateki kainyû*) with human embryos "comes very close to the manipulation of the life of a person (*hito*)." Therefore, there is "anxiety" (*osore*) and "irritation" (*fuan*) concerning their use, which is in danger of infringing on human dignity. It goes on to note: "On the other hand, the envisioned massive benefits of reproductive medicine correspond to the human right to the pursuit of happiness, which follows from 'human dignity'".[49] In other words, the production, use, and destruction of human embryos for medical research is thought to both potentially enhance and potentially infringe on human dignity. Consequently, the ethical problems of such research, as seen by the report, call for a solution through an evaluation and calculation of dangers and benefits to human dignity. This relates to the concept of "human dignity" that informs the report.

C. The Moral Status of Human Embryos

An essential part of the evaluative process in the *Interim Report* is its assessment of the moral and legal status of human embryos. In the discussion of this problem, the report makes a primary differentiation between "human fertilised embryos" (*hito jusei hai*) and "specified embryos" (*tokutei hai*), "human cloned embryos" (*hito kurôn hai*) being subsumed under the latter.[50] The moral status of a human cloned embryo is judged to be close to, but not identical with, that of a human fertilised embryo.[51]

The status of the latter is determined in the light of a survey of 1) "philosophical/ethical," 2) "religious," and 3) "legal/administrative" viewpoints.[52]

Concerning the philosophical and ethical aspects, the report presents as a majority opinion the view that human embryos are neither "things" (*mono*) nor "persons" (*hito*). They are accorded an intermediate status, expressed by the term "sprout of human/personal life" (*hito no seimei no hôga*).[53] As R. Ida mentions in his opinion statement,[54] this term, which is already used in the *Law on Human Cloning* (Supplementary Provisions, Art. 2), was first coined in an earlier report on *Fundamental Thoughts on Human Embryo Research, focussing on Human Embryo Stem Cells* (*Hito hai seikan saibô o chûshin to shita hito hai kenkyû ni kansuru kihonteki kangaekata*) by the Subcommittee for Research on Human Embryos of the Council for Science and Technology Bioethics Committee.[55] The problem of giving the term a positive meaning beyond the double negation of "thing" and "person" is explicitly mentioned. However, the only further qualification made is that it is meant to express a moral status that requires a certain amount of respectful behaviour, but not absolute protection from manipulation, lesion or destruction, thus permitting its use for medical research.[56] The reasons for this position, which is the official position of the report, are stated in a later section and will be discussed below.[57] It should be noted that in its preliminary survey of philosophical and ethical aspects, the report does not quote, or even mention, any alternative views, nor does it refer to philosophical or ethical literature on the subject. Accordingly, there is also no evaluation of arguments.

This also holds for the religious views mentioned in the report, which states that no official positions were available from Buddhist, Shintô, and Protestant Christian organisations.[58] In its quotation of individual anonymous opinions, the report is careful to balance statements from each religion, from the view that human life begins with the moment of fertilisation to others which de-emphasise the moral import of this position.[59] Finally, it sums up its survey of philosophical, ethical, and religious views with the statement:

> "There are a variety of positions concerning the status of the human embryo in the philosophical and ethical debate. However, in our current review it is necessary to establish a position that will determine the actual way human fertilised embryos will be handled. Therefore, we will first summarise the views in the current laws, administrative guidelines, and regulations of the relevant medical associations,"[60]

thus displaying a disinterest in discursive moral argumentation and implicitly opting for a legalist-positivist stance.

From its summary of current Japanese laws and regulations relating to the status of human fertilised embryos, the report derives the position that although they are not subjects endowed with dignity and the right to full legal protection, they are still to be protected as the initial stage of an individual human life with the potential to evolve into human personal life. It is assumed that reification and careless treatment of these embryos would in some way have a detrimental effect on the protection of human dignity.[61] Thus, as a rule, the use of embryos in scientific research is not allowed. However, if such research promises "many and great benefits" (*tadai no onkei*), it can be permitted within strictly regulated limits that should ensure a certain standard of respectful behaviour not further specified in the report.

On the status of human cloned embryos, the report mentions the position that these have a different quality from human fertilised embryos and should consequently be accorded a lower moral status. Cloned embryos produced through nuclear cell transfer are the result of a completely artificial process non-existent in nature, and thus not part of the process of sexual reproduction. In addition, because cloning for reproduction is prohibited by law and penalised in Japan, they do not come into existence in the context of human reproduction. Thus, a human cloned embryo cannot be seen as the initial stage of an individual personal life, and its production, manipulation and destruction have no bearing on human dignity.[62] However, the majority opinion quoted in the report is that, although human cloned embryos are created in an "artificial process non-existent in the world of nature" and therefore differ somewhat from human fertilised embryos, this difference does not merit a lower moral or legal status, because they still have the inherent potential to grow into a human being with a personal life of its own.[63]

The status of "sprout of human/personal life" was formulated with the idea of according respectful behaviour and legal protection, while at the same time opening a door for research that promises help to severely afflicted human beings. The *Interim Report* proposes that the production and use of human fertilised embryos should be allowed for both further research in assisted reproduction and research on therapies for "severe diseases" (*nanbyô*).[64] It does not elaborate on criteria for "severe diseases" or provide a list of them, a fact that is reportedly criticised by some of the Commission's members.[65] The production of human embryos for the creation of stem cell lines is assumed to be unnecessary due to the existence of "excess embryos" from in-vitro fertilisation, and the report therefore recommends that it be prohibited.[66] The same principles apply to the production and use of cloned human embryos. In other words, the *Interim Report* recommends that production and manipulation of cloned human embryos should be prohibited as a rule, but could be allowed for medical research, if the results promise "extremely great benefits for the health and welfare of the people" (the inclusion of "welfare" (*fukushi*) in this formulation deserves special attention).[67] However, no consensus could be reached on the benefits of such research in the present situation. Thus, the report stops short at recommending that the existing ban on human cloning for purposes of medical research be lifted.

D. "Human Dignity" and "Ethics" in the *Interim Report*

It is the explicit aim of the *Interim Report* to formulate a basic way of thinking with respect to research on human embryos that ensures the protection of "human dignity" (*ningen no songen, hito no songen*). This is regarded as a fundamental ethical idea or principle (*rinen*) of Japanese society that has to be "firmly upheld" (*kenji*).[68] Consequently, all actions that directly infringe on human dignity are unanimously regarded as impermissible, and research on human embryos, including the use of cloning techniques, is considered viable insofar as it is in accordance with this fundamental ethical principle.

However, while human dignity is considered the highest ideal orienting moral and legal regulation, it is not regarded as a concept that implies any strict, or absolute, rules. The report itself states that "the content of 'human dignity' can hardly be clearly defined, but its most basic point is that human beings have their own substantial value as individuals (*sorezore kojin to shite koyû no kachi o yû shi*) and are respected (*sonchô sareru*)."[69] This definition implies a valorisation of individual human dignity, which is consequently regarded as something relative. This process of valorisation becomes evident when the report refers to requests for research that promote the "the value of life and health and the pursuit of happiness of the people" as equally "stemming from 'human dignity'",[70] and consequently balances these requests with the possible detrimental effects the use of human embryos may have on the protection of human dignity.

It should be noted that, in the opinion of the report, the "respect" (*sonchô*) of human embryos can be maintained, while at the same time producing and destroying them in the process of medical research. In principle, the way of thinking adopted in the report therefore opens the door to the instrumentalisation of human beings, a fact deplored by Shimazono Susumu in his opinion statement attached to the report.[71] Still, the report does not consider this to be the case, because it carefully distinguishes between human persons who are subjects of "human dignity," and human embryos as "sprouts of human/personal life." Thus, the specific value of human embryos in the opinion of the report does not require their absolute protection from manipulation and destruction in order for "respect" to be maintained, but the value of an individual human person may.

This again may be reading too much reflective consistency into the report, which displays a rather cavalier attitude towards close philosophical argumentation. The reason for this is the report's concept of "ethics" (*rinri*). "Ethics" to the authors clearly does not mean an elaborated system of principles or values. Instead, it refers to the mores and accepted behaviour within a given society. Thus, the report's concluding remarks (*musubi*) state:

> "The difficulty of contemporary bioethics as an 'ethics' lies in the fact that our 'ethics' have not followed up with the explosive progress of the life sciences. The problem is a deep one, as there is no 'ethics' people can share with self-assurance, when expert researchers still do not understand at what point a life becomes a 'human being,' an existence of its own to be addressed as

> 'someone,' or are unclear even about the standards for life and death."[72]

The general concept of ethics as a set of communally shared values and rules, implicit in this statement, informs the report's approach to conflicting ethical opinions. Instead of analysing and weighing arguments, it opts for a quantitative approach, the ideal solution being a consensus or at least a clear majority. It is thus only consistent with this approach that the decision on the final draft of the report was reached by a majority vote, a procedure that met with criticism from major Japanese newspapers.[73] A related feature of this attitude is the positivism or legalism adopted by the report's main authors. As we have seen above, the moral status of the embryo was mainly assessed by reviewing existing laws and guidelines, and the concluding remarks emphasise the role of laws and guidelines in shaping the bioethical mores and convictions of the future: "[...] until bioethics firmly take roots among the people, there will probably be no other way than to indicate it in the form of rules like 'guidelines' and 'laws'".[74] The *Fudamental Principles'* text follows the same line of thinking.

E. The Position of the *Interim Report* within the Japanese Debate

The positions taken by the report reflect to a large extent general notions about Japanese or East Asian attitudes to bioethical questions, as indicated by ascriptions like "pragmatic" (as opposed to "principled"), "community oriented" (as opposed to "individualistic"), or "consensual" (as opposed to "critical/conflictual"). Still, they should not be taken to represent "the" Japanese attitude towards these matters. First of all, members of the Expert Research Commission itself have voiced strong dissatisfaction with the report, especially with its lax attitude towards close argumentation.[75] Ida, Katsuki, and Washida each consider that the reasons given for allowing the production of human embryos for use and destruction in research to be insufficient.[76] Shimazono and Washida also argue that the report is inconsistent in its handling of the concept of dignity.[77] The above named have also formulated a minority opinion statement concerning the *Fundamental Principles.*

Apart from dissent by members of the Commission, some arguments and positions adopted by the report have already been criticised within the Japanese bioethical literature. Exploring the concept of the "sprout of human life," Yahata Hideyuki has argued against the objectivist approach that is employed by traditional Angloamerican bioethics (and the *Interim Report*) in assessing the human embryo's moral status.[78] To Yahata, the objectivist search for properties of an individual from which to infer dignity inevitably leads to the exclusion of ever more spheres of human existence from the protected sphere of the person; it is therefore not an adequate foundation for the protection of human dignity.[79] In contrast, Yahata opts for a deontological approach which would accept unilateral responsibility for human embryos as a responsibility for the future existence of autonomous beings.[80]

In a discussion of the ethics of cloning for reproductive purposes, Yamaguchi Okitomo warns against the identification of legal and ethical rules that informs much of the discussion on human cloning, again including the *Interim Report*. Yamaguchi gives a critical assessment of three main arguments brought forth in favour of a general ban on human reproduction by cloning, that is, 1) the argument based on the violation of personal integrity, 2) the argument based on insecurity, and 3) the argument based on commodification. Ad 1), Yamaguchi says cloning does not lead to a violation of personal integrity. This would only occur if the individual personality was determined by the DNA, a hypothesis which is untenable in the light of the different personalities to be found in uniovular twins.[81] Ad 2), the uncertainties currently accompanying reproduction through cloning cannot legitimise a general ban on the technique, but only a ban to be lifted once the current difficulties have been overcome.[82] Ad 3), as commodification of human gametes is already accepted, Yamaguchi claims that it would be inconsistent to prohibit human cloning for this reason.[83] He thus challenges the current Japanese legislation, which is one of the presuppositions of the *Interim Report.* At the same time, he insists that even if reproduction through cloning was legally permitted, it would not make it ethically acceptable, because ethical requirements can be more strict than legal rules.[84] He goes on to argue for an ethics of self-restriction with respect to the "unnatural desires" that lead to the demand for reproduction through cloning (including a reflective passage on the different meanings of "nature" (*shizen*) and "naturalness" (*shizensa*) in the ethical debate).[85]

Yamaguchi is not the only one to argue against the legal ban on human reproduction through cloning. As part of his sweeping critique of traditional bioethics, Okamoto Yûichirô denounces legal restrictions on cloning as "paternalistic" (*patânaristikku*), "discriminative" (*sabetsu-shugiteki*), and based on "male centrism" (*dansei chûshin shugi*).[86] In his opinion, cloning technology is a perfect match to the lifestyle of the present, and its prohibition goes against the principles of liberalism on which current society is founded:

> "It is entirely up to the individual, to choose what kind of child she wants to give birth to. The kind of technology she wants to use in giving birth, is probably equally her individual choice. [...] If this is the case, why would a completely unrelated stranger oppose the cloning of a human being?"[87]

Okamoto has a point when he challenges the validity of arguments against cloning based on technical difficulties or on the "unnaturalness" of the technique (the criticism does not apply to the points raised by Yamaguchi).[88,89] Still, in terms of validity, his argument is too sweeping to be convincing.

With his staunch individualistic brand of liberalism, Okamoto is rather on the margin of the Japanese debate. Still, his doubts about the legal ban on reproductive cloning are shared by others who are much more careful in arguing their cause. In 2001, Muramatsu Akira published one of the most systematic Japanese explorations of the problems related to advanced reproductive medicine, including human cloning for reproduction and research. In his book, titled "When does a human being become a human person?" (*Hito wa itsu hito ni naru no ka*), he strongly criticises the naturalistic reductionism that is amply present in the bioethical debate. He then describes the ways in which human beings make their appearance as persons in the world, and how they are grasped by others, analysing the potential for consciousness,[90] the various dimensions of corporeal appearance and activity,[91] and social interaction and recognition as three overlapping sources of the human experience of personality.[92] Returning to the issue of cloning, he argues, like Yamaguchi, that producing a cloned human being would not infringe on this individual's personality: "Even if the genetical information is identical, the human person is not identical. A human clone is not a replica or copy of a human being."[93] Although there may be issues of security or of social acceptability, these do not warrant a general ban on cloning, but should be addressed individually.[94]

Still, Muramatsu is far from being enthusiastic about reproduction through cloning, which he regards as a rather outlandish technique. The more important problem to him is cloning for medical research. "This is a much more severe question. The birth of a human clone would in any case be the birth of a human person, while the use of a human embryo is an instance of 'destruction.'"[95] Referring to his earlier discussion on the sources of our experience of personality, Muramatsu cites the following arguments against the production and use of human embryos in medical research: 1) A human embryo is a body in its entirety. If the manipulation of such a body becomes an accepted practice, this will endanger our understanding of the human body as constituting a human person.[96] 2) The recognition of a human person does not depend solely on cognition. "We accept someone as a human person, before we know him as a person."[97] Conversely, the constitution of a person depends to a large extent on being accepted - spoken to, addressed - as "someone," before displaying distinctly personal behaviour.[98] There may be no objective characteristic which forces us to accept human embryos as persons. However, in order for persons to come into existence, it is important that they be recognised as such, before they can prove themselves. Muramatsu admits that 1) and 2) do not force us to accept, and protect, human embryos as persons. But, he adds 3), as it is not irrational to treat human embryos as persons, and as there are people who regard embryos in this manner, it would be good to share in their attitude, as a matter of recognition and respect for their personality.[99]

In comparison to the literature quoted above, the *Interim Report* misses out on important methodological and ethical arguments that were raised in the Japanese debate prior to its formulation. In this respect, it should not be seen as representing the results of a Japanese discourse on cloning. Still, there is one feature common to the report and most of the Japanese literature in the field: A comprehensive criminalisation of all forms of human cloning is hardly ever to be found. While Okamoto, in his all-out enthusiasm for reproductive cloning, is rather an exception, almost all authors seem to accept the possibility that cloning in one way or another could conform with the idea of human dignity. Although there are authors who argue that all forms of cloning are in some way immoral, a fundamentalist stance that would exclude all manipulation of human gametes for cloning as a crime will hardly be accepted in Japan. It is generally dismissed as the product of a monotheistic, creationist religion that fails to convince the Japanese public as well as Japanese bioethicists. Still, many oppose the unconcerned attitude of the administration towards the ethical issued involved in the manipulation of human gametes, embryos and fetusses, and wish to make sure that human dignity is protected in all stages of research and application of advanced reproductive medicine.

5. Conclusion

Our survey of the Japanese discussion on human cloning has revealed that, not surprisingly, there is no unified "Japanese opinion" on the subject. Public opinion in Japan, as in the EU, is generally favourable towards biotechnology and most interested in the medical benefits promised by reproductive medicine. The hope of new cures for "difficult" diseases overrides all doubts concerning cloning technology, be they ethical or pragmatic. The Japanese administration takes a similar, but not identical position. Like the public, it is not concerned about the niceties of ethical reasoning; it endorses biotechnology for its expected benefits. However, the government is well informed about the international debate on human cloning. This is probably the reason for its strict stance on cloning for reproduction, which while not rejected by the public, is penalised by law. Attempts at producing a "cloned baby" would draw Japan into the international limelight, and lead to criticism from abroad.

Within Japanese society, scientists are most positive towards biotechnology in general, and research into human cloning in particular. In contrast, experts from fields other than science are most critical in this matter. Thus, a greater degree of knowledge about human cloning seems to lead to greater acceptance among researchers in the field, while it has the opposite effect on those not directly involved in scientific research. Still, it should be noted that the arguments of experts in bioethics have not found their way into public opinion, nor have they had an impact on policy making so far. While opinions on cloning vary to a considerable degree, the spectre is not all-encompassing. Arguments in favour of a comprehensive criminalisation of all forms of human cloning are virtually absent in Japan. Even critics of reproductive cloning or research involving the destruction of human embryos consider these acts to be unethical rather than criminal. As far as our materials suggest, this is the only point on which the situation in Japan differs from that in Europe.

The variations of opinion on a large scale conform to the differences of involvement with cloning and its regulation. As such forms of involvement - as a lay person and prospective medical patient, as a scientific researcher, as an administrative regulator, as a professional ethicist, etc. - are to a large extent similar at least in highly industrialised societies, the parallels that are to be detected between Europe and Japan come as no surprise. On the other hand, our materials did not suggest a decisive influence of any national tradition on attitudes towards human cloning, unless the complete irrelevance of religious arguments is seen as a specifically Japanese trait.

Notes

1. The following statistics of Japanese attitudes toward biotechnology and the international comparison should be read with the awareness that they are based on different opinion polls which did not use the same wording and the same sampling procedures. Therefore the results are only partly comparable, but they should suffice at least to show trends in the dynamics of public and/or experts' attitudes.
2. Ng, 2000, Table 5.
3. cf. Macer, 2002, 723.
4. The sample consisted of 2,700 respondents from all over Japan, among them academics, professionals, researchers, and administrators in various numbers. The survey method was by mail and the response rate 78.2%; Sôrifu Kôhôshitsu, 1998a).
5. Sôrifu Kôhôshitsu, 1998a, 4 -5.
6. The translation of the questions are taken from NISTEP, 2001, 207; to this question, too, multiple answers were possible.
7. Ibid., 19.
8. Ibid., 24.
9. Ibid., 26, 28.
10. Eurobarometer, 2003, 14.
11. Ibid., 15.
12. Macer, 1998, Summary.
13. Surveys in Europe and in Japan have shown that knowledge concerning biology and biotechnology is unevenly distributed among the respondents according to the problem involved but also to nationality - which seems to reflect differences in science education. In no country surveyed were the respondents able to correctly answer more than 6.35 out of nine questions concerning biology and biotechnology on average; see, e.g., Eurobarometer, 2003, 20-22; Macer, 1998, chapter "Background Knowledge and Understanding"
14. This is one of the points which were stressed at two round tables hosted by the German Federal Ministry of Education and Research in 1995 and 1996; for the results see VDI-Technologiezentrum, 1996.
15. Thus the title of a chapter in "Roads to Reform," *Asia-Pacific Perspectives*, 1.1, (2003), 12.
16. These data were taken from the Public Opinion Polls on the Future of

Science and Technology - as the official English title of these polls reads - conducted in 1981, 1987, 1990, and 1995. As a time series, these polls offer invaluable material for the study of the dynamics of public opinion concerning science and technology in Japan, comparable to the Eurobarometer studies conducted by the European Union in its member states.

17. see Table 4/1, 4/2, and 5.
18. see several contributions to Jaufmann and Kistler, 1991.
19. NISTEP, 1995, Fig. 7-2-1.
20. cf. NISTEP, 2001, 3, 4.
21. Kagaku Gijutsu Seisaku Kenkyûjo, 2002; see also Kagaku Gijutsu Seisaku Kenkyûjo, 1992.
22. Macer, 1998, Table 5.
23. cf. Morioka, in this volume.
24. Sôgô Kagaku Gijutsu Kaigi Seimei Rinri Senmon Chôsakai, 2003.
25. see section 4 of this paper.
26. GSK, 1998.
27. CST 1998.
28. e.g., STA, 1998, 5.
29. Kagaku Gijutsu Seisaku Kenkyûjo, 1999.
30. Ibid., p. 149.
31. Ibid., 1999, 34-42.
32. Itô, o 1999, 121-132; Kiba, 1999, 132-147.
33. cf. Morioka, in this volume.
34. Saegusa, 2000, 581.
35. see his paper in this volume.
36. *Fundamental Principles,* 1.
37. Ibid., 4.
38. Ibid., 5.
39. Ibid., 6; see also page 12.
40. Ibid., 6.
41. Ibid., 7.
42. Ibid., 7, 13.
43. Ibid., 13-14.
44. Ibid., 15.
45. Ibid., 14, 16-19.
46. Ibid., 4-5.
47. Ibid., 12-13.
48. *Asahi shinbun* 2004, *Mainichi shinbun* 2004, *Sankei shinbun* 2004, *Nihon keizai shinbun* 2004.

49. *Interim Report*, 1.
50. The classification system of "specified embryos" is taken from the *Law on Human Cloning,* Art. 2; see also Morioka.
51. *Interim Report*, 31.
52. Ibid., 11-15.
53. Ibid., 11.
54. Ibid., 46.
55. *Kagaku Gijutsu Kaigi Seimei Rinri Iinkai/Hito Hai Kenkyû Sho Iinkai*, 2000, 8.
56. *Interim Report*, 12.
57. Ibid., 15-18.
58. Ibid., 12-13.
59. Ibid., 12-14.
60. Ibid., 14.
61. Ibid., 17.
62. Ibid., 31.
63. Ibid., 30, 31.
64. Ibid., 20.
65. Ibid., 21.
66. Ibid., 23.
67. Ibid., 32.
68. Ibid., 1, 17-18; the term *songen* occurs 35 times in the main text of the report and 15 times in the individual opinion statements.
69. Ibid., 17.
70. Ibid., 17
71. Ibid., 51.
72. Ibid., 42.
73. *Mainichi shinbun* 2004, *Nihon keizai shinbun* 2004, *Sankei shinbun* 2004.
74. *Interim Report,* 42.
75. see above, section 1, and the statements of Ida Ryûichi, Katsuki Motoya, Shimazono Susumo, and Washida Kiyokazu, *Interim Report*, 46-47, 49-50, 51-52, 60.
76. *Interim Report,* 46, 49, 60.
77. *Ibid.*,51, 60.
78. Yahata, 2002, 117-120.
79. Ibid., 120.
80. Ibid., 2002, 126-127.
81. Yamaguchi, 2002, 44.
82. Ibid., 44.

83. Ibid., 44-45.
84. Ibid., 45.
85. Ibid., 51-53.
86. Okamoto, 2002, 128, 133-135.
87. Ibid., 128.
88. Ibid., 125-128.
89. Ibid., 129-130.
90. Muramatsu, 2001, 123-152.
91. Ibid., 153-202.
92. Ibid., 203-228.
93. Ibid., 230.
94. Ibid., 231.
95. Ibid., 232.
96. Ibid., 233.
97. Ibid., 234.
98. Ibid., 234-235.
99. Ibid., 235.

References

Asahi shinbun, "Kurôn hai - kenkyû wa mitomete ii ga [Cloned embryos it's ok to allow research, but ...]." (Tôkyô, 6.25 2004).

CST (= Kagaku Gijutsu Shingikai Seimei Rinri Iinkai Kurôn Shoiinkai). 1998 *Kurôn gijutsu ni kansuru kihonteki kangaekata ni tsuite (chûkan hôkoku): Heisei 10/06/15* [Fundamental principles concerning cloning technology (Interim report)]. Tôkyô: CST, 1998. (Also online: http://www.mext.go.jp/b_menu/houdou/10/06/980618h.htm.)

Eurobarometer, *Europeans and Biotechnology in 2002.* 2nd edition March 21st 2003.

GSK (= Gakujutsu Shingikai Tokutei Kenkyû Ryôiki Suishin Bunkakai Baiosaiensu Bukai). 1998 *Daigaku ni okeru kurôn kenkyû ni tsuite (hôkoku); Heisei 10/07/03* [Concerning university research on cloning, 1998/07/03]. Tôkyô: GSK, 1998 (Also online: http://www.mext.go.jp/b_menu/shingi/12/gakujutu/toushin/980702.htm#1-4).

Guidelines for Derivation and Utilization of Human Embryonic Stem Cells/Hito ES saibô no jûritsu oyobi shiyô ni kansuru shishin. (Also online: http:// www3.kmu.ac.jp/ legalmed/ethics/wadai2.html).

Hito ni kansuru kurôn gijutsu nado no kisei ni kansuru hôritsu. (Also online: http://www3.kmu.ac.jp/legalmed/ethics/wadai3.html).
Itô, Kosuke. "Dôi keisei no tame ni hitsuyô na tôjisha [Parties involved in consensus development]." In *Sentan kagaku gijutsu to hôteki kisei (Seimei kagaku gijutsu no kisei o chûshin ni)* [Legal regulations on advanced science and technology - Regulations on life science)], NISTEP (ed.) (= Kagaku Gijutsu Seisaku Kenkyûjo). Tôkyô: NISTEP (=Policy study; 1), 1999, 121-132.

Jaufmann, Dieter and Ernst Kistler. *Einstellungen zum technischen Fortschritt: Technikakzeptanz im nationalen und internationalen Vergleich.* Frankfurt/M.: Campus, 1991.

Kagaku Gijutsu Kaigi Seimei Rinri Iinkai / Hito Hai Kenkyû Sho Iinkai. *Hito hai seikan saibô o chûshin to shita hito hai kenkyû ni kansuru kihon tekikangaekata* [Fundamental principles concerning human embryo research: Focusing on embryonic stem cells]. Printed matter, dated 6 March 2000.

Kagaku Gijutsu Seisaku Kenkyûjo. *Nichi-Bei-Ô ni okeru kagaku gijutsu ni taisuru shakai ishiki ni kansuru hikaku chôsa* [A comparative survey of social attitudes toward science and technology in Japan, the United States, and Europe]. Tôkyô: Kagaku Gijutsu Seisaku Kenkyûjo, 1992.

Kagaku Gijutsu Seisaku Kenkyûjo. *Sentan kagaku gijutsu to hôteki kisei (Seimei kagaku gijutsu no kisei o chûshin ni)* [Legal regulations on advanced science and technology - Regulations on life science]. Tôkyô: Kagaku Gijutsu Seisaku Kenkyûjo (= Policy study; 1), 1999.

Kagaku Gijutsu Seisaku Kenkyûjo. *Seimei to hô: Kurôn kenkyû wa doko made jiyû no ka* [Live and law: How free is research on clones]. Tôkyô: Ôkurashô insatsukyoku, 2000.

Kagaku Gijutsu Seisaku Kenkyûjo. *Kagaku gijutsu ni kansuru ishiki chôsa* [The 2001 Survey of Public Attitudes Toward and Understanding of Science and Technology in Japan], (NISTEP Report; 72). Tôkyô: Kagaku Gijutsu Seisaku Kenkyûjo, 2002. (Also online: http://www.nistep.go.jp/ achiev/abs/jpn/rep072j/rep072aj.html; downloaded June 18, 2003, 9.45).

Kiba, Takao. "Dôi keisei shuhô [Approach to consensus development]." In *Sentan kagaku gijutsu to hôteki kisei (Seimei kagaku gijutsu no kisei o chûshin ni)* [Legal regulations on the advanced science and technology - Regulations on life science], NISTEP (ed.) (= Kagaku Gijutsu Seisaku Kenkyûjo). Tôkyô: NISTEP (=Policy study; 1), 1999, 132-147.

Macer, Darryl R.J. *Attitudes to Biotechnology in New Zealand and Japan in 1997 (Eurobarometer Survey).* Tsukuba: Eubios Ethics Institute, 1998. (Also online: http://www.biol.tsukuba.ac.jpzj.html; downloaded Feb. 25, 2004, 17.03.).

Macer, Darryl R.J. "International Aspects: National profiles, Japan." In *Encyclopedia of Ethical, Legal, and Policy Issues in Biotechnology,* Murray, T.J. and M.J. Mehlman (eds.). New York: Wiley, 2002, 722-731.

Mainichi shinbun. "Shasetsu: Hito kurôn hai - hôkoku isogazu wakugumi no seibi o [Editorial: Human cloned embryos - don't rush with the report and construct a framework]." (Tôkyô ed., 6.25 2004).

Muramatsu, Akira. *Hito wa itsu hito ni naru no ka: Seimei rinri kara jinkaku e* [When does a human being become a human person?]. Tôkyô: Nihonhyôronsha, 2001.

Ng, Mary Ann Chen, Chika Takeda, Tomoyuki Watanabe and Darryl Macer. "Attitudes of the Public and Scientists to Biotechnology in Japan at the Start of 2000." *Eubios Journal of Asian and International Bioethics,* 10 (2000), 106-113.
(Also online: http://www.boil.tsukuba.ac.jp/~macer/ bioJ2000.html; downloaded May 27, 2003, 10.39).

Nihon keizai shinbun. "Shasetsu 2: Kurônhai, isogaba maware [Editorial 2: Cloned Embryos - if you're in a hurry, take a detour]." (Tôkyô ed., 6.25 2004).

NISTEP. *Science and Technology Indicators: 1994.* Tôkyô: NISTEP, 1995.

NISTEP. *Science and Technology Indicators: 2000*. Tôkyô: NISTEP, 2001. (Also online: http://www.nistep.go.jp/achiev/ftx/eng/rep066e/idx066e.html; downloaded Feb. 26, 2004, 15.00).

Okamoto, Yûichirô. *Igi ari!: Seimei kankyô rinri gaku* [Here's a different view! - Bioethics and environmental ethics]. Kyôto: Nakanishiya, 2002. "Roads to Reform," *Asia-Pacific Perspectives*, 1.1, (2003), 6-25.

Saegusa, Asako. "Japan's Cloning Law Fails." *Nature Biotechnology*, 18, 6 (2000), 581.

Sankei shinbun. "Shutchô: Hito kurôn hai - bosô fusegu hadome ga zentei da [Opinion: Human cloned embryos - a working break against abu precondition]" (6.25 2004).

Sôgô Kagaku Gijutsu Kaigi. *Hitohai no toriatsukai ni kansuru kihonteki kangaekata (an)*. [Fundamental principles for the handling of human embryos (draft)]. Printed matter, dated 23 July 2004. Tôkyô: Sôgô kagaku gijutsu kaigi.

Sôgô Kagaku Gijutsu Kaigi Seimei Rinri Senmon Chôsakai. *Hito hai no toriatsukai ni kansuru kihonteki kangaekata: chûkan hôkoku* [Fundamental principles concerning the handling of human embryos: Interim report]. Printed matter, dated 26 December 2003. Tôkyô: Sôgô Kagaku Gijutsu Kaigi Seimei Rinri Senmon Chôsakai.

Sôrifu Kôhôshitsu. *Kagaku gijutsu ni kansuru yoron chōsa* [Opinion poll on science and technology]. Tôkyô: Sôrifu Kôhôshitsu, 1982.

Sôrifu Kôhôshitsu. *Kagaku gijutsu to shakai ni kansuru yoron chōsa* [Opinion poll on science and technology in society]. Tôkyô: Sôrifu Kôhôshitsu, 1987, 1990, and 1995.

Sôrifu Kôhôshitsu. *Kurôn ni kansuru yûshokusha ankêto chôsa* [Opinion poll among experts on cloning]. Tôkyô: Sôrifu Kôhôshitsu, 1998a.
Sôrifu Kôhôshitsu. *Shôrai no kagaku gijutsu ni kansuru yoron chôsa* [Opinion poll on the future of science and technology]. Tôkyô: Sôrifu Kôhôshitsu, 1998b.

STA. "Interim Report by CST Subcommittee on Cloning: Distinguishes Between Human and Animal Cloning," *STA Today* 10, 7 (July 1998), 5.

VDI-Technologiezentrum. *Statusbericht zur Akzeptanz der Bio- und Gentechnologie in der deutschen Öffentlichkeit.* Düsseldorf: VDI Technologiezentrum (= Technologie-Monitoring; 15), 1998b.

Yahata, Hideyuki. "Hito no 'seimei no hōga' wa 'songen' o motsu ka [Does the 'sprout of human/personal life' possess 'dignity'?]." In *Hito no seimei to ningen no songen* [Human/personal life and the dignity of human beings], Takao Takahashi (ed.). Kyôto: Nakanishiya, 2002, 105-132.

Yamaguchi, Okitomo. "Kurôn gijutsu to seishoku iryô no zehi o tou [The pros and cons of cloning technology and reproductive medicine]." In *Sei to shi no rinrigaku* [The ethics of life and death], Shun'ichirô Shinohara and Takahiko Hatae (eds.). Kyôto: Nakanishiya, 2002, 33-57.

Cloning Issues in China

Qiu Renzong

Abstract: The paper elaborates the various factors which influence the debate on cloning issues in China: cultural dimensions, scientism, market force, and legal/regulatory factors, and describes the intellectual basis for the cloning debate and the development of bioethics in mainland China. It then sets out the arguments for and against human reproductive cloning and the arguments for human therapeutic cloning. It concludes that the cloning debate needs cross-cultural dialogue and international collaboration.

Key Words: Cloning culture, Confucianism, Taoism, Person, Embryo, Fetus, Scientism, Human reproductive cloning, Human therapeutic cloning.

1. Introduction

Since news of Dolly the sheep was first published in *Nature* in 1997, human cloning has been the topic of ongoing debate in China,[1] both in scientific circles and among the general public. Hundreds of articles on the subject have been published in academic journals, popular magazines, and newspapers, and it has been featured in a considerable number of TV programmes. Decision makers in science and medicine began to address the ethical, legal and social issues involved in human cloning, and for the first time in its history the Minister of Health convened a meeting of experts to discuss the implications of nucleus transfer technology. The Chinese government has joined France and German in their efforts to persuade the United Nations to endorse a resolution prohibiting human reproductive cloning, despite the fact that some Chinese scientists and philosophers are in favour of it. Since then bioethics as a discipline has been very much in the limelight.

The cloning debate in China is taking place at a time when the country is in transition from a monolithic society to a more or less pluralistic civil society. The factors influencing the debate are complicated. China has had a characteristic culture for thousands of years, which is built on three teachings: Confucianism, Taoism and Buddhism. Of them, Confucianism is the dominant one and is firmly rooted in many Chinese minds and in many social conventions. In the decades after 1949 the chief ideology was Marxism-Leninism in its Chinese version. Since the new policy in 1987, China has been gradually integrated into the international community, and the influence of Western culture has become increasingly pervasive. Liberalism, feminism and communitarianism became new conceptual resources. With the government's commitment to market economy and the rule by law, economic and legal factors have also become very important during the last two decades.

2. Cultural Dimensions of Cloning Issues

Chinese Cosmology

For Confucians and Taoists the universe and all of its inhabitants are composed of *qi*, a physico-psychological entity with the capacity for creation, change and transformation, which has two basic forms, *yin* and *yang* (or *yin-yang*). The generation, change and transformation of all things are produced by their interaction. The universe and its inhabitants are participants in an organic whole that interacts in a self-generating cosmos. There is no ultimate cause, no creator, no external will, and no judge. The universe has always existed, there was never a time when it did not exist, and therefore the question of a creative act simply could not logically arise.

According to this unique cosmology, all forms of life are composed of *qi* and its two basic forms, *yin* and *yang*. As classical writers claimed, "Life is the condensation of *qi*, and death is the dispersion of *qi*." (*Zhuang Zi* 22) "*Yin-Yang* is the *tao* of Heaven and Earth, the law of all things, the cause of changes and the basis of life and death." (*Interior Classic of Huangdi: Questions and Answers about Living Matter*) "*Yin-yang* causes the beginning and end of all things, and is the root of life and death. If you run counter against them, you will suffer catastrophe, and if you follow them, you will avoid disease." (*Interior Classic of Huangdi: Questions and Answers about Living Matter*).

In the evolution of *qi*, different forms of *qi* assembling produce things in different kinds, and different degrees of *qi* refinement form things in different degree. The human being is composed of the most refined form of *yin-yang*.

Peitai - Embryo

Embryo in Chinese is *peitai*. In ancient books, *pei* and *tai* mean "beginning," "condensation of blood," "anything not formed," "in a woman's womb," "pregnant for one month" or "pregnant for three months."[2] Since the study and dissection of the aborted foetus or embryo as well as the dissection of the adult corpse were prohibited until modern-day China, the ancient Chinese had only an ambiguous conception of the embryo.

Human Embryo

The human embryo is a product of *ying* and *yang* interaction unlike the clone.[3] There are many transformations in the process from embryo to newborn. The transformation from embryo/foetus to new-born child is something like the transformation from bud to leaf/flower or from seed to plant. There is a continuum between bud and flower or from seed to plant, but bud and flower or seed and plant are different, both in kind and in degree. In the same vein, the human embryo will develop or transform into a human person, but it is not yet a human person, though it has the inherent teleology or self-actualisation to develop into a human person.

Concept of Person: Substantialist vs. Relational

Western classical philosophers think of being and person in terms of substance: to exist in itself, and not to be a part of any other being, is what causes a being to be a unity, identity and whole. A substantialist view of person may identify a person with the human genome, with a soul, or with rationality, as in Descartes' and Kant's understanding. Against the rationalist view of person are a number of intractable counter-examples: What about the marginal human beings, such as neonates, infants, the senile, and seriously retarded individuals?

The Confucian concept of personhood is a relational one. For Confucians a person is a psycho-somatic-social unity which has body (*ti*) or shape (*xing*) and a psyche (*shen*), and has rational (*li*), emotional (*qing*) and social-relationship (*qun*) capacities. All these elements derive from *qi* and are related to each other. "Being" is always in the process of "becoming." When a human being is born, he or she is imperfect both physically and morally. He or she is born as a human being only biologically, but not morally. A moral human, i.e., a person *per se*, has to develop in a process of "becoming" called person-making.

Many classical Confucians argued that human beings are distinguishable from animals because of their relational/social capacity. "Men are not as strong as oxen, nor do they run as fast as horses, yet how is it that oxen and horses are mastered by men? It is because men are capable of social organisation and animals are not." (*Xun Zi* 9) A person is never seen as an isolated individual but is always conceived of as part of a network of relations. The self-cultivation of a biological human being to become a moral person is a process that is carried out both in and through the social context and for the purpose of fulfilling social responsibility rather than self-actualisation per se. Relations constitute and complete personhood.[4]

Concept of Ren

One of the implications of this relational concept of personhood is that a person should not only consider him- or herself. The Confucian paramount concept of *ren* is love,[5] compassion and care, a moral capacity to feel for others. We love others and are compassionate towards others not because we benefit from doing so, but because we have a heart which cannot bear the sufferings of others, as Mencius wrote:

> When men suddenly see a child about to fall into a well, they all have a feeling of alarm and distress, not to gain the friendship of the child's parents, nor to seek the praise of their neighbours and friends, nor because they dislike the bad reputation of not rescuing the child. (*Meng Zi* 2a:6)

Ren is an extension of the natural compassion that everyone feels in view of the hardship and misfortune of others. *Ren* can be and has to be cultivated and developed in interpersonal and social relationships, first and foremost in the family: love, compassion and care for parents (*xiao*, filial piety) and siblings (*ti*, fraternity) are the root of *ren*. (Confucius, *Lun Yu* 1.2). The affection that is assumed to grow naturally in the family should become the model for the way we treat others and society at large.

For Confucianism, medicine as well as science and genetics should be the art of *ren*. Physicians, scientists, geneticists, and other professionals should cultivate *ren* and complete the process of person-making. The golden rules of *ren* are: (1) Don't do to others as you would not have them do to you; (2) Establish yourself and help others to establish themselves, develop yourself and help others to develop. (*Lun Yu* 5.12, 15.24, 6.28). Scientists and physicians should do good (*gong*) to others (patients, clients etc.) and benefit as many people as possible (*bo shi yu min, er neng ji zhong*, *Lun Yu* 6.2);[6] they should not do harm to the health and life of others, as Mencius said "To do no harm is the art of *ren*" (*Meng Zi* 1a:7), and they should not make money by taking advantage of patients' vulnerability.[7]

Xun Zi argued that *ren* entails respect (*jing*) for human dignity, because humans "are the most precious beings on earth" (*zui wei tianxia gui*, *Xun Zi* 9). They should not be harmed, enslaved, exploited, oppressed, innocently killed, stigmatised, discriminated, battered, manipulated, treated as mere means, sold and bought, and manufactured etc.

In Confucianism there are two fundamental requirements for professionals. The one is that professionals should not do harm. Confucius says: "Don't do to others as you would not have them do to you." Mencius says: "To do no harm is the art of *ren*." Do you wish to harm yourself? You don't. So you ought not to do harm to others. The other requirement is respect for human dignity. Confucius says: "It is possible to deprive a whole army of its commander. But it is not possible to deprive even a common person of his will." (*Lun Yu* 9.26). So from the Confucian perspective, what is wrong with Nazi doctors and Japanese Unit 731? The first wrong is that they have done tremendous harm to victims, the second is they have violated human dignity.

3. Scientism and Market Force

The debate on cloning issues is also influenced by historical, social, and economic factors as well as by traditional culture. Since 1919 Chinese intellectuals have believed that Mr. Sai (science) and Mr. De (democracy) can save China from exploitation and humiliation by the powers. Thus, scientism is very popular: science and technology are thought to be capable of solving all issues facing China. During my lectures to scientists the points they often raised were "Science is not a double-edged sword, but a single edged one," "All unsolved issues are caused by inadequate investment in science and technology and will be solved by more investment," "We know better than anyone else how to protect human subjects," or "Ethical values will change with the advances of science and technology and accommodate themselves to these advances" etc. Accordingly, many scientists are reluctant for their research to be regulated.

Following the shift from planned to market economy, market fetishism has been prevalent in China. A couple of years ago a prominent scientist said in an interview with a journalist of *Beijing Youth*, one of the most popular newspaper in Beijing: "Genes are money." Since the proposition "Science and technology is the first productive force" was put forward by Deng Xiaoping, cooperation or integration between research institutes and business enterprises is becoming increasingly common. For example, some human genome research centers, such as Beijing Human Genome Research Centre, Shanghai Human Genome Research Centre, and Genome Research Centre of Chinese Academy of Science, are all integrated with a company which may produce something for profit on the basis of genetic research. However, the possible conflict of interest has largely been ignored or underestimated.

4. Legal and Regulatory Factors

Legal and regulatory factors also play an important role in the debate on cloning issues:

In Article 9, Chapter II Citizens (Natural Persons), *General Principles of the Civil Law*, it is stipulated that "A citizen shall have civil rights from birth to death."
In Article 101, Section 4 Personal Rights, *General Principles of the Civil Law*, it is stipulated that "The citizen's personal dignity shall be protected by law."
In Article 28, *Law of Succession*, it is stipulated that "At the time of the partitioning of the estate, reservation shall be made for the share of an unborn child."
And in Article 302, *Criminal Law*, it is stipulated that "Whoever steals from or insults a corpse shall be sentenced to a fixed-term imprisonment of up to three years."

The implications of these articles are that a natural person comes into being at the moment of birth. A human embryo or foetus is not a person. Neither is a corpse. However, there is some transitivity: both the human embryo or foetus and the corpse can be seen as having a certain interest. A person enjoys full respect of his or her personal dignity, and a "pre-person" (embryo or foetus) or "post-person" (corpse) enjoys due (not full, but modest) respect.

Since 1998 both Parliament (National People's Congress) and Government (State Council) have promulgated some important laws and regulations. The *Law on Practising Doctors* enacted by the National People's Congress and enforced on May 1, 1999 contains a chapter on legal accountabilities. Specifically, Article 8 prohibits experimental clinical treatment without the consent of the patient or a member of his or her family, and Article 9 prohibits the disclosure of a patient's privacy.

In 1998 the *Interim Regulations for Ethical Review of Biomedical Research Involving Human Subjects* were promulgated by the Ministry of Health. In Chapter 1 Article 1 it is stated that "This regulation is for safeguarding human dignity, protecting human life and health, and ensuring adherence to basic ethical principles." Other chapters contain stipulations on: informed consent (Chapter 3), responsibilities of the investigators (Chapter 4), rights of human subjects (Chapter 5), and IRB (Chapter 6). In Chapter 8 Article 25 the following practices are prohibited:

- All scientific experiments related to human asexual reproduction;
- All research using human embryos and aborted foetuses;
- Import and export of aborted foetuses and organs,
- Sale or/and purchase of human cells, tissues and organs.

The *Interim Measures for the Administration of Human Genetic Resources* were promulgated jointly by the Ministry of Science and Technology and the Ministry of Health and enforced on June 10, 1998. In Article 13 (6) it is stipulated that "Applications will not be approved where there is no evidence to confirm that informed consent has been obtained from the donor of the human genetic sample and from a member of his/her family."

On September 1, 1999 Zheng Xiaoyu, Director of the State Food and Drugs Administration (SFDA) promulgated the *Clinical Drug Trial Guidelines*. The following articles set out some of the most important guidelines :

> *Article 4*: All research involving human subjects must comply with the ethical principles elaborated in the Helsinki Declaration and in CIOMS's (Council of International Organisation of Medical Sciences / World Health Organisation) International Ethical Guidelines on Biomedical Research Involving Human Subjects, i.e. justice, respect, maximum benefits to human subjects, and avoidance of harm as far as possible.
> *Article 8*: In the process of clinical drug trials the individual rights and interests of human subjects must be safeguarded, and the research must be scientific and reliable.

5. Bioethics in Mainland China: Intellectual Basis for Cloning Debate

The cloning debate in mainland China is based on the intellectual infrastructure which has been established since 1979.

Although there is a long history of traditional medical ethics, bioethics began in modern China with a series of events during the 1970s and 1980s. (The following paragraph relies heavily upon Zhai 2004.)

In December 1979 a Conference on the Philosophy of Medicine was held in Guangzhou at which a major report focused on the ethical issues raised by advanced biomedical technologies, such as, e.g., life-sustaining technology, assisted reproductive technology and organ transplant technology. In 1980 the journal *Medicine and Philosophy* launched, and its first issue contained an article on brain death and euthanasia. In 1986 a workshop on bioethics was held for young teaching staff in medical schools in Nanjing City organised by Southeast University and the Railway School of Medicine. Many of the participants later became the backbone of bioethics in China. In 1987 the book *Bioethics* was published.[8] The volume was extremely popular in China, and 50,000 copies were sold in a very short space of time.

Among the most important events of the 1980s we have to mention two conferences in 1988 and two legal cases. In July 1988 the first National Conference on Social, Ethical and Legal Issues in Euthanasia was held in Shanghai. The Conference concluded with a statement on the right of the terminally ill to choose how they wished to die; all except two participants had endorsed the statement. The other conference was the National Conference on Social, Ethical and Legal Issues in Human Artificial Reproduction which was held in Yueyang, Hunan Province. This Conference ended with a policy recommendation to the Ministry of Health and the State Commission on Family Planning on the regulation of artificial insemination and sperm banks. In 1986 and 1987 two legal cases on euthanasia and artificial insemination by donor respectively were publicised in the mass media and sparked off a debate among professionals and the general public.

Case 1: A 59-year old woman who had suffered from cirrhosis for many years was admitted to a hospital in the city of Han Zhong, Shaanxi Province, on June 23, 1986. Her diagnosis was: (1) cirrhosis and ascites, (2) coma and liver-kidney syndrome, and (3) second and third degree bedsores. By June 27, the patient's condition had worsened. The patient's son and youngest daughter asked Dr. P. to help them to let her die. At first he refused but was then persuaded to prescribe a 100 mg dose of a painkiller which the Chinese Ministry of Health permits only in doses of 25-50 mgs. However, when the head nurse refused to administer the injection, the doctor requested a student nurse to do so. Not daring to disobey his order, and against her own better judgement, she injected only half the dose. At midnight the patient was still alive and her two children asked the doctor on duty, Dr. L., to inject the drug according to the directions Dr. P. had left. Dr. L. agreed and the nurse on duty injected it. The patient died at 5 a.m. on June 29. The patient's other two daughters then sued the two doctors, claiming that they had murdered their mother. The two doctors, the son, and the youngest daughter were arrested, taken into custody, and released pending trial. In 1991 the middle court of Han Zhong City ruled that the doctors were guilty of hastening the patient's death, but the guilt was insignificant and they were treated as not guilty. The case triggered widespread discussion of euthanasia among physicians, nurses, medical scientists, philosophers, ethicists, and lawyers in newspapers, radio, and television.[9]

Case 2: In five years of marriage, a young couple had not been able to bear a child so they went to a hospital in Shanghai City for help. The doctor diagnosed the problem as the husband's inability to ejaculate. The couple knew that this doctor performed artificial insemination by donor (AID) and asked him to try it with them. They also decided not to let the husband's family of the husband know. The doctor agreed to use the sperm of a healthy donor with the same blood type as the husband. He did not charge them and did not ask them to complete an AID application form. The insemination was successful and the wife gave birth to a baby boy in 1987. But the husband's mother and other members of his family came to feel that the child did not resemble his father and suspected that the wife had been involved in affairs with other men. The husband's family asked the wife to take the baby and leave the family because the boy did not come from their family line. The wife left the family and sued her husband. The husband told the judge he did not know the truth and sought a divorce in court, but this was denied because under Chinese law a husband cannot divorce his wife until one year after childbirth. The court ruled that the husband must pay the wife per month for support. One year after the child was born, the court awarded the divorce. The child lives with his mother.[10]

Since 1988 almost each year there has been a national conference on bioethics in addition to local conferences. Every two years a national conference has been organised by the Chinese Society for Medical Ethics and another by the Committee on the Philosophy of Medicine affiliated with the Chinese Society for Philosophy of Nature, Science and Technology. These national conferences have covered almost all topics relating to bioethics. Academic exchanges between the Mainland, Hong Kong, Taiwan and Macao have become increasingly frequent. Apart from these, there have also been bilateral or international conferences between China and other countries.

After the publicity of Dolly the sheep, bioethics began to be institutionalised on the Mainland and it became a focus not only of academics but also of the government and the legislature, the public and the mass media. The regulations or laws which have been promulgated since 1998 include:

MOH: Interim Guidance on the Ethical Review of Biomedical Research Involving Human Subject (1998)
Ministry of Science and Technology (MOST) and MOH: Interim Measures for the Administration of Human Genetic Resources (1998)
National People's Congress (NPA): Law on Practising Doctors (1999)
SFDA: Clinical Drug Trial Guidelines (2000)
MOH: Regulations on Human Assisted Reproductive Technologies (2001)
MOH: Regulations on Compulsory Labelling of GMO (2002)
MOH: Ethical Principles of Human-Assisted Reproductive Technologies (2003)
MOH: Guidelines on Human-Assisted Reproductive Technologies (2003)
MOST & MOH: Ethical Guidance on Human Embryonic Stem Cell Research (2004)

6. Arguments for Human Reproductive Cloning

The debate on cloning issues has been in two discourses, the one is the traditional discourse, the other the modern discourse.[11] Sometimes there is a degree of overlap between the two. For example, in the traditional discourse one argument for human reproductive cloning (HRC) was provided by Lee, SC who argues that HRC can make up for the inadequacy of nature which violates the *tao* of nature. This argument is reminscent of Confucius' idea that "Humans can carry forward the *tao*." (*Lun Yu* 15.2)[12] This argument can be called the New-Confucian argument. The Neo-Taoist argument was provided Chen, DH who argues that HRC does not run counter to the law of nature. He supports this with a quotation from *Dao De Jing* 77, "It is the *tao* of nature that the excessive be reduced and the deficient supplemented."[13] Because these arguments rely heavily on the Confucian and Taoist classics, they could be called quasi-religious arguments.

There are a number of arguments for HRC in the modern discourse or quasi-secular discourse. The main ones are as follows:

- People have the right to reproduction, they can choose any form of reproduction available including HRC.
- HRC is no different from *in vitro* fertilisation (IVF). IVF is now widely accepted ethically, so HRC should and will be ethically acceptable.
- It is impossible to prohibit a technology once it is invented and applied.

The counter-arguments against the above arguments for HRC are as follows:

- In the case of HRC the *tao* of *yin-yang* or the law of nature is violated. *Yin-yang* is the core of Confucian/Taoist cosmology. According to the *Yi Jing*, only *yin* and *yang* together can generate something, "One *yin* and one *yang*, this is called the *tao.*" (*Yi Jing, Xici xia*)
- HRC is substantially different from IVF. It does not involve gene recombination (*yin-yang*).[14]
- HRC is not human reproduction, but a kind of producing or manufacturing. Thus, the right to reproduction is not applicable.
- "Impossible to prohibit" is not an argument for what is ethically permissive. There are many things in today's world which seem impossible to prohibit, such as corruption, drug trafficking, women and children trafficking. The impossibility of prohibiting these practices is not a reason to make them ethically permissible.

7. Arguments Against Human Reproductive Cloning

The arguments in China against HRC are mainly based on the 'do no harm' and the 'dignity' argument.

HRC will cause serious harm to the possible cloned child. For Confucian or Taoist medicine, health is the balance between *yin* and *yang*, and illness is caused by a *yin-yang* imbalance. Whether the somatic cell is derived from a male or a female, the imbalance between *yin* and *yang* is obvious. In modern scientific phraseology, HRC is a kind of asexual reproduction. Errors and defects in reprogramming are unavoidable. And many foetal defects cannot be detected before delivery. This is somatic harm. The psychological and social harm to the cloned child is that he/she cannot play the same role as his/her prototype. The indeterminate status and role of the cloned child will cause it great distress and suffering.[15]

HRC will also cause serious harm to others apart from the possible cloned child. The first of these is the women who donates eggs: a successful HRC may need 50-100 oocytes for one cloned embryo. Psychological anxiety will also be caused to the person who donates the somatic cell: what will the clone be when he or she grows up as a copy of the person in family and social relationships. And last but not least are the possible security problems to society: if there are many cloned people, they might possibly abuse their likeness with someone else.

Many Chinese scholars are concerned that HRC might jeopardise orderly familial relationships (*ren lun*), which in turn are the foundation of the state.[16] For Confucians "Marriage is the union of two families into one in order to serve ancestors and extend life to future generations. Man and woman are different, so there is affection between husband and wife, then there is kinship between father and son, then there is political relations between king and his subjects." (*Book of Rites*, *Hun yi*)

The other main argument against HRC is that it violates human dignity.[17] According to the Confucian classics, "Only the human being is the most intelligent in the myriad of things" (*Shang Shu*, *Taishi shang*), and "The human being is the most precious between heaven and earth" (*Book of Filial Piety* 9). Accordingly, the human being should not be objectified or manufactured in the way a material product is manufactured. HRC will lead to the manufacture and objectivisation of human beings. HRC will also make it impossible to prevent other types of cloning in which the human is treated as a mere means, e.g., a child may be cloned simply to provide organs for transplant or just out of curiosity. HRC will create a moral slippery slope along which human dignity will be increasingly neglected: It will lead to a veritable "Brave New World."

8. Arguments for Human Therapeutic Cloning

Human therapeutic cloning (HTC) is the use of nucleus transfer technology to create a human embryo from which totipotent stem cells can be derived for therapeutic purposes. There are few, if any, arguments against HTC in Mainland China. Confucians may argue that the human embryo created by cloning is not viable because of the imbalance between *yin* and *yang*. There are two answers to the question of whether there is a *yin-yang* imbalance in a human embryo created by cloning: radical Confucians may answer yes, so HTC should be rejected; but moderate Confucians may argue that if the cell or tissue or organ derived from stem cells were to be implanted in a body with *yin* and *yang*, the imbalance could be corrected.

The argument for therapeutic cloning is based mainly on beneficence. It is argued that HTC may create a possibility to cure a great number of fatal, currently incurable diseases and relieve the suffering of millions of people. According to current assessment, the probable benefits outweigh any possible risks. Thus, although many of the factors involved are uncertain, at least therapeutic cloning should be permitted ethically.

However, arguments for therapeutic cloning depend upon the position taken towards embryo's moral status. As I argued above, according to the accepted Confucian view personhood begins at birth. The human embryo is not a person, not a human personal life. The destruction of an embryo, like abortion, should not be taken as "killing a person." However, the human embryo is a human biological life, a precursor of person, not merely a thing like the placenta. Accordingly, it deserves due respect. Without sufficient reason, the manipulation or destruction of an embryo is not ethically permissible. Saving millions of human personal lives could be sufficient reason. It should also be disposed of with due procedure. Undue disposal will violate human dignity.

Due respect must be given to the entities which are between persons who enjoy a higher moral status and those (e.g., a piece of stone) which have no moral status. A person enjoys full respect, a stone has no moral status. In between these, the non-human animals, human embryos and foetuses, corpses etc. deserve modest but not full respect. When we say the human embryo should enjoy due respect, it means that:

- The human embryo may be used in research only extra-corporeally and only when it is no more than 14 days old;
- The human embryo may be used only in cases where the research goal cannot be achieved with other methods;
- The human embryo should not be treated as a commodity or as property; Scientists should in some way acknowledge the human embryo's contribution to human health and life.
- Due procedure means that after use, the human embryo should be buried or cremated with a ceremony.

If there are alternative sources from which to derive totipotent stem cells for research, such as the use of aborted foetal primordial germ cells or the use of spare human embryos after the success of IVF, or making adult stem cells totipotent, we have no reason to create a human embryo. However, where alternatives are not available for research in the near future (e.g., making adult stem cells totipotent), or for therapeutic purposes, the use of nucleus transfer to create embryo from which to derive stem cells is not absolutely prohibited. Rather, it should be deemed permissible, or even necessary, but it should be subject to strict regulation.

Case 3: A Controversial Report

On September 7, 2001 a report was published in *Beijing Youth Daily*: Professor Chen, Zhongshan Medical University, has transferred a skin cell nucleus from a 7-year old boy to the denucleated egg of a rabbit, and created a "human" embryo, from which he then derived totipotent "human" stem cells that "can be used for therapeutic purposes." The reason he did so is "the difficulty of obtaining a human egg."

This report raised many questions: Is this a "human" embryo? Are these "human" stem cells? Can they be used to treat human disease? If it is difficult to obtain a human egg, why did Prof. Chen not use germ cells from an aborted female foetus or eggs from a dead female body with informed consent? Can the stem cells with rabbit mitochondria be used to treat human diseases? A further question is: Is it permissible to create a chimera between human and animal?

Based on the human dignity argument, the creation of a chimera by the fusion of human and animal embryo, or the hybrid between human and animal gametes should be prohibited. However, the chimera between human nucleus and animal mitochondria is permissible if it is for research within 14 days after fertilisation or nuclear transfer and only if there is sufficient reason to do so. But Professor Chen did not have sufficient reason. Now he has discontinued his research.

Case 4: Dr. Zhuang from Zhongshan University transferred the nucleus of a fertilised egg (he used the husband's sperm and the wife's egg) to the denucleated egg of a donor, because the wife suffers from mitochondria DNA disease. Some argue that this should be prohibited because it is nucleus transfer for reproductive purposes. But others argue that it is the treatment or prevention of mitochondria DNA disease rather than human reproductive cloning. Accordingly, it should be permitted with informed consent.

9. Guidelines

On September 13, 2001 the Centre for Applied Ethics of the Chinese Academy of Social Sciences, the Centre for Bioethics of the Peking Union Medical College, the Committee on the Philosophy of Science of the Chinese Society for the Philosophy of Nature, Science and Technology, and the ELSI Committee of the Human Genome Project China jointly organised a workshop to discuss the ethical and regulatory issues relating to human embryonic stem cell research. On the basis of the workshop, a draft of the Proposal of Ethical Principles and Regulations on Human Embryonic Stem Cell Research (Ethical Principles and Regulatory Recommendations on Human Embryonic Stem Cell Research) was submitted to the Ethics Committee in the Ministry of Health and discussed at a meeting of this Committee in November 2001. At about the same time, the Ethics Committee of the National Human Genome Center (Southern) drafted Ethical Guidelines for Human Embryonic Stem Cell Research in October 2001 and revised them in August 2002 (Ethical Guidelines for Human Embryonic Stem Cell Research). Both documents recommended that HRC should be prohibited and that HTC should be allowed. The ethical principles of respect, informed consent, non-maleficence and beneficence, and non-commercialisation are all taken into consideration in the guidelines. It is also suggested to prohibit the re-implantation of the embryo into the human or animal uterus for stem cell research, to mix a human and animal gamete or embryo to make a chimera, to use the embryo after more than 14 days, to add any external gene into the embryo, or replace the nucleus of the embryo with any other human or animal nucleus, to force the donor to become pregnant and undergo abortion under duress, or to manipulate the method and time of abortion, and, finally, to sell and buy human gametes, embryos or foetal tissue. The need for ethical reviews, monitoring, inspection, and ethical training is emphasised.

Meanwhile, the MOH promulgated *Regulations on Human-Assisted Reproductive Technology* and enforced them on August 1, 2001. Article 22 prohibits the selling or/and buying of human gametes, zygotes and embryos, surrogate motherhood, and unauthorised sex selection.

In July 2003 the MOH promulgated the *Guidelines on Human-Assisted Reproductive Technology* which stipulated that the voluntary principle of informed consent or informed choice must be strictly observed and the patient's privacy respected.

The following acts are prohibited:

- Sex selection without medical indications;
- Surrogate motherhood technology;
- Donation of human embryo;
- Human egg plasma and nucleus transfer technology for the purpose of reproduction;
- Creation of a hybrid using a human gametes and a gametes of other species; likewise the implantation of gametes, zygotes and embryos of other species into the human body;
- Implantation of human gametes, zygotes or embryos into the body of other species;
- Manipulation of the gene in human gametes, zygotes or embryos for the purpose of reproduction;
- Fusion of sperm and egg of close relatives; in the same period of treatment the gamete or zygote must be derived from same male and female;
- Transfer of gamete, zygote or embryo to other people;
- Research without the patient being informed and without her or his free consent;
- Research on human chimera embryo;
- Human reproductive cloning.

In the same month the MOH also promulgated the *Ethical Principles of Human Assisted Reproductive Technology* which include:

Benefit to the patient,
Informed consent,
Protection of children,
Social good,
Privacy and confidentiality,
Non-commercialisation, and
Ethical reviews and surveillance.

On December 24, 2003 the MOST and the MOH jointly promulgated the Ethical Guiding Principles for Research on Human Embryonic Stem Cells. (MOST/MOH) The following is an authorised translation.

1. The Ethical Guiding Principles for Research on Human Embryonic Stem Cell (hereinafter referred to as the Guiding Principle) are formu-

lated for the purpose of bringing human embryonic stem cell research in biomedical domains conducted in the People's Republic of China to accord with bioethical norms, to ensure internationally recognized bioethical guidelines and domestic related regulations to be respected and complied with, and to promote a healthy development of human embryonic stem cell research.

2. Human embryonic stem cells described in the Guiding Principles include stem cells derived from donated human embryos, those originated from germ cells and those obtained from somatic cell nuclear transfer technology.

3. Any research activity related to human embryonic stem cells conducted in the territory of the People's Republic of China shall abide by the Guiding Principle.

4. Any research aiming at human reproductive cloning shall be prohibited.

5. Human embryonic stem cells used for research purpose can only be derived from the following means with voluntary agreement:

1) Spared gamete or embryos after *in vitro* fertilization (IVF);
2) Fetal cells from accidental spontaneous or voluntarily selected abortions;
3) Embryos obtained by somatic cell nuclear transfer technology or parthenogenetic split embryos; and
4) Germ cells voluntarily donated.

6. All research activities related to human embryonic stem cells shall comply with the following norms:

1) Embryos obtained from IVF, human somatic cell nuclear transfer, parthenogenesis or genetic modification techniques, its *in vitro* culture period shall not exceed 14 days starting from the day when fertilization or nuclear transfer is performed.
2) It shall be prohibited to implant embryos created by means described above into the genital organ of human beings or any other species.
3) It shall be prohibited to hybridize human germ cells with germ cells of any other species.

7. It shall be prohibited to buy or sell human gametes, fertilized eggs, embryos and fetal tissues.

8. The principle of informed consent and informed choice shall be complied with , the form of informed consent shall be signed, and sub-

jects' privacy shall be protected in all research activities related to human embryonic stem cells.
The informed consent and informed choice mentioned above refer to that the researchers shall use accurate, clear and popular expressions to tell the subjects the expected aim of the experiment as well as the potential consequences and risks and to obtain their consent by signing on a form of informed consent.

9. Research institutions engaged in human embryonic stem cell shall establish an ethical committee, which consists of research and administrative expert in biology, medicine, law and sociology with the responsibilities for providing scientific and ethical review, consultation and supervision of the research activities related to human embryonic stem cells.
10. Research institutions engaged in research related to human embryonic stem cells shall formulate corresponding detailed measures and regulatory rules in compliance with the Guiding Principles.
11. The Ministry of Science & Technology and the Ministry of Health of the People's Republic of China shall be responsible for the interpretation of the Guiding Principles.
12. The Guiding Principles shall go into effect as of the date of its promulgation.

Chinese bioethicists have commented on this Guiding Principles,[18] pointing out that this is a first ethical guideline for research in a specific field which was promulgated jointly by the Ministry of Science & Technology (which is the department of Central Government that has main responsibility to administer the research) with MOH after the MOH promulgated the *Interim Regulations for Ethical Review of Biomedical Research Involving Human Subjects* in 1998 and the SFDA promulgated *Clinical Drug Trial Guidelines* in 1999. The document first points out the consistency between compliance with ethical norms and international and national ethical guidelines on the one hand and smooth development of scientific research on the other. It explicitly stipulates the prohibition of human reproductive cloning, the prohibition of human embryonic research *in vitro* no more than 14 days, the prohibition of implanting human embryos used for research into human or other animal's reproductive system, the prohibition of hybrid forms combining human germ cells and germ cells of other species, and finally the prohibition of the sale and purchase of human gametes, fertilised eggs, embryos and foetal tissue. It emphasises that the principles of informed consent and informed choice must be adhered to in order to protect the subject's privacy, and the need to establish ethics committees etc. All these are very important and should be given a positive evaluation.

However, at the same time bioethicists have criticised that there are a number of problems in this Guidance. The definition of terminologies, for instance, is inadequate. For example, what is the difference between blastula and parthenogenetic split blastula? What do they refer to respectively? In the same vein, what is the difference between the somatic cell nucleus transfer technique and the parthenogenetic reproduction technique? These terms should be defined. Secondly, there are several inconsistencies in the document. For example, in Article 5 the sources of human embryonic stem cell are listed, but the techniques, such as parthenogenetic reproduction and genetic modification are not mentioned. However, these two techniques are mentioned in Article 6. Thirdly, the document makes no mention of penalties for failure to comply. If somebody, e.g., stubbornly carries out human reproductive cloning, how should he/she be dealt with? Last but not the least, the words "ethics" and "ethical" occur many times in this document, likewise the need to establish ethics committees. However, with regard to the make-up of such committees the inclusion of professional ethicists is not mentioned. Is this an oversight on the part of the decision-maker, or is ethics not considered to be a specialist discipline? And in fact we have seen members of Ethics Committees who have no background or training in ethics but who have merely attended a workshop on the subject abroad. As a result, they were incapable of understanding and applying major ethical theories and basic ethical principles to ethical review practice, they applied a double standard in their ethical reviews of protocols, and they violated international and national ethical guidelines. Furthermore, some ethics committees only ever conducted scientific reviews and not ethical reviews. They were ethics committees in name only. This situation must be changed and capacity building enhanced.

10. Cross-Cultural Dialogue and International Collaboration Is a Must for the Cloning Debate

Cross-cultural and international dialogue on cloning or any other bioethical issue is urgent and very important, because all these issues are global in nature. However, cross-cultural and international dialogue can only be fruitful if there is mutual understanding and mutual respect. I am saddened that even in bioethical publications there are seriously distorted articles about China and other countries.[19]

Cloning issue is a ongoing topic of debate and many issues remain to be addressed during human embryonic stem cell research or therapeutic cloning, the international collaboration both in science and ethics is necessary and important. We are sincerely grateful to all colleagues who have come to China to help us improve and develop bioethics. However, some Western colleagues exaggerated the importance of their own contribution and ignored the work that has done by Chinese colleagues.[20] Western colleagues who engage in dialogues or collaborations in developing country should be culture sensitive. For example, in China there are different approaches and there has been much heated debate. Some Western colleagues have misunderstand this as a "big fight" in China's bioethics. It is misleading: Chinese favor the Confucian maxims: "Harmony is precious" and "Harmony is nor becoming identical." (*Lun Yu* 1.12 and 13.23)

Our country has been humiliated by Western powers for hundreds of years. It is our sincere hope today that those colleagues who sincerely want to help us will not take an arrogant attitudes towards us and will show us that they will not impose anything on us, exaggerate or widen the difference within us, interfere with our internal affairs, and especially that they would not make groundless allegations to the international forum.[21] We are still a country which is poor in resource and in which the young especially are easily induced, even with small reward. Most Western colleagues are very sincere about wishing to help us, but there are some, albeit very few exceptions.[22] We truly hope that the collaborative relationship between us would be just, fair, equal and equitable.

As for coping with the differences between various cultures in the field of bioethics, the following principle may be appropriate: Seek common ground while preserving the differences. It may be wise to endorse an international convention on the banning of human reproductive cloning as a first step; there is still sufficient time to negotiate on the issue of human embryonic stem cell research/therapeutic cloning.

Notes

1. In this article the term China refers to mainland China, the situation in the remaining part of China, such as Hong Kong, Taiwan and Macao is different to a greater or lesser degree.
2. Zong *et al.*, 2003, 1853-1857.

3. According to the Confucian viewpoint, a cloned embryo is not the product of *yin-yang* interaction, so it is not viable.
4. Hui, 2004, 29-44; Ip, 2004, 49-56.
5. The term *ren* in Chinese has different translations in English, such as kindheartedness, benevolence, or humanity. But I think the translations humanness and humaneness may be better.
6. Xue, 1999, 51.
7. Ibid., 53.
8. Qiu, 1987.
9. Qiu, 1991, 16-27; Zhai, 2002.
10. Qiu, 1991.
11. This section is cited from Qiu, 2002, 71-88.
12. Lee, 1998, 105-122.
13. Chen, 1998, 73-86.
14. In gene recombination there is one half set of genes from the father and another half set from the mother.
15. Ip, 1998, 18-30.
16. Shen, 1998, 125-144; Kuang, 1998, 135-149; Fan 1998a; Liao, 1997.
17. Fan 1998b.
18. Qiu, 2004, 1.
19. In 1998 Mr. Carl Becker presented some untrue facts in his presentation at a session of the World Congress of Bioethics in Tokyo. Although his claims were immediately refuted by Chinese participants, Mr. Becker's article titled "Money Talks, Money Kills? - The Economics of Transplantation in Japan and China" was published in *Bioethics* vol. 13, no. 3/4, 1999. In this article the author accused China of increasing the number of the prisoners sentenced to the death penalty for the sole purpose of selling their organs for transplantation. Mr. Becker knew very little about China and had no evidence for his statements. At the Beijing International Conference on Bioethics, which was held in January 2004, Professor Cao Nanyan from Tsinghua University argued that, as a responsible publication of the International Association of Bioethics, the journal *Bioethics* was duty-bound to ensure that all its articles were based on reliable sources and unprejudiced (www.chinaphs.org/biothics). It was said that before publishing the article was sent to a reviewer and rejected. However, the editor of *Bioethics* persisted in publishing it regardless of the outcome of the review procedure. Why was this? Chinese bioethicists have published articles in journals arguing against the use of the organs of executed prisoners (e.g. Qiu, 1999, 22-24). Although the gov-

ernment approved the practice on condition that prisoners had given their informed consent, it is the physician and the hospital manager who take the organs. How can they bring about an increase in the number of executions? We have criminal law and three independent authorities, i.e. the police, the public prosecutor and the law courts to determine whether a suspect is guilty and whether he/she should be sentenced to capital punishment. The principle "innocent until proven guilt" is fundamental in judiciary procedure and a suspect is entitled to have a defense lawyer. One can disagree with the Chinese system of capital punishment, but to draw such a ridiculous conclusion is to go too far. The selling and buying of organs is illegal in mainland China. If there is no mutual understanding, how can we have fruitful dialogue on cloning issues or any issue in bioethics?

20. For example, someone was called "the father of bioethics in China" in his home country. There are no fathers and sons in bioethics: everybody is equal.
21. For example, when we worked together on a project, the Western colleague did not inform us about the whole truth of the project including the real budget. On completion of the project, even though the Chinese side had made a substantial contribution, the Western colleague made no mention of it at an international forum; also the Western colleague often quarreled with Chinese quarter on insignificant matters.
22. It is unusual that one side of a controversial case was invited to attend a meeting in Germany (2003) and another meeting in France (2004), and the invitee raised groundless allegations against the other side with notorious means of propaganda, while the other side was not invited and accordingly had no opportunity to defend itself. This violates the principle of due procedure in a so-called democratic society.

References

Book of Change, cited from Xue, GC. (ed.). *Confucianism, Taoism and Buddhism*. Beijing: Chinese Medicine Classics Press, 1999.

Chan, W.T. *A Source Book in Chinese Philosophy*. Princeton, NJ: Princeton University Press, 1963.

Chen, D.H. "Bioethical Dimension of Secular Daoism: A Reflection on Cloning and Genetic Engineering." *Proceedings of International Sympo-*

sium on Bioethics, vol. 1. Tsung Li and Jia Yi, Taiwan, June 16-19 (1998), 73-86.
Ethics Committee of the National Human Genome Center (Southern). "Proposed Ethical Guidelines for Human Embryonic Stem Cell Research." In *Medicine and Philosophy* (6) (2001), 8-9.

Ethical Principles and Regulatory Recommendations on Human Embryonic Stem Cell Research, *Newsletter*, 2001, no. 1, Research Center for Bioethics, PUMC, 3.

Fan, R.P. "Human Cloning and Human Dignity: Pluralist Society and the Confucian Moral Community." In *Chinese & International Philosophy of Medicine*, vol. 1, no. 2 (1998a), 73-94.

Fan, R.P. "Genetic Intervention and the Confucian Tradition." *Proceedings of International Symposium on Bioethics*, vol. 2, Tsung Li and Jia Yi, Taiwan, June 16-19 (1998b) , 1-14.

Hui, E. "Personhood and bioethics." In *Bioethics: Asian Perspectives - A Quest for Moral Diversity*, Qiu Renzong (ed.). Dordrecht: Kluwer, 2004, 29-44.

Interior Classics of Huangdi: Questions and Answers about Living Matter, vol. 2 *Plain Questions*. Beijing: Chinese Medicine and Pharmacy Press of Science and Technology, 1998, 27.

Ip, P.K. "Human Cloning, Ethics and Asian Values." *Proceedings of International Symposium on Bioethics*, vol. 1. Tsung Li and Jia Yi, Taiwan, June 16-19 (1998), 18-30.

Ip, P.K. "Confucian Personhood and Bioethics: A Critical Appraisal." In *Bioethics: Asian Perspectives - A Quest for Moral Diversity*, Qiu Renzong (ed.). Dordrecht: Kluwer, 2004, 49-56.

Kang, P.S. "To Clone or Not to Clone: the Moral Challenges of Human Cloning." In *Chinese & International Philosophy of Medicine*, vol. 1, no. 2 (1998), 95-124.

Kang, P.S. "Cloning Humans? Some Moral Considerations." In *Bioethics: Asian Perspectives - A Quest for Moral Diversity*, Qiu Renzong (ed.). Dordrecht: Kluwer, 2004, 115-128.

Kuang, T.J. "Value Dispute on Human Cloning and Genetic Engineering." *Proceedings of International Symposium on Bioethics*, vol. 1. Tsung Li and Jia Yi, Taiwan, June 16-19 (1998), 135-149.

Lee, S.C. "On the Social and Ethical Puzzles in Human Cloning: An Analysis of Applied Ethics." In *Ethics and Life-Death: Collected Papers of Asian Symposium on Applied Ethics*, Lee, S.C. (ed.). Tsung Li: The National Central University Press, 1998, 105-122.

Liao, S.B. et al. "On Human Cloning." *Guangming Daily*, March 14 (1997).

Lun Yu, Analects of Confucius. Shanghai: The Commercial Press, 1926.

Meng Zi. Shanghai: World Bookstore, 1932.

MOST/MOH: Ethical Guiding Principles for Research on Human Embryonic Stem Cells. *Health News*, January 14, 2004.

Qiu, Renzong. *Bioethics*. Shanghai: Shanghai People's Press, 1987.

Qiu, Renzong. "Morality in Flux: Medical Ethics Dilemmas in PR China." *Kennedy Institute of Ethics Journal*, 1 (1991), 16-27.

Qiu, Renzong. "Can the Use of Organs from Executed Prisoners Be Justified?" In *Medicine & Philosophy*, no. 3 (1999), 22-24.

Qiu, Renzong. "The Tension between Biomedical Technology and Confucian Values." In *Cross-Cultural Perspectives on the (Im)possibility of Global Bioethics*, Julia Tao Lai Po-wah (ed.). Dordrecht: Kluwer, 2002, 71-88.

Qiu, Renzong. "Comments on MOST/MOH's Ethical Guidance for Research on Human Embryonic Stem Cell." In *Medicine & Philosophy*, no. 25(4) (2004), 1.

Shen, V. "Is Human Cloning Supported by Any Ethical Argument?" In *Chinese & International Philosophy of Medicine*, vol. 1, no. 2 (1998), 125-144.

Xue, G.C. (ed.). *Confucianism, Taoism and Buddhism*. Beijing: Chinese Medicine Classics Press, 1999.

Xun Zi, in Zhang, ST. *Commentaries on* Xun Zi. Shanghai: Shanghai People's Press, 1992.

Zhai, X.M. *Dying with Dignity*. Beijing: Beijing Capital University Press, 2002.

Zhai, X.M. "ABA Country Report for China 2003." *Eubios Journal of Asian and International Bioethics*, 14 (1) (2004), 5-9.

Zhuang Zi, cited from Xue, GC (ed.). *Confucianism, Taoism and Buddhism*. Beijing: Chinese Medicine Classics Press, 1999.

Zong, F.B. et al. (eds.). *Ancient Interpretations of Words* (Gu Xun Hui Zhuan). Shanghai: The Commerce Press, 2003.

Culture and Bioethics in the Debate on the Ethics of Human Cloning in China

Ole Döring

Abstract: This article scrutinizes the debate on human cloning at the interface between bioethics and policy making in China. It is argued that the culture-orientated study of bioethics depends on a well-calibrated balance of empirical groundwork and sound theoretical reflection and methodology, conducted over an extended period of time. At present, the debate, for its limited scope, cannot symbolize genuine moral features that may be validated as straightforward manifestations of cultural life. However, statements from the debate still convey an array of culturally and ethically relevant information. Material issues relating to the morality of cloning are considered, as are abstract and cultural theoretical reflections. It is claimed that even highly specialized and culturally specific questions such as human cloning can be discussed in relation to the conceptual puzzles of culture and cultural understanding. This article applies a „thin" theoretic axiomatic assumption that a „good life" is a reasonable aspiration, considering the methodological and theoretical implications of „reason." This combined ethical and cultural investigation emphasizes the ultimately discursive character of bioethics, as distinct from prudential or power-related measures.

Key Words: China, Culture and ethics, Human cloning, Theory of practical reason, Bioethics.

1. Introduction: Cloning as a Cultural and Ethical Issue

A global perspective on bioethics, with a particular interest in culture, opens up a wide spectrum of themes. It includes fundamental conceptual issues, such as human dignity, as well as methodical questions concerning requirements for an ethical discourse and proper procedures for argumentation and for decision making and moral matters relating to justice and the avoidance of harm. Generally speaking, based on the hermeneutic principle of acknowledgement of the originator of moral claims, every discussion that addresses moral and ethical problems seriously can be regarded as culturally interesting, with no subject-specific bias, regional preference or culturalist reservations.[1]

For all its significance human cloning is certainly not the most urgent issue for a cross-cultural or culture-transcending bioethics to address. For instance, China, as a developing country faces immediate moral and ethical problems on a huge scale, with enormous tasks related to the building of the health care sector, the introduction of a basic social security system, and emerging patterns of a civil society and a state of law.[2]

In particular, in the absence of an established intellectual reservoir for critical studies on modernisation and in the absence of a pluralistic public debate in China, the relationship between science and society is a largely unchartered field, with many culturally relevant questions not yet adequately articulated. As a consequence, the moral and ethical opinions expressed in national law, professional regulations and statements from bioethics experts can reflect moral sentiments in Chinese society and the cultural characteristics of China only to a very limited and probably insufficient extent.[3] The culture-orientated study of bioethics depends on a well-calibrated balance of empirical groundwork and sound theoretical reflection and methodology, conducted over an extended period of time.[4]

On the other hand, ethical and sociological studies about research in the life sciences and medical practice have the potential to trigger the relevant research on all levels, empirical, methodological, theoretical, and ethical. Such enquiries yield obvious implications for issues of cloning, such as what defines a human being in terms of worthiness of protection, what are the priorities regarding multiple interests or clusters of stakes, whose concerns should receive ultimate attention, or how should medical practice be organised in clinics and research laboratories so that principles are respected and concerns are adequately addressed? The rationale in arguments related to human cloning can also be explored on a more abstract level of reflection: how are claims about justice or criteria to assess the moral meaning of a technique conceptualised in view of the foundations of understanding about practical matters, across barriers of languages and cultures?

According to such an approach, even highly specialised and culturally specific issues, such as human cloning, can be discussed in relation to the conceptual puzzles of culture and cultural understanding. This includes consideration of why cloning is or is not regarded as morally problematic. Notably, such an approach, on the deep-seated level of philosophical reflection, connects material moral judgement with ethical concepts and principles and, beyond that, with an interest all humans share by virtue of being humans, namely, realising a "good life."

Thus, a comprehensive venture is proposed that would systematically integrate moral diversity and cultural richness, within the bounds of sustainability.

Accordingly, cultural and ethical limitations can be discussed in terms of a "thin" axiomatic assumption that a "good life" is a reasonable aspiration, or as expressions of negative freedom, together with the methodological and theoretical implications of "reason" (*Vernunft*, as distinct from rationality, *Verstand*). Hence, a combined ethical and cultural investigation emphasises the ultimately discursive character of bioethics, as different from rational, prudential or power-related measures.

Whereas this endeavour is enshrined in the conceptual horizon of the meaning of humanity, or the human being, the methodological point of this approach is to focus on given arguments, purposes of action and criteria for moral deliberation, as they are put forward in the respective region, that is, China. The first empirical step, therefore, is to identify stakeholders, key players, relevant institutions, interests and modes of discourse and decision-making in China.[5]

2. On Cultural Assessment in Ethics

In secular societies, *culture* denotes a problematic area for our intellectual endeavours to understand humanity as being greater than a sum of parts (such as genes) and functions (such as social roles), or an expression of the circumstances of upbringing (such as education). Culture delineates the horizon for critical deliberation about the meaning of humanity. It marks a conceptual point of reference for our creative attempts to explore the questions of humankind. Thus, the conceptual structure of culture is intrinsically heterogeneous.[6] Moreover, empirically, there is no cultural entity that could be adequately addressed as, for example, "China." We have to identify the subjects of cultural studies in ways that make them definable, accountable and traceable in order to use the term culture in meaningful ways.

The relationship between culture, morals and ethics is notoriously difficult to analyse. Conceptually, the genetic or historic fact that a moral claim is rooted or embedded in, or has indeed become, a tradition of a certain culture does not indicate whether a claim is justified. Culture is not directly normative. An accurate definition of culture describes *how it is* (or has been) and not how it *should* be.

However, many functions of culture, such as moral and social creativity, feed into the process of embroidering the fabric of moral life and ethical reflection. When we look at how cultures generate moral claims (e.g., in their attempts to argue that "we should/should not apply the dignity criterion to an early embryo"), this merely *describes* a prescriptive assertion and the related processes of deliberation. It does not inform us in itself whether or not the respective assertion is tenable. It gains a *prescriptive* quality only when someone actually accounts for it as being his/her moral view, for example, by saying, "I believe that we should/ should not apply the dignity criterion to an early embryo."

Furthermore, culture is *creative*. Human activities in general change the factual world in ways that are sometimes normatively significant. Hence, the second major pattern in the concept of culture is that it refers to something created by humans. We are particularly interested in the human inventions that contain prescriptive ideas.

Frequently, such ideas are framed in descriptive language. For example, China's leading bioethicist, Qiu Renzong, has stated that, "Anatomy and surgery have never fully developed in traditional China, because dissection of a dead body and operation on a body will violate the principle of filial piety."[7]

This statement is typical of a "cultural argument." It seems to express a common sense among many people in China that correlates filial piety and attitudes with anatomy and, in this case, organ transplantation. It is difficult to say exactly what this tells us about Chinese culture. For example, on the descriptive level, amplification of the validity of the general historical assertion is needed, whether or not the author approves of the described reasoning, in terms of accepting it as a prescriptive claim. It is evident from the same article and his other writings that Qiu does not regard traditional Chinese culture as being *directly* normative for contemporary ethics in China.[8] In this instance, he expressly argues that medical sciences ought to include transplantation studies. This conclusion stands in no explained relation to the factual assumption cited above. In fact, Qiu intends to argue for a *modification* of Chinese values, inasmuch as they are deemed *unsuitable* for modern China.

The reference to *xiao* (filial piety) in this argument is not entirely logical, since the researcher is not actually required to dissect a relative (in particular). In fact, historically, there *have* been anatomic dissections.[9]

On the other hand, it is true that many people in contemporary China hesitate to donate body parts, even after death. There is evidence that Chinese who wished to donate their body after death for research and who had signed a document to that effect, are frequently overruled by their relatives, who claim the body for customary burial. The reasons for this behaviour and the driving motivations or moral considerations should be the subject of empirical studies. An exploration of the past does not appear to be the most meaningful strategy to achieve such clarification. It is not valid to draw generalising conclusions from the past culture to explain the present.

Cultural reflection provides an apperceptive reference frame for the general condition of humanity. It manifests the ways we understand and aspire to a "good life," as signified, e.g., in the works of literature, in the institutions and processes of social life, and in the outcomes of science and technology. Accordingly, medicine constitutes a scientifically supported approach to understanding, helping and healing people, offering means to improve their present condition on different physiological and psycho-social levels. Ethics in particular attempts to reflect upon the meaning of a "good life" and how our lives might be improved in general terms, under conditions of moral and anthropological plurality. This includes giving thought to the means and techniques that can be acceptably applied in biomedicine.

3. Interacting With Key Players

From this assessment there follow some methodological and theoretical features of an interactive approach to studies of morals and ethics in different cultures. On the heuristic basis of a peer-level approach (Roetz, 2002), it is challenging and meaningful to discuss cultural and related differences in terms of *quality*, much more so than in consideration of less controversial terms, such as diversity of opinion or tolerance, although this is clearly in the interest of advancing ethics and understanding.

A. Assessing another Culture

For example, as bioethical standards, including terminology, principles and norms or protocols, become increasingly consistent internationally, how can we avoid a "copycat modernisation"? Since the beginning of the Deng Xiaoping era, the overall political rationale in China has increasingly become a pragmatic one. Regarding science and technology management, it has been observed that, "China's new standards regime, (..., as a strategy is) best understood in terms of a 'neo-techno-nationalism,' in which technological development in support of national economic and security interests is pursued through leveraging the opportunities presented by globalisation for national advantage."[10]

The funds earmarked for research and development in the Tenth Plan (2000-2005) focus, among other things, on functional gene chips and bio chips, new medicines, modernisation of the production of traditional Chinese medicines, and the establishment of key technical standards.

In such a political climate, it can be difficult to sensitise scientists and bioethicists to inherent problems of modernisation, namely scientism or technicism, bioreductionism, genetification or medification, and comercialisation, when it is their task, at the same time, to facilitate practices that are to some extent based upon these very problems.

As a consequence, it is a shared concern for ethicists from different cultural backgrounds to constructively discuss alternative visions, for example, in the pursuit of sustainable development in biomedicine. Such self-awareness in bioethics presupposes an interest in and the capacity for self-critical development as well as the willingness to accept reciprocated responsibilities on all sides. It also reflects the problem of a mere "copycat modernisation" or biotech imperialism. Thus, the role of bioethicists in their respective social and national entrenchment needs to be considered in their respective social and national entrenchment.

Whose Culture?

A related problem is one of commitment and "cultural authenticity." In culture-transcending bioethics, the distinction between actual *commitment* (as a professional, an expert or a stakeholder) and assertion of social and/or cultural *representation* is not trivial. It is different from enquiries about the quality or legitimacy of a "cultural argument" in that it emphasises a logical and descriptive situation, but not a prescriptive judgement. Mindful of this difference, cultural studies should avoid to place unreasonable expectations on the author's *cultural* significance, and instead regard him or her as a source of certain (moral, ethical) judgements that can be interpreted from various angles. It is less interesting to discuss to whom a given statement can be attributed: either to ethics, as a programmatic discipline, or to a cultural tradition. The decisive measure is the *validity* and, to some degree, the intelligibility of the argument. This hermeneutics does justice to the author both as an individual and as a cultural person.

Together with the elucidation of expectations, the significance of particular contributions to bioethics should be considered in the proper relational setting, such as by experts or professionals, if we wish to address something as a "culture." On the basis of a peer-level approach, that is, without an intellectual bias, the significance of a *plurality* of contributions is highlighted. In terms of humanity the genuine moral claim of every individual is of interest. The horizon of the viability of ethics is the limit.

On these grounds, in preparation for such a discourse, study of and discussions with key players who represent shared concerns, for example China, can be helpful in outlining a hypothetical sketch of a "culture" in bioethics. This covers a broad spectrum of moral opinions, ethical styles, interests, and conceptual languages as well as exemplary casuistry. In particular, such an outline indicates the ethical issues and their characteristic systemic priority. However, it is difficult to generalise from sets of key players, since the examples are not necessarily representative of a culture.

A similar approach, which can be only mentioned in passing here, would be to focus on institutions rather than individuals. In the case of China, as a country undergoing substantial transformation in all relevant areas, and in the absence of strong religious communities, such an institutional approach can deliver preliminary indicators and an empirical or theoretical basis for a fully-fledged research programme.

Without anticipating the course of such a research programme, it appears likely that such an endeavour will produce methodological insights that reach beyond interaction with key players and institutional analysis, pertaining to the art of understanding and the foundations of normativity.

First of all, it may help in making every instance of communication more meaningful. The scientific relevance of cultural samples depends on their empirical accuracy, which, in turn, requires the study of and engagement in a proper discourse. In its present form, the discourse is far from ideal, though in pluralistic democratic and civil societies probably less so than in authoritarian countries. In this regard, Qiu Renzong's mentioning of China as being in "transition from a monolithic society to a more or less pluralistic civil society" is remarkable. In principle, however, it is a *human* enterprise. Thus, it is not only an issue of aided capacity building, but rather a methodical scientific requirement aimed to strengthen every human engagement in bioethics in terms of a discourse, from the "grass-roots" upwards. It stands to reason that the legitimacy of such engagement is bound by ethical principles.

"Chinese Ways"

Many Chinese bioethicists disregard the problem of cultural authenticity and tackle ethical issues in an intuitive manner, either as "citizens of the world" (Weltbürger) or with little related reflection. Some, however, discuss it expressly for genuine ethical reasons and others refer to culture for mere instrumental purposes.[11] In substance, all agree that bioethics should observe ways to respect the *individual*, albeit *according to Chinese ways*. What is particular about *Chinese* ways to respect a human being, permeating practice and reflection, and distinct from other, *non*-Chinese ways, and at the same time relevant for bioethics? Qiu Renzong explains,

> The distinctive character of medical ethics in China lies in seeking the balance between the individualistic and the collectivistic approach, and seeking the balance between rights and duties. Underlying this distinctive understanding is the conception of personhood. A person is taken as an individual who is in close relation with others: he/she is neither an independent atom nor a drop of water, bound to lose her/his identity, neither in anonymity nor in a community. (...) Any extremely individualistic or extremely collectivistic approach is not appropriate.[12]

In other words, both collectivism and individualism are ill suited to culturally sensitive ethical reasoning. The latter, individualism, because of its encouragement of selfish attitudes and its lack of social responsibility; and the former, collectivism, because of its simplistic and naive understanding of the normative status of a community and because of its affinity with political or ideological interests. Both are one-sided and without real ethical merit.

B. Examples

Human Cloning Research

In China, human cloning techniques are developed and used in basic research, whereby application is restricted to therapeutic purposes. Two outstanding individuals engaged in human cloning are the professors Lu Guangxiu (Changsha) and Sheng Huizhen (Shanghai). They are working in different research environments. Lu Guangxiu is a specialist in clinical reproductive medicine and fertility treatment.[13] Her research areas include basic research, applied medicine, psycho-social studies, and ethical regulations on institutional, provincial and national levels. Sheng Huizhen is an embryologist working on human embryonic stem cells,[14] with no immediate clinical affiliation and engaged in ethical regulatory work.[15]

Clinical Fertility Research

According to Lu,[16] the early human embryo cannot be distinguished in moral terms from other body tissue cells during the first 14 days of gestation. She argues that the moral status of an embryo is qualified by its being part of the woman's (or the prospective mother's) organism and by its carrying the parent's genetic heritance, with the potential to become a full human being. Hence, it may not be utilised at liberty for medical or research purposes. Full ethical attention should be directed to the protection of the adult. Lu maintains the priority of preventing harm, respecting self-determination, and protecting the integrity of women and couples. Tentatively, the Nestor of reproductive medicine in China, who has spent her entire academic career inside her home country, appears to share the moral rationale of many libertarian ethicists and researchers in Europe and North America.

Based on her long experience in reproductive medicine and cloning research, Lu Guangxiu is sceptical about the future of experiments that would depend on the supply of the "resource egg." She believes that the duty to protect donors will ultimately lead to a permanent shortage of biomaterial supply, if the existing ethical regulations are properly observed.[17] Given the high priority of patient protection that is expressed in and framed around obtaining informed consent, this is an unsolvable problem. Lu expressly criticises a "clinical and research culture" that is reliant on the procurement of egg donors. Commenting on the cloning experiments conducted in South Korea,[18] Lu argues that ethical standards, which actively support or call for egg donation, are not suitable for China.[19]

Fundamental Research

In Shanghai, embryo researcher Sheng Huizhen adopts a different moral approach. She argues that, on the biochemical level, the fusion of sperm and egg marks the most plausible reference point for the beginning of a human's life. Accordingly, protection of the human being should be afforded right from this initial moment of development.

She is working on a hybridisation technique that fuses human nuclei and denucleated rabbit cells. According to Sheng, the entity resulting from this process is technically not a human being, owing to the mitochondrial rabbit matter. Although a different order of species which may not be able to develop beyond the blastula stage, these embryonic cells are hoped to bear all the functional capabilities required for modelling and deriving stem cell lines that would eventually be employed for therapeutic use on humans.[20] Developing such stem cell lines is meant to satisfy both ethical and medical requirements. In the long run, Sheng hopes that her research will enable researchers to avoid the use of human embryos in future experiments.[21] With her position, Sheng Huizhen seems to be conceptually close to conservative or restrictive moral standpoints.

The moral views of leading Chinese scientists differ quite significantly between these two positions, according to their special research focus. They cannot be identified with any particular characteristic of *Chinese* culture, but relate to the respective researcher's sense of his or her medical or scientific mission.

Who or What Shall Be Protected and Why?

Opinions also diverge about the fundamental rationale of protection. When, how and why a human being is believed to achieve the status of *worthiness of protection* determines the grounds for ethical, legal and moral guidance. Generally, the term "dignity" (*zunyan*) is used to express the quality that denotes the worthiness of protection. Does dignity come gradually, or is it endowed at a certain moment during the process of gestation? Is it an objectively conferred quality independent of our judgement, or is it associated with reasoning or the quality thereof? Is it attached to individual persons, or does it reflect trans-personal ideas of humanity? These puzzles challenge Chinese and European ethicists alike.

Consider the beginning of human life. If dignity, or an equivalent property, emerges gradually, how does the degree of protection grow in proportion? How can the respective stage of worthiness of protection be formalised and verified?

On the other hand, is it possible or indeed desirable, in ethics, as distinct from legalistic thinking, to limit our conceptual capacities to the confinements of a functional paradigm, or to analytical and instrumental rationality? The Hong Kong physician and medical ethicist, Au Kit-sing, has drawn a poetical analogy of what it might mean to destroy an embryo for medical research: "Do we seriously believe that it is justifiable to kill a caterpillar because by definition it is only 'potentially' a butterfly?"[22] This alludes to the fact that an embryo might be more than we can hope to understand by virtue of rational assessment, going far beyond the biological meaning of potentiality, for example, through critical reason or in teleological terms.[23] Like the beauty of the butterfly, the value of an embryo might be intrinsically a matter of intuitive moral knowledge, such as is known in Chinese as the "good knowledge" (*liang zhi*).[24]

Consequences for Humanity

At present, however, the actual moral status of an early embryo does not receive much attention from Chinese bioethicists. The emphasis is rather on consequentialist or utilitarian concerns. In view of the practical bearing of human cloning, many Chinese intellectuals believe that it would upset family traditions. For example, "It would exacerbate the diversity of family forms and would cause fundamental changes in family relations and kinship criteria."[25]

This general concern gives scope for differing moral emphasis. For example, the Taiwanese philosopher Shen Qingsong (Vincent Shen) maintains, "Many of today's arguments against human cloning are based upon a social interpretation of the criterion of harmony. For example, that human cloning might undermine the actual purpose of the human sexual relationship, parenthood and the family system. (...) For me the most important consideration here is the interest of the child."[26]

The Shanghai philosopher Chen Rongxia offers a more fundamental assessment. Referring to Daoist and Confucian concepts, she claims that "nature possesses an intrinsic value which is independent of humans."[27] She criticises that "since the advent of modern technology, natural entities including humans first have been degraded to simple matter so that their natural essence is lost. Second, the desire of man to conquer nature (including himself) has been released without restraint after the attitude of awe of nature has vanished."[28] To tackle these problems, she highlights "the doctrine of the Dao that focuses on the relationships between humanity and technology." "From the viewpoint of the doctrine of Dao, human cloning is immoral because it violates nature. In the history of evolution, it is sexual reproduction that has resulted in diversity. The process of evolution depends on the variance of beings. Cloning—asexual reproduction—suppresses the appearance of variation. It thus violates the nature of life."[29]

Traditionally, Chinese medical ethicists refer to the moral attitude towards any form of human life, as represented by Sun Simiao from the Tang dynasty. Sun warned that "Whoever destroys life in order to save life places life at an even greater distance," and maintained, "this is my good reason for the fact that I do not suggest the use of any living being for medical purposes."[30]

Today, some Chinese ethicists attempt to identify a conservative or humanistic trait in Chinese medical ethics that would challenge biased views on Confucianism as a school of thought disinterested in issues related to the moral status of the early human being and other forms of life. These works call for a greater sense of awe for the "miracle of life" in the face of practices such as human cloning, and argue that there has in fact been a viable Confucian tradition in support of such views.[31]

Such an interpretation contradicts the views of other Chinese bioethicists. For example, according to the Taiwanese New-Confucian scholar, Lee Shui-Chuen (Li Ruiquan), there is a *categorical* moral difference between foetuses and embryos and infants. Infants "are under the guardianship not only of their parents, but also of all members of the community, which is usually represented by public authority."[32] Hence, society and the state have the duty to protect every born human being. Lee adds that in cases of immaturity or incapability to protect themselves, the efforts to protect them from harm should be reinforced. On the other hand,

> As to the moral status of foetuses, embryos and zygotes, their situation is more problematic. According to the notion of the moral community, foetuses and embryos could procure membership of the community with the approval of their expectant mothers. Here the mother has the ultimate authority, as she is the only person who can have direct interaction with her foetus, not to mention her role in its gestation. (…) In other words, nobody else has a right that overrides hers. (…) As to zygotes and blastocysts, it seems to be of no avail to confer membership to them, as there seems, first, no sensible communication between them and any person; and second there is a lack of individuality that would justify them to be treated as a member. They should be treated as the property of their owner, though with limiting moral obligations for proper use in experiments or disposal. Experiments with embryos and blastocysts have to be on guard as they could be future persons and for the erosive effect of such experimentation on the concept of moral personhood - there is a slippery ethical slope there.[33]

For Lee, the human embryo as such is not regarded as a morally distinguished entity. He does not envisage any Confucian support for the argument against human cloning for reasons of "playing God," or the hubris problem. He argues,

> through our sincere moral practice, we could help beings on Earth to manifest their mandate without any distortion, that is, ideally no being is subjected to unnecessary harm or evil. The agent (...) becomes a being with infinite value, an incarnated being and as great as Heaven and Earth and is thus regarded as making a trinity with Heaven and Earth. (...) Thus, to play the same role as Heaven and Earth is all and this is what a moral act means. If biotechnology could help man to achieve

> this purpose, it is morally good and fine. There is no problem of a human being playing God's role in his or her efforts to 'make good' defects of all kinds of being (...). Confucianism basically supports the employment of biotechnology in the relieving of human defects, whether they are inborn or man made.[34]

Hence, Lee is ready to endorse human cloning for therapeutic purposes.[35] Moreover, besides the frequently expressed alertness to the risks currently associated with the technology of mammalian cloning,[36] there is no fundamental argument against creating human beings by means of reproductive cloning, because

> Confucians tend to regard them as parent and offspring, not as 'delayed twins.' Furthermore, there is no fear of child abuse such as treating the clone as a substitute or instrumental for the original, as they are distinct individuals. Each has his or her personal identity, though the clone might have a certain psychological crisis as being a 'copy' of the original. (…) As long as the clone is an individual, he or she has all the protection of the moral and legal laws as any child born in other ways.[37]

This view challenges the appropriateness of making *generalising* value judgements about moral matters which by definition depend on and are grounded in experience, such as the moral status of unborn humans. Here, the mother-to-be has the highest moral authority in determining the status of the particular undeveloped offspring inside her womb. Above all, Lee emphasises the morally constitutive role of the experience of relationship. It refers, first, to the individual interpersonal relationship between the mother-to-be and her embryo or foetus and, later on, to the entire "moral community."

What Makes an Embryo Human?

In comparison with the cloning researchers Lu and Sheng introduced above, bioethicist Qiu Renzong inhabits a middle position. According to him, "a human embryo is a human biological life, a precursor of a person, not merely a stuff like the placenta, nor a thing having a price. So it deserves due respect. If there is not sufficient reason, it won't be permissive to manipulate or destroy it. Saving millions of human lives could be a sufficient reason."[38]

Thus, in principle, the embryo is at the discretion of utilitarian calculation. "Sufficient reason" to destroy an embryo means that, in principle, not all humans are equal in their fundamental worthiness of protection. However, the embryo is appreciated as a human being in biological terms. Accordingly, there are "degrees of personhood" that matter ethically.[39] In line with the cloning researchers, Qiu adopts a pragmatic and benevolent attitude, acknowledging the protection of (adult) humans and vulnerable people, including patients and donors, in particular, as a chief concern. Hence, the "difficulty to obtain human eggs" is quoted as a justification of (hybrid) cloning for research purposes.[40]

Moreover, Qiu Renzong emphasises the fundamental importance of social relations for morality.[41]

> The biological and psychological dimensions are necessary conditions for an entity to be a person, the social dimension is a sufficient condition for it to be a person.[42]

In probably the most distinct and concise description of the beginning of human life in moral terms by a Chinese bioethicist, Qiu offers the following summary:

> For the Chinese the newborn begins his/her *sheng huo* and not only his/her *sheng ming* at birth. Although *sheng huo* and *sheng ming* are both translated into English as 'life,' their meaning is quite different. *Sheng ming* means life in the biological sense, but *sheng huo* means life in the social sense. An individual's *sheng huo* must be unfolded in interpersonal relationships. There is no *sheng huo* if there is no relationship between the individual and others. Confucians pay more attention to such interpersonal relationships and formulate ethical norms for them (…). In Chinese, the term *lun li xue* (ethics) means norms of interpersonal relationships. The other two Chinese terms, *xing ming* and *ren sheng*, also show something important. *Xing ming* is the biological life, all forms of life have their own *xing ming*, but only a person has his/her *ren sheng* (personal life). *Ren sheng* begins with life after birth.[43]

Whereas it is not clear how the embryo and the foetus should categorically not have the property of *shengming* before birth, and how the latter differs ethically from *xingming*,[44] the peculiar and almost exclusive interest in human life after birth is re-emphasised and defended. With the focus on the morally constitutive meaning of life in a social sense as an extension and integral aspect of personal life, Qiu defies radically individualistic concepts of personhood and connects the substance of morality with the human being as a *zoon politicon*. Still, not much is contributed to the assessment of the early stages of human development, or to enlighten us about what makes a human being worthy of protection in his or her own right. Further exploration is encouraged as to the possibility to integrate rather than set individual and social aspects against one another.

Qiu presents an overview of the ethical positions adopted by his colleagues in the light of traditional Chinese ethics and his vision of the task of bioethics.

> Some scholars (...) argue for the abandoning of all research into the nucleus of the cell and the exploration of the secret of life. Others try to reinterpret Confucianism and Taoism in order to adapt it to modernity. (... He argues) that Confucianism and Taoism have to be reinterpreted in a way in which they can adapt to modernity, but that there is a hard core that cannot be interpreted in different ways or reinterpreted away. A proper interpretation or reinterpretation of Confucianism will not impede the application or development of biomedical technology, and meanwhile it can play a very important normative role.[45]

In his ontological view, Qiu provides an interpretation of the origin of the value of human life that goes beyond Lee's (and others') undivided focus on endowment by relationship by highlighting the concept of "filial piety" (*xiao*).[46] He explains that a Confucian standpoint holds,

> the body is a conjugation of the two finest energies, called *yin* and *yang* (...). Keeping personal bodily integrity is a value, for each body is endowed from her/his parents and cannot be harmed in any way, otherwise it would violate a primary ethical principle called *xiao*, or filial piety. (...) the body or any part of the body can be damaged only in the following two circumstances: 1. To

> meet her/his parent's needs. (...) 2. To sacrifice himself for practising *ren*.[47]

He confides that

> "manipulation of a gene or cell will not conflict with a reinterpreted Confucianism if it is for the treatment or prevention of disease," as long as it is guided by *ren* (humanity) as love, care, and doing good to others. As a general rule, "human action should follow the *dao* of nature. Humans carry forward the *dao*, but cannot transcend the *dao*. (...) The human genome will change for various reasons, but its 'identity' or 'integrity,' which makes *Homo sapiens* as it is, has no change. The same can be said of nature."[48]

In summary,

> Confucianism would not generally object to the use of technology. But it would object to its use when it would cause such a serious adverse impact on nature that it would become a totally different nature that no longer could sustain the survival and reproduction of most present and future species. (...) Genetically-enhanced or cloned humans may constitute a new species of human beings, which would affect the harmony between nature and humans.[49]

The Primacy of the Born Human

Hence, the general principles of moral concern in the area of human cloning, according to China's leading bioethicist, are established. They mean to prevent harm from or provide benefit for born human beings, and can be characterised in terms of a paternalistic social ethics of pragmatic conservatism. It delineates tentatively the range of legitimate interference as it is defined by the integrity of the nature of the "*Homo sapiens*, as it is."[50] In this vein, the student of China's bioethics would appreciate the elaboration of criteria to distinguish a modified nature that would be accepted from a "totally different nature," that would be deemed unacceptable in moral terms. As to the issue of belonging to the "moral community," we are not encouraged to conceive of early human life in terms independent from the judgements of others. Inasmuch as this means to grant discretion to the would-be mother, there is a remarkable grain of liberal confidence in the individual to be accounted for.

However, given the ambivalence and heteronomy of human psychology and moral experience, some consideration to counter-balance individual judgement and in particular to safeguard the early human being in its own right appear desirable, especially in the setting of the culture-transcending debate. That is to say, the main critical point of philosophical argument against human cloning, which is not so much related to hubris (or "playing God") or metaphysical dogmatism, is still waiting to be addressed. How does this approach respond to a transcendental and deontologically inspired hesitation to make a judgement (here: about the early embryo) that would exclude individuals from the moral community of humankind, with reference to any positive or empirical attribution?

The philosopher Cheng Chung-ying has contributed significantly to the development of contemporary Chinese philosophy and our understanding of the meaning of Confucianism today. In a systematic argument about the values and norms required to guide our biomedical and biogenic skills, he addresses the power issue. Cheng demands that "biomedical technological inventions should not be used as instruments of domination and political power, but must instead contribute to the cause of justice."[51] Seeking a "deeper understanding of humanity," Cheng at the same time warns against bio-reductionism and scientism.[52] Accordingly, he believes that "even a human foetus could be recognised as a human individual, albeit a potential one."[53] Therefore, "As human beings or potential human beings (human beings in-becoming), they deserve human care and human love as much as we are able to extend."[54] This assertion hints at the notion that early human beings have some intrinsic value that is independent from the meaningfulness generated through relationships. One wonders, however, to what degree and according to which moral standards our care must be extended to the early human being, whether this is our *duty* or depends on our *ability* to extend it. Cheng has not elaborated this further.

It is noteworthy that bio-reductionism was identified as a problem in China earlier. Investigations in the area of eugenics studies have shown that modern China has produced both reductionist concepts and non-reductionist concepts.[55] In bioethics, theories of social relatedness and narrative approaches bear a strong, albeit not fully explored, potential to reject reductionism by reminding us of the complexity of the related issues.[56]

Cheng's close affiliation with Confucian ideas as described earlier becomes clear when he discusses the challenges and adverse consequences of human cloning. "When we move to the area of manufacturing birth by test tubes, contract pregnancies, and cloning, (...t)he real problem is the consideration of the biological individual involved in contract pregnancy with regard to his or her becoming a moral human being. How would this individual feel in relation to three women who could claim to be his mother?"[57] The soundness of natural human relations is at stake. Unfortunately, Cheng does not discuss the creation of human embryos doomed for research purposes, that is, he avoids the issue that would help to clarify his notion of potentiality and the related worthiness of protection. It could also help to make sense of Qiu's declaration about the fusion of gametes being the ethically "central factor of human reproduction," the absence of which forbids human reproductive cloning.[58] This speculative notion could be interpreted in a sense that rejects bio-reductionistic views about procreation, or it could express a peculiar, normatively relevant appreciation of sexuality. Neither Cheng nor Qiu provides sufficient clarification about these fundamental matters. However, in order to gain more plausibility, their arguments would benefit from defining in some detail what it is, or what kind of quality it is that dignifies human life through such special acts of bonding, beyond proper moral character.

The Taiwanese medical ethicist, Tsai Fu-chang (Cai Fuchang), proposes a "two-dimensional concept of Confucian personhood."[59] Like Qiu and others, he acknowledges the crucial importance of social duties and relations. Moreover, he urges us to be critical about potential imbalances of power, especially in the doctor-patient relationship, in order to safeguard the individual. He argues that it is important in medical practice to "protect patients from being manipulated or coerced by the collectivist pressure (mostly from their families) and promote their autonomy without despising their traditional family values."[60] Again, alertness to the fair treatment of adults prevails over the interest in the particular situation of the early human life that is at stake in the practice of research cloning.

Beyond Functionality

In his analysis of personhood, in a different context of medicine, Au Kit-Sing comments that "evaluating personhood simply by assessing the patient's functional capacity is inadequate."[61] He adds another facet to the notion of social relatedness. The value we find in a human being depends on the way we *choose* to see it. For example, there can be a "powerful effect of the fatalistic labelling" on self-esteem and esteem of others. The impact of taking "a different perspective" may be vital,[62] and it is always substantial.

Whereas direct correlations cannot be established between the situation where capacities degenerate (the one to which Au refers) and the situation where capacities unfold, some ethical conclusions can be drawn from the moral reasoning invoked and from observations about moral judgement in general, without claiming these to be conclusive or fully applicable to all kinds of practice. One obvious conclusion is to look further than merely *functional* concepts in our attempts to assess any human being, including early development stages, in moral terms. The second apparent point is that we should pay utmost attention to what Au calls "labelling." We cannot pretend to understand an issue just because we have given it a conceptual *designation*, but we ought to be aware of implicit moral messages or rationales. Au by implication also hints at the danger of instrumentalising language, especially regarding notoriously difficult matters such as the moral meaning of a human life which cannot speak on its own behalf. A culturally interested bioethics appreciates the intricate and ambiguous characteristic of normative language.

We can appreciate that the observed ethical bias in favour of human beings *after* birth, which exceeds the issue of early human development, does not necessarily - logically or in principle - refute the possibility of ascribing inherent value and unrestricted worthiness of protection to the unborn. It would be perfectly sound for any of the positions described above to develop a more sympathetic attitude towards early human life, either for ethical reasons similar to those appealed to, or for independent considerations. Such a shift of focus would appear more likely if work on bioethics in China was carried out under less competitive conditions, provided that time was taken to further refine arguments and that more philosophers and critical social scientists were involved.[63]

If we take for granted a genuine moral interest in the entirety of human life, then the very early stages could be assessed, for example, in terms of Qiu's metaphysical postulate of *Yin* and *Yang* as a development programme, or in terms of intuitive bonding according to Lee's paradigm of relationship, or along consequential lines of reasoning, or the high esteem for *humanity* in general. Such a scenario, however, hinges not only on moral concern and ethical attention, but also on the definition of an early human being as a morally fully dignified entity (as suggested, for example, by embryologist Sheng Huizhen).

4. Concluding Observations

As argued above, in terms of cultural significance, the analysis of key players in bioethics is of limited merit, just as the relevant legal documents and ethical guidelines cannot tell us very much about Chinese culture.[64] None of them can claim to represent the majority of the Chinese people, a sufficiently long period in time, or a majority of certain groups in society, let alone the entire moral spectrum of a culture. First of all, laws are not produced in a transparent manner in the framework of biomedicine-related policy making, and key players are individuals with their own agenda. Therefore, they and/or their statements cannot symbolise genuine moral features that may be validated as straightforward manifestations of cultural life.

However, their statements still convey an array of culturally and ethically relevant information, even when they are appreciated as political testimony. They help the student to assess the mindset and preferences of a selected target group, although in the case of laws or regulations it can be difficult to identify the members of a particular group and their respective interests or values. They also inform us tentatively about a culture of politics, power and decision making that resides in the "black box," the core of power in a political system that is not transparent.

At the interface between bioethics and policy making in China, there is a tendency to argue that the destruction of an early human life form needs to be justified by ambitious medical objectives, which are otherwise unattainable. Cheng Chung-ying's notion that "if we grant ethics a realistic-ontological basis, there is common ground (...) for universal agreement and commensurable understanding" appears optimistic.[65] However, in the case of bioethics in China, a pragmatic *raison d'état* is indicated. Thus, even when disregarding the lack of political or cultural legitimacy, major compartments within the fabric of culture can be understood more clearly, owing to the factual normative import of such institutions and individuals.

As to the general philosophical concern about the sustainability of cultural diversity in an area of ethics, the early state of the relevant debates suggests patience and continued efforts to study carefully the intricacies of culture and bioethics in China, with special respect to the different levels of scholarship and discourse. However, if we take it for granted that the Chinese way, according to Qiu Renzong, postulates the fallacy of extremely individualistic and collectivistic approaches, it does not seem difficult to conceive of a cross-cultural consensus and common ground about respecting the individual, with its limiting qualifications.

Still, even a moderate perception of humanity requires deeper philosophical analysis. Ethically, the more disconcerting challenge seems to be the difficulty to identify, within any culture, who does exactly count as an individual, with the associated status of being worthy of protection, and why.

Nonetheless, the fundamental intuition that inspires the common-sensual statement of "seeking common grounds while reserving differences" corroborates that there are some values and principles we all must respect, or in fact do respect in practice, by virtue of arguing our differences peacefully and constructively, as citizens of the world.[66] Otherwise, such a slogan would appear as a newly redressed echo of the political human-rights rhetoric of the 1980s and 1990s.[67]

It is a matter of consistency and sincerity to call these values and principles, in their regulative form, *universal*. Philosophy's Eros compels bioethics to reflect on their normative implications and logical preconditions of an ethics with a regulative structure, and to explore them in content and range. The programmatic motif of a culturally alert bioethics accordingly is, how can moral diversity flourish and what kind of difference must be tolerated, among cultures and within.

As regards the particular issue of human cloning, the major question and challenge seems to be: What are the implications of the notion of "endowment through relations" for the role of agents who are in no way related to an early human life, such as researchers in the life sciences? If only bonding qualifies for moral judgement, for example, will an unrelated researcher be required to abstain from any ethical assessment of the embryo or embryonic material he wishes to use for research? Or, adversely, will a researcher have to build meaningful relations with the donors and the social background of the embryo before being allowed to act?

Both of these obvious implications appear to be pretty far fetched in light of feasibility and the research practice in China. However, what else can be the point of the bonding argument, as far as the researchers are concerned? It seems that, in addition to the greatly emphasised merit of a relational approach, there is no way to avoid highlighting the moral meaning of the early embryo in terms independent of or preceding all forms of bonding. In this regard, additional arguments from Chinese ethicists and embryologists are urgently called for.

Notes

1. Döring and Chen, 2002.
2. Döring, 2003a.
3. Döring, 2004a.
4. Döring, 2001a.

5. Döring, 2004d.
6. Simmel 1919 and 1998.
7. Qiu, 2002b.
8. Qiu, 2000, 2002a.
9. Unschuld, 2003.
10. Suttmeier and Yao, 2004.
11. Döring, 2004d.
12. Qiu, 1999.
13. Mann, 2003.
14. Chen, 2003.
15. Ethics Committee, 2004.
16. Döring, 2004b.
17. Leggett, 2003.
18. Cyranoski, 2004.
19. Personal communication with Lu Guangxiu in Changsha, May 22, 2004.
20. Denker, 2004.
21. Personal communication with Sheng Huizhen in Shanghai, May 25, 2004.
22. Döring, 2004c.
23. Systematic and teleological modes of argument in ethics can be found in classical Chinese philosophy, for example in *Mengzi* 6.1.8, where the intrinsic "goodness of human nature" is explained with reference to an eroded and barren mountain, which had originally been and is meant to be bursting with vitality (Döring, 2003a).
24. Roetz, 1993, 168-172; Fung, 1973, 603-605.
25. (Rance) Lee, 1999.
26. Shen, 1998.
27. Chen (Rongxia), 2002, 222.
28. Ibid., 221.
29. Ibid., 222.
30. Wang, 2000, 2002.
31. Nie, 1999, 2000a.
32. Lee (Shui-chuen), 2002, 175.
33. Ibid., 175f.
34. Lee (Shui-chuen), 1999, 192.
35. Compare the more radical views in Ip, 2002.
36. Rhind, 2003.
37. Lee (Shui-chuen), 1999, 197.
38. Qiu, 2003, 675f..

39. Qiu, 2000, 143.
40. Ibid., 143.
41. In terms of comparative cross-cultural studies, it is noteworthy that a German bioethicist, Dieter Birnbacher, albeit without explicit reference to Chinese or other Asian authors, has recently argued along quite similar lines of reasoning, refering to the core Confucian concept of "piety." (Birnbacher, 2004). Earlier, libertarian German ethics had argued from a view of rationalism, emphasising that the quality of dignity is conveyed through our capacity to moral reasoning (Markl, 2001).
42. Qiu, 2000, 141.
43. Ibid., 138f.
44. Ibid., 142. The formulation, "the newborn begins his/her *sheng huo* and not only his/her *sheng ming* at birth," might be flawed. It contradicts Qiu's clear notion that "the foetus is not a person, but still a form of human biological life" (Qiu, 2000, 142) and the logic of the gradual emergence of personhood. Moreover, the foetus is (presumably in a figurative sense) compared with the respected human corpse (Ibid., 142), although it is obviously a living human entity the destruction of which stands for justification. Still, the distinctions between embryo, foetus and early infant, or between *shengming* and *xingming*, in moral, terms remain unclear.
45. Qiu, 2002b, 72.
46. Ibid., 73.
47. Ibid., 76, with reference to the *Analects.*
48. Ibid., 78.
49. Ibid., 79.
50. Shi and Lin, 1996.
51. Cheng, 2002, 354.
52. Ibid., 335.
53. Ibid., 344.
54. Ibid., 342.
55. Dikötter, 1995, 1998.
56. Tao, 2002; Au, 2000; Nie, 2000b.
57. Cheng, 2002, 350.
58. Qiu, 2003, 672.
59. Tsai, 2002.
60. Ibid., 211.
61. Au, 2000, p. 214.
62. Ibid., 215.

63. Döring, 2004d, 220-224.
64. Döring, 2003c, 238.
65. Cheng, 2002, 346.
66. Qiu, 2003, 677.
67. Roetz, 2004.

References

Au, Derrick KS (Ou Jiecheng). "Brain Injury, Brain Degeneration, and Loss of Personhood." In *The Moral Status of Persons: Perspective on Bioethics*, Gerhold K. Becker (ed.). Amsterdam and Atlanta: Rodopi, 2000, 209-217.

Becker, Gerhold K. *Changing Nature's Course. The Ethical Challenge of Biotechnology*. Hong Kong: Hong Kong University Press, 1996.

Becker, Gerhold K. *The Moral Status of Persons: Perspective on Bioethics*. Amsterdam and Atlanta: Rodopi, 2000.

Birnbacher, Dieter. "Prinzip der ‚Pietät - Begründung der (begrenzten) Schutzwürdigkeit früher Embryonen." *Ethik in der Medizin* (16), 2004, 155-159.

Cheng Chung-ying. "Bioethics and Philosophy of Bioethics: A New Orientation." In *Cross-Cultural Perspectives on the (Im)Possibility of Global Bioethics*, Julia Lai Po-wah Tao (ed.). Dordrecht: Kluwer, 2002, 335-358.

Chen Rongxia. "Religious Emotions and Bioethics." In *Advances in Chinese Medical Ethics*, Döring and Chen (eds.). Hamburg: Mitteilungen des Instituts für Asienkunde, 2002, 214-222.

Chen, Ying; He, Zhi Xu; Lu, Ailian; et al. "Embryonic Stem Cells Generated by Nuclear Transfer of Human Somatic Nuclei into Rabbit Oocytes." *Cell Research* 13 (2003), 251-64. (see also http://www.cell-research. com/20034/2003-116/ 2003-4-05-ShengHZ.htm)

Cyranoski, David. "Korean bioethicists call for inquiry into stem-cell work." *Nature* 429 (03 June 2004), 490.

Denker, Hans-Werner. "Early human development: New data raise important embryological and ethical questions relevant for stem cell research." *Naturwissenschaften*, 91 (2004), 1-21; (see also www. springerlink.com/openurl.asp?genre=article&id=doi10.1007/s00114-003-0490-8)

Dennis, Carina. "Stem cells rise in the East." *Nature* 419 (2002), 334-336.

Dikötter, Frank. *Sex, Culture and Modernity in China*. London: Hurst, 1995.

Dikötter, Frank. *Imperfect Conceptions*. London: Hurst, 1998.

Döring, Ole and Chen Renbiao. *Advances in Chinese Medical Ethics*. Hamburg: Mitteilungen des Instituts für Asienkunde, 2002.

Döring, Ole. *Chinese Scientists and Responsibility: Ethical Issues of Human Genetics in Chinese and International Contexts*. Hamburg: Mitteilungen des Instituts für Asienkunde Nr. 314, 1999.

Döring, Ole. "Introducing a 'Thin Theory' for Cross-Cultural Hermeneutics in Medical Ethics. Reflections from the Research Project 'Biomedicine and Ethics in China'." *Eubios Journal of Asian and International Bioethics* Vol. 11 (2001a), 146-152.

Döring, Ole. "Verstehen als Anerkennen. Überlegungen zu einer zeitgemäßen Kulturhermeneutik am Beispiel der Medizinethik im heutigen China." *Bochumer Jahrbuch zur Ostasienforschung* Band 25, 2001. Bochum: Iudicium (2001b) 9-52.

Döring, Ole. "Global Governance, National State and Health System Reform: Assessing the Case of China." In *Globalization, Global Health Governance and National Health Politics in Developing Countries. An exploration into the dynamics of interface*, Wolfgang Hein/ Lars Kohlmorgen (eds.). Hamburg: Schriften des Deutschen Übersee-Instituts 60 (2003a), 269-285.

Döring, Ole. "Maßstab im Wandel. Anmerkungen über das Gute und die Medizinethik in China." In *Zukünftiges Menschsein: Ethik zwischen Ost und West* (Schriftenreihe des ZEI, Bd. 55), Ralf Elm und Mamoru Takayama (Hrsg.). Baden-Baden: Nomos-Verlagsgesellschaft (2003b), 319-353.

Döring, Ole. "China's struggle for practical regulations in medical ethics." *Nature Reviews Genetics* 4 (2003c), 233-239.

Döring, Ole. "Chinese Researchers Promote Biomedical Regulations: What Are the Motives of the Biopolitical Dawn in China and Where Are They Heading?" *Kennedy Institute of Ethics Journal* Vol. 14, No. 1 (2004a), 39-46.

Döring, Ole. "Zwischen moralischem Rubikon und rechtlichem Limes: Chinas bioethisches Selbstverständnis nimmt Gestalt an." *China aktuell* 7/04 (July 2004) (2004b).

Döring, Ole. "Was bedeutet ‚ethische Verständigung zwischen Kulturen'? Ein philosophischer Problemzugang am Beispiel der Auseinandersetzung mit der Forschung an menschlichen Embryonen in China." In *Weltanschauliche Offenheit in der Bioethik*, Eva Baumann; Alexander Brink; Arnd May; Peter Schröder; Corinna Schutzeichel (Hrsg.). Berlin: Duncker und Humbolt (2004c), 179-212.

Döring, Ole. *Chinas Bioethik verstehen. Ergebnisse, Analysen und Überlegungen aus einem Forschungsprojekt zur kulturell aufgeklärten Bioethik.* Hamburg: Abera (2004d).

Ethics Committee of the Chinese National Human Genome Center at Shanghai. "Ethical Guidelines forHuman Embryonic Stem CellResearch." *Kennedy Institute of Ethics Journal* Vol. 14, No. 1 (2004), 47-54.

Fung Yu-Lan. *A History of Chinese Philosophy*, Translated by Derk Bodde. Princeton (Princeton University Press), Vol. 2, 1973, 603-605.
Ip Po-keung (Ye Baoqiang). "The Ethics of Human Enhancement Gene Therapy." In *Advances in Chinese Medical Ethics*. Hamburg: Mitteilungen des Instituts für Asienkunde, Ole Döring and Chen Renbiao (eds.). Hamburg: Mitteilungen des Instituts für Asienkunde, 2002, 119-132.

Lee, Rance. "Ethical, Legal, and Social Implications of Human Genetics. Views of Chinese Intellectuals on Human Genetic Engineering." In *Chinese Scientists and Responsibility: Ethical Issues of Human Genetics in Chinese and International Contexts*, Ole Döring (ed.). Hamburg: Mitteilungen des Instituts für Asienkunde Nr. 314, 1999, 66-81.

Lee Shui-chuen. "A Confucian Perspective on Human Genetics." In *Chinese Scientists and Responsibility: Ethical Issues of Human Genetics in Chinese and International Contexts*, Ole Döring (ed.). Hamburg: Mitteilungen des Instituts für Asienkunde Nr. 314, 1999, 187-198.

Lee Shui-chuen. "A Confucian Assessment of 'Personhood'." In *Advances in Chinese Medical Ethics*, Ole Döring and Chen Renbiao (eds.). Hamburg: Mitteilungen des Instituts für Asienkunde, 2002, 167-177.

Leggett, Karby. "China Has Tightened Genetics Regulation - Rules Ban Human Cloning. Moves Could Quiet Critics of Freewheeling Research." *Asian Wall Street Journal* (13 October 2003): A1.

Mann, Charles C. "The first cloning superpower. Inside China's race to become the clone capital of the world." *Wired Magazine* (11 January 2003).

Markl, Hubert. "Der Mensch ist moralisch großzügig geschneidert und leidet an den Folgen seiner Doppelnatur." *Süddeutsche Zeitung*, 2.11.2001 (see also http://www.sueddeutsche.de/aktuell/sz/artikel91702.php).

Nie Jing-Bao. "'Human Drugs' in Chinese Medicine and the Confucian View: An Interpretative Study." In *Confucian Bioethics*, Fan Ruiping (ed.), 1999, 167-208.

Nie Jing-Bao. "Abortion in Confucianism: A Conservative View." *Proceedings of the Second International Conference of Bioethics*, Chungli. Taiwan (2000a), 130-155.

Nie Jing-Bao. "The Plurality of Chinese and American Medical Moralities: Toward an Interpretative Cross-Cultural Bioethics." *Kennedy Institute of Ethics Journal* 10 (3) (2000b), 239-260.

Qiu Renzong. "Medical Ethics in China. Status Quo and Main Issues." In *Chinese Scientists and Responsibility: Ethical Issues of Human Genetics in Chinese and International Contexts*, Ole Döring (ed.). Hamburg: Mitteilungen des Instituts für Asienkunde Nr. 314, 1999, 24-32.

Qiu Renzong. "Reshaping the Concept of Personhood: A Chinese Perspective." In *The Moral Status of Persons: Perspective on Bioethics*, Gerhold K. Becker (ed.). Amsterdam and Atlanta: Rodopi, 2000.

Qiu Renzong. "A Vision of the Role Medical Ethics Could Play in the Transforming of Chinese Society." In *Advances in Chinese Medical Ethics*, Ole Döring and Chen Renbiao (eds.). Hamburg: Mitteilungen des Instituts für Asienkunde (2002a), 3-11.

Qiu Renzong. "The Tension between Biomedical Technology and Confucian Values." In *Cross-Cultural Perspectives on the (Im)Possibility of Global Bioethics*, Julia Lai Po-wah Tao (ed.). Dordrecht: Kluwer (2002b), 71-88.

Qiu Renzong. "Cloning in biomedical research and reproduction: Ethical and legal constraints - A Chinese perspective." In *Cloning in Biomedical Research and Reproduction. Scientific Aspects - Ethical, Legal and Social Limits*, Ludger Honnefelder and Dirk Lanzerath (eds.). Bonn: Bonn University Press, 2003, 667-678.

Rhind, Susan M.; Taylor, Jane E. et.al. "Human Cloning: Can it be Made Safe?" *Nature Genetics Reviews* Vol. 4 (Nov. 2003), 855-864.

Roetz, Heiner. *Confucian Ethics of the Axial Age*. New York: SUNY, 1993.

Roetz, Heiner. "Philologie und Öffentlichkeit. Überlegungen zur sinologischen Hermeneutik." *Bochumer Jahrbuch zur Ostasienforschung*, Vol. 26 (2002), 89-111.

Roetz, Heiner. "Bioethik und Kultur. Zu einem schwierigen Verhältnis." *Technikfolgenabschätzung*, Nr. 1, 13. Jahrgang - März 2004, 85-89.

Shen Qingsong. "Youhe lunli panduan zhichi funong ren?" ("Are there ethical arguments to support human cloning?"), *Zhongwai yixue zhexue* (3), 1998, 125-144.

Shi Da-Pu (Shi Dapu) and Lin Yu. "The Conflict between the Advancement of Medical Science and Technology and Traditional Chinese Medical Ethics." In *Changing Nature's Course. The Ethical Challenge of Bio-*

technology, Gerhold K. Becker (ed.). Hong Kong: Hong Kong University Press, 1996, 112-118.

Simmel, Georg. "Der Begriff und die Tragödie der Kultur." In *Philosophische Kultur*, Georg Simmel (Hrsg.). Leipzig: Alfred Kröner, 1919 (2. Auflage), 223-253. Compare Georg Simmel. *Philosophische Kultur*. Berlin: Wagenbach, 1998, 195-219.

Suttmeier, Richard P. and Yao Xiangkui. "China's Post-WTO Technology Policy: Standards, Software, and the Changing Nature of Techno-Nationalism." *Special Report* No. 7, May 2004, The National Bureau of Asian Research (ed.).

Tao, Julia Lai Po-wah. *Cross-Cultural Perspectives on the (Im)Possibility of Global Bioethics*. Dordrecht: Kluwer, 2002.

Tsai, Daniel Fu-chang. "The Two-Dimensional Concept of Confucian Personhood in Biomedical Practice." In *Advances in Chinese Medical Ethics*, Ole Döring and Chen Renbiao (eds.). Hamburg: Mitteilungen des Instituts für Asienkunde, 2002, 195-213. (This is a revised version of his earlier publication: Tsai Fu-chang. "How should doctors approach patients? A Confucian reflection on Personhood." *Journal for Medical Ethics*; 27, 44-50).

Unschuld, Paul U. *Was ist Medizin? Westliche und östliche Wege der Heilkunst.* München: Beck, 2003.

Wang Yifang. *Jingwei Shengming: Guangyu shengming, yixue he renwen guanghuai de duihua* (Awed by Life: Dialogues on life, medicine and humanistic concerns). Nanjing: Jiangshu People's Press, 2000.

Wang Yifang. "Ideals and Ethics: the Concept of 'The Art of Humaneness' is not Reliable." (revised translation of "Kaobuju de renshu" from Wang, 2000) In *Advances in Chinese Medical Ethics*, Ole Döring and Chen Renbiao (eds.). Hamburg: Mitteilungen des Instituts für Asienkunde, 2002, 266-275.

Chinese Ethics and Human Cloning: A View from Hong Kong

Gerhold K. Becker

Abstract: The paper addresses the issue of human cloning from a Hong Kong perspective within the broader context of considerations about the implications of cultural differences for global bioethics. The tendency, particularly in contemporary ethical discourse in Hong Kong, to interpret cultural issues in bioethics as a vindication of ethical pluralism and of the need to replace bioethics in the singular with culture-based alternatives is examined and traced back to the formation of what has become the dominant bioethical model, principlism. It is argued that the alternative versions of bioethics under discussion fail due to conceptual inconsistencies and an impoverished understanding of culture at the intersection between traditional forms of life and the conditions of modern biomedicine. Recovering the rich texture of cultural diversity in Hong Kong as the framework for ethical discourse, the paper explores four major lines of argument for a ban on human cloning that are largely endorsed across Hong Kong society.

Key Words: Cloning, Bioethics, Personhood, Human dignity, Principlism, Culture, Chinese ethics, Confucianism, Hong Kong, Reproductive technology.

The search for commonalities in cross-cultural bioethics starts from the assumption of substantive cultural differences. Investigations of such differences have produced an expanding body of research that raises some stimulating questions about the cultural basis of bioethics and its implications for universal moral norms. Unless explorations in moral diversity are to serve mainly historical or anthropological interests, the discovery of cultural bases of divergent moralities seems to question the universality of ethical standards. Linking up with the larger philosophical question about the role of reason in post-modern times, cultural issues in bioethics have been taken as vindication of ethical pluralism and the need to substitute culture-based alternatives for bioethics in the singular, or at least for what is widely regarded as its dominant version in the West.

In such a context, I have been asked to contribute to this symposium specifically from a Hong Kong perspective. The expectations behind this request seem to be, firstly, that Hong Kong as a socio-political entity can claim a sufficiently recognisable cultural identity that warrants this special treatment alongside China; secondly, that there exists some sort of ethical and, in particular, bioethical discourse that differs somehow, and presumably for cultural reasons, from what is going on elsewhere; and, thirdly, that this discourse has something specific to say about the issue of human cloning that merits attention. All three expectations may be flattering to a representative from Hong Kong, but they are problematic. Before I can look at them in some detail, however, another and much broader concern needs to be addressed.

1. The Search for (Cultural) Alternatives to Bioethics

This concern relates to the background assumption that bioethics may be more intimately linked to culture than previously thought and that its cultural roots must be clarified before questions about a universally shared common morality and the universal validity of moral standards can be answered. The usual test case in the debate about a common morality is the universal normativity of human rights, their application and their enactment in international law. In the case of human cloning, such clarification seems all the more urgent, since its practical implications reach far beyond specific cultural traditions and may affect humanity itself. Whereas at the international level, at least at present, unanimity seems to exist across the various cultures in support of a global ban on human reproductive cloning, the underlying moral reasons in support of such legislation may well differ.

Needless to say, the issue concerning the normative claims of individual cultures on ethics is much too large for my present purpose, yet a few remarks may be in order so as to avoid misunderstandings.

I believe the recent debate about cross-cultural issues in bioethics has been spurred by two different but closely related developments with international (or cross-cultural) implications: the increased interest in the implementation of human rights in international, regional (conventions, covenants) and national law, and the development of an ethics with a specific focus on biomedical practice and research. Both developments have been charged with cultural bias and even cultural imperialism, and their alleged "Western" (or Eurocentric, or indeed liberalist) ideology prompted the search for alternatives.

It is certainly a historical fact that bioethics as a distinct subdiscipline of ethics originated in the North American debate in the early decades after the Second World War about the moral constraints of medical research. The publication of the so-called Belmont Report in 1978 by the *US National Commission for the Protection of Human Subjects of Biomedical and Behavioral Research* is seen as a milestone in this development, since it recommended a simple triad of moral principles which were thought universally valid and highly applicable in medicine and health care: respect for persons, beneficence, justice.

Yet it was the specific version of "principlism" that gave bioethics its distinct profile and thus galvanised the critique. By unfolding the second of the Belmont-report principles into two separate moral principles, Tom L. Beauchamp and James F. Childress arrived at their full set of four bioethical principles: autonomy, beneficence, nonmaleficence, justice. Ever since the invention of the "Georgetown mantra" (Jonathan Moreno) and its authors' explicit claim that it represents the core set of principles of a "common morality" as distinct from "customary morality") to which all "morally serious persons" supposedly subscribe, it has become highly fashionable to denounce this claim as culturally biased, and to offer alternatives to the model of so-called "Western" bioethics.[1]

If nothing else, principlism has put bioethics into sharp profile. As it appears to offer a shortcut to ethical decision-making even in complex contexts of medical research and health care scenarios, it holds great appeal to physicians, clinicians and institutional ethics committees; it has also been highly influential in the drafting of medical codes of conduct.[2] It has highlighted at least some of the most fundamental moral values nobody can ignore in today's biomedical practice. Or so it seems.

To its critics it is obvious that the principlist version of bioethics has overdone its job and put too sharp a focus on issues of autonomy, equality and justice at the expense of similarly important other values. Or, to put it differently: the significance of these values derives from their place within a system of cultural parameters that cannot claim universality.

Assuming a systematic point of view from within the "Western" bioethical paradigm, the critique is concentrated on the claim that "principlism" has failed to provide a "well-developed unified theory" of rules that would express the "unity and universality of morality."[3] Instead it is seen as suffering from the "anthology syndrome," which is characteristic of bioethics in general, by simply parading a variety of principles whose place in various unconnected and competing moral theories remains obscure.

From a cross-cultural perspective it has been argued that the four principles are incapable of serving as a reference points for the solution of ethical problems within the specific parameters of non-Western cultures that do not share their value basis. In particular, it is the insistence on the centrality of individual autonomy, informed consent, and truth telling that is not shared by bioethicists from Asia. To the critics, they resemble more an ad-hoc checklist of alien, unrelated, and abstract principles or moral directives than the fundamental norms of substantive ethics, which can only be the ethics of a particular culture.

Two things should be noted along the way. Firstly, Asian advocates of alternative bioethics tend to overlook cultural differences *within* the West and fail to recognise the critique of principlism from European critics and their contribution towards a more inclusive value basis of bioethics.[4] Secondly, it is possible to see in principlism above all not an empirical but a conceptual or transcendental claim about the four principles that is largely immune to objections on the ground that a particular moral tradition or culture lacks one or more of these principles.[5] Although Beauchamp in particular has pointed to support for principlism in the alleged factual convergence of different cultures towards the four principles,[6] another and more challenging reading finds in these fundamental principles the necessary condition for cross-cultural moral judgement: "These fundamental precepts alone make it possible for a person to make cross-temporal and cross-cultural judgments and to assert firmly that not all practices in all cultural groups are morally acceptable."[7] "What justifies the norms of the common morality is that they are the norms best suited to achieve the objectives of morality."[8]

The recent search for alternatives to Western bioethics then is, to a large extent, a direct response to "principlism" and its claim to offer a moral language and a set of moral commitments that can be shared with everyone regardless of individual philosophy, political inclinations or religious beliefs.

2. Ambiguities of Alternative Bioethics

Whereas attempts at the redefinition of human rights along cultural lines have become notorious for political overtones that found their clearest expression yet in the *Bangkok Declaration on Human Rights* (1993), they share with critics of Western bioethics the focus on issues of autonomy, personhood and the very concept and basis of moral rights. In contrast, culture-based alternative models tend to organise their guiding values around interpersonal and communal concerns within the thick fabrics of substantive society-based moralities. In both areas, the Chinese cultural tradition has been promoted as one of the strongest contenders for a fully-fledged alternative to Western models of both human rights and bioethics.

It is, of course, not the only contender. Before I turn to one version of such alternatives that plays a particular role in what may be called the Hong Kong bioethical discourse, I will briefly consider issues arising from a related approach towards an alternative model of Western bioethics.

In their book *Beyond a Western Bioethics: Voices from the Developing World*, the authors from the Philippines challenge Western bioethics by what they claim is "the radically different character that bioethics takes in the developing world."[9] They buttress their claim that Western bioethics is inherently incapable of capturing the different "spirit" of Filipino bioethics by contrasting the Western focus on "anonymous formal principles," on conceptual analysis, on the individual, and above all on autonomy with a moral phenomenology of socially lived values revolving around the family. On this account, the deficit of Western bioethics is its neglect of the experiential basis of ethics in a concrete and socially centred way of life. Steeped in strong emotions, religious beliefs and the inescapability of the social web, the authors portray Filipino (bio)ethics as structured by the forces of traditional society. The medical life world that provides the culture-specific parameters within which the alternative model of bioethics unfolds is a direct extension of the larger social world and its unquestioned hierarchy of traditional authorities. The social institution most important for the grounding as well as the transmission of moral values and thus for the formation of bioethics is not the state but the family. As the locus of support and assistance in daily living, the familial community holds final authority in decision making for its members and thus becomes the prime model in the organisation of all areas of social life. Grounded in the socio-moral responsibilities of familial authority and its overriding concern for harmony, Filipino bioethics assumes an unabashed paternalistic attitude that leaves little space for individual choice and dissenting views. Moral agents are expected to have fully internalised the dominant social order and to submit to authority. As the authority is represented through individuals and extends in form of concentric circles from the parents and elders to employers, the state and the Church, Filipino culture is seen as person-oriented in as far as persons "take precedence over abstract, impersonal issues or ideas" or "concerns for social justice or honesty." In consequence, Filipinos are said to be "very respectful," "reluctant to speak their minds openly," and willing to "let those in authority make decisions" on their behalf. "Abstract philosophical and political issues such as justice or fairness usually are not considered as significant as personal allegiances."[10]

The impression one gets from this account of Filipino bioethics is highly ambiguous. While the interest in an alternative to Western bioethics is obvious, it is neither obvious that this account of bioethics is truly representative of *contemporary* Filipino culture nor that it is the attractive alternative to Western bioethics it is supposed to be. Only two points can be mentioned here.

Firstly, the authors seem to take a myopic view of their own culture without realising that, while traditional Filipino culture still exerts its influence in contemporary society, the dominant cultural factor in public life is, on all accounts, the Catholic Church and her social and moral teachings. Yet this cultural force is clearly Western in origin or at least alien to traditional Filipino culture and in robust opposition to traditional Filipino religion, which is mainly animistic. The influence of the Catholic Church is, of course, a typically Filipino phenomenon and not characteristic of any other Asian country, although similarities may be found in the dominance of the non-indigenous cultural force of Islam in the societies of Malaysia and Indonesia. The question that needs to be addressed, then, concerns the role of religion in contemporary bioethics.

Secondly, and more importantly, the account of the Filipino life world and its fundamental values is the ahistorical description of the status quo of a static society without any major socio-economic tensions and without doubts about the legitimacy of the authority of the powers that be. It is doubtful that major normative bioethical conclusions can or should be drawn from the socio-cultural facts the authors have described. Assuming that the existing structures of authority and the channels of decision-making have been accurately identified, the reasons for the apparent neglect or even the absence of individual autonomy in such processes may nevertheless be misrepresented. Even if they applied to the closed traditional Filipino society of the past, they may be a reflection of the dominant influence of socio-economic forces rather than a genuine expression of Filipino culture. The issue then is to what extent social models of the past can still claim legitimate moral authority to define and resolve bioethical concerns of the present.

Apparently, it is not simply a methodological matter to delineate an alternative model of bioethics along the lines of traditional culture and structures of power and authority but an issue of prime concern that raises questions about the normativity such models may claim even within the confines of their own cultures. Whether or not the principle of autonomy that within the bioethical context unfolds into matters of informed consent and truth telling, among others, can find its genuine place in a specifically Filipino bioethics cannot be answered with reference to the apparent disinterest of poor people with little or no formal education to burden themselves with responsible decision making, particularly in a medical environment that is alien to their own life world. The respect these people have traditionally shown to authority may be more indicative of influences of socio-economic factors than of alternative moral values or their ranking. As differences between rural and urban Filipino society become currently more entrenched and conflicting interests emerge, the supremacy of the traditional model of bioethics is being challenged from within Filipino society itself, and the specific features of an alternative Filipino bioethics in the singular become even more elusive.

The recent legislative initiative in the Philippines towards instituting a *Magna Charta* of Patient Rights not only confirms tensions within Filipino society that may run most clearly along the rural and urban divide but also challenges the traditional model of Filipino bioethics (as portrayed by Alora et al.) and its claim to truly represent the Filipino alternative to Western bioethics.[11] The prime focus of the Magna Charta on issues of autonomy and rights (including the rights to privacy and confidentiality, to disclosure of and access to information, to choose one's physician, to refuse participation in medical research and, above all, the right to be informed of one's rights and obligations as a patient) reflects the need for the review of traditional family structures and their bioethical implications and addresses shortcomings that can no longer be remedied from within. While such a shift may appear to some as a betrayal of traditional morality and a yielding to alien, presumably Western, influence, it is part and parcel of the public discourse on the values of contemporary Filipino society that also defines its bioethics.

Presumably, culturally based alternative models to Western bioethics can be most readily identified within homogeneous societies with dominant background institutions. The increasing complexity and diverging social interests of traditional societies generate moral conflicts. Traditional, culture-based moralities may be able to address these conflicts only up to a point, beyond which they would require new resources. Such new resources will neither be entirely alien nor exclusively indigenous, but they may well have a basis in traditional morality whose significance for the evolution of ethics may have been ignored or underestimated. If it is not simply the traditional that people are seeking, but the good (Aristotle), one should expect that no version of ethics, and in particular bioethics, can be construed as a window-less, closed system revolving around nothing else but its own centre.

In drawing some provisional conclusions from these deliberations, some light may fall on the cultural basis of bioethics in Hong Kong.

It seems that the search for cultural alternatives to Western bioethics would imply more than a haphazard collection of certain alternative values whose bioethical significance is taken for granted but whose socio-cultural moorings remain dubious. The inner complexities of cultures and cultural traditions must not be underestimated, particularly since they apparently gain identity only through intimate interactions with rival forces within the same cultural tradition. Moral dynamics cannot be reduced to static and fixed snapshots without becoming unappealing and lifeless relics of the past. Their value will be similar to that of historical artefacts which hold a fascination for those collecting them, but whose influence on contemporary society is rather limited. I believe that the search for cultural alternatives to bioethics requires a comprehensive and public moral discourse that involves all the moral resources and actively engages all the major stakeholders. Whether such discourse will indeed result in a full-fledged alternative bioethics or whether it will produce only a minimalist set of values reflecting the smallest common denominator, may be of less significance than the insight it will facilitate into the shared foundations of morality.

3. Some Problems with Confucian Bioethics

Hong Kong is certainly no closed society, and it is therefore doubtful that we should find a specifically Hong Kong bioethics. It is, however, a predominantly Chinese society, and that has suggested to some Hong Kong bioethicists that a specifically "Chinese" ethics must prevail and that the leading candidate of such ethics is Confucianism.[12]

Yet such a conclusion may be a little rash, and this for several reasons. While it is true that Confucianism has made an indelible mark on Chinese society, it has neither been a monolithic doctrine nor was its influence exclusive. Not only can "Confucianism" stand for many things, including "a world view, a social ethic, a political ideology, a scholarly tradition, and a way of life,"[13] there are also significant and bioethically relevant differences in its perception at the scholarly and the grass roots levels, and between the actual influence it exerts in rural and in urban areas. Furthermore, various factors suggest that the Confucian tradition in

Hong Kong is less pronounced than in rural areas on the Chinese mainland. Such differences manifest themselves more readily when one leaves behind the usual grand constructs of a dominant version of classical Confucianism and examines its actual and measurable influence in people's daily lives at the micro-level.

Secondly, the Chinese moral tradition in general and that of Hong Kong in particular are notoriously complex and by no means confined to Confucianism. The usual concentration on ancient or medieval Confucianism ignores not only the great philosophical rivals to Confucianism (Daoism, Mohism, Legalism) as well as the Buddhist tradition but also the variety within Confucianism itself. Yet more importantly, it neglects the complexity of contemporary Chinese culture and morality. It makes, of course, good sense to ask whether e.g. Mencius' or Xunzi's representation of Confucianism offer a conceptual basis for modern issues such as autonomy and rights, or whether such concerns remain outside their ambit. Much has been made of the fact that the Confucian tradition is primarily focused on relationships and reciprocal responsibilities instead of prioritising the role of individuals, whose duties can be independently alt: assessed, and that this focus is even reflected in the Chinese language, which originally had no single term corresponding to the English term "right." Yet it is much less philosophically appreciated that the Chinese tradition, in spite of its reputed isolation from the West, has been actively engaged in a rights discourse for more than hundred years. A discourse *within* Confucianism began to emerge that roughly coincided with the modern rights debate in Europe and squarely addressed Confucian inadequacies in the recognition and protection of legitimate interests of the individual. Drawing on historical and contemporary Chinese resources, the case has been made for the existence of a distinctive discourse about rights with its own concepts, motivation and trajectory.[14] It has even been claimed that the contemporary human rights discourse in China is closer to the original intention of the Universal Declaration than some of the solemn proclamations in the West.[15] This development shows not only the "dynamic, interactive, and internally contested nature of Chinese rights discourse," but also its significance in the search for common moral ground across cultures.[16] Similarly, Chad Hansen has reminded us that China's history of contact with Western ideas is as long as that of North America, and that Hong Kong "now has a tradition of abiding by the rule of law that is nearly as old as the abolition of slavery in America."[17]

All this suggests that the Chinese moral tradition available for the construction of a contemporary bioethics is not limited to the model of classical Confucianism on which it is usually exclusively based. As for Hong Kong in particular, the Chinese moral tradition in its full complexity is but one, albeit a powerful factor in the development of a bioethics with Hong Kong characteristics.

Lastly, it is still not clear to what extent key bioethical concepts such as the concept of the person can be reconstructed on specifically Confucian premises without endorsing what may be deemed "certain unattractive features" deriving from its socio-historical basis with its rigidly hierarchical relationships. Jiwei Ci has argued that the construal of moral relationships exclusively in terms of kinship (family) and hierarchy-reciprocity (state), could be successful only in a traditional agrarian society without significant geographical mobility and the ensuing need to establish moral relationships outside the boundaries of family and clan. To people in modern societies and impersonal political and economic orders, Confucianism has little to say as to how to fill the moral space that defines the relationship between relative strangers: "As a result, those who have absorbed the Confucian concept of human relations would be socially and ethically at sea if they were to enter into relations with strangers, where the conjunction of hierarchical-reciprocal relations and kinship ties simply does not exist."[18]

This suggests it is doubtful that a comprehensive "Confucian bioethics" can be construed which "stands out as a significant communitarian bioethics that offers a coherent way of engaging the good life."[19] The attempt to press Confucian bioethics into the matrix of a debate defined by liberalism or cosmopolitanism on the one side and communitarianism on the other seems problematic. It is also doubtful that a Confucian bioethics could be construed solely on the basis of the classical model if it were to transcend both versions of moral theory and be able to respond persuasively to the same set of contemporary moral issues the rival theories are supposed to resolve. The doubts result from the intrinsic relationship between Confucianism and the parameters of a particular cultural tradition and a particular society at a specific historical juncture to which Confucianism responded and by which it was shaped. Attempts to counter issues of autonomy and human rights with reference to a relational concept of the person would have to show that Confucianism indeed has the moral resources to reconstruct the foundations of contemporary Hong Kong society without reverting to a "Confucian civilisation" revolving around a state-sanctioned ideology of rigidly hierarchic human relationships. The ambiguity of the affirmation of the Confucian tradition has been noted not only in bioethics but also in policy analyses of the current

Hong Kong government. The government's reluctance to tackle the issue of democracy as stipulated in the Basic Law and its reliance instead on top-down decision making has been interpreted as a deliberate reversion to the political values of a Confucian past with its focus on authority, obedience and social stability, which would appear quite out of tune with the political and social aspirations of the Hong Kong public.[20]

It seems then that a Confucian bioethics faces a dilemma: Provided that Hong Kong constitutes a pluralist "public sphere" of diverse moral communities where Protestants and Catholics, Muslims, Buddhists, and Taoists live alongside Confucians and where the overwhelming majority belongs to none of these but "keeps an open mind" about any traditional values of particular communities, ethical rules exclusively based on and determined by Confucianism would have little appeal outside the small circle of committed Confucians. Yet, "if the ethical rules and guidance in question have nothing, or too little, to do with the traditions of Chinese philosophy, then Chinese bioethics will lose its distinctiveness."[21]

4. Cultural Space and Public Moral Discourse

The search for Hong Kong bioethics has to start from the assumption of a plurality of comprehensive moral (and religious) doctrines (John Rawls), which cannot be reduced without distortion to a unified Chinese morality. Attempts to construe Chinese medical morality and culture as monolithic require an explanation. In the opening part of this paper I suggested one such explanation with reference to the very birth of bioethics and its perceived value imperialism, which provoked the search for cultural alternatives and indigenous sets of moral values and norms. Similarly, Jing-bao Nie points out: "The assumption of a monolithic and unified Chinese culture in general, and a singly medical ethics in particular is a myth." Nie suggests a further explanation for the search for Chinese bioethics by linking it up with attempts of earlier anthropologists (e.g. Lucien Levy-Bruhl) to discover the specific mentalities of primitive societies. In a similar vein, "twentieth-century sinologists have attempted to identify the unique way in which the Chinese people see, think about, and approach the world." The question is, of course, "is there really a unique Chinese mentality? And if there is, does that mentality truly dominate each individual and pervade every aspect of Chinese life?" For Nie, "the answer to both these questions is 'no'."[22]

As far as Hong Kong is concerned, one can safely assume that it is not and never has been a homogenous society with one specific mindset. While it is usually taken for granted that Hong Kong has developed a rather unique blend of a dual socio-cultural identity comprising the values of traditional Chinese society and its Western colonial masters, even this description has been challenged as superficial. The Chinese population of Hong Kong is made up of various ethnic groups that include native villagers descending from "the five great clans" that moved into the area from Canton, the Hakka, who had to settle on the hilly and infertile land not claimed by the clans, and the Tanka (or "boat people") communities as well as Chinese immigrants.[23] While the latter group represents the overwhelming majority of Hong Kong people, it is not homogeneous either but comprises people of different dialects and geographic origin who arrived in Hong Kong at different times. Until recently, a sense of the temporary and transient (attached to the various waves of immigrants and to the migrants' search for further destinations) was one of the characteristics of Hong Kong identity that channeled its fabled energy and vitality toward the economic sphere.[24] It also fostered a materialistic mind-set with a pragmatic and "generally amoral attitude towards success and failure, hopes and frustrations."[25]

Instead of postulating a fixed and unitary cultural identity, the identity of Hong Kong people seems neither stable nor free of tension. It is constantly (re-)negotiated between the local cultures of diversified ethnic groups in the rural areas (or what is still left of them) and the population in the urban centres integrated into the fabric of an expanding metropolis. Yet this negotiation has been driven by a variety of forces, including the forces of colonialism, industrialisation and urbanisation, the traditions of British common law and Chinese customary law, and the forces of modern science, technology and medicine. Yet the struggle for identity takes place against the backdrop of traditional popular beliefs and customs, which provide a sense of continuity and stability in the midst of change and uncertainty. The cultural uniqueness of Hong Kong is encapsulated in the following verses called *Eggplant* by Ping-kwan Leung, a Cantonese-writing poet living in Hong Kong:

> "With what mixed feelings, I wonder, your parents
> had followed the flux of emigrants and crossed the wide seas
> in time, hybrid fruit and new vegetables slipped into their vocabulary
> their tongues slowly got used to foreign seasonings
> Like many of their generation, people began to drift

away
from the center, their appearance changed
But now and then from shreds of something here and bits of
something elsewhere we discovered a vaguely familiar taste
like meat and skin cooked to a mush, gone apart
back together again: that taste of ourselves, extinct, distinct."[26]

The complexity of Hong Kong culture is clearly reflected in the emerging bioethics discourse and most evident in the following three areas: high-tech medical intervention, health care (and its reform) and the evolution of the law.

The socio-cultural parameters of Hong Kong bioethics are particularly noticeable at the intersection of modern medical technology and traditional Chinese beliefs. One such example is transplantation medicine, which has challenged the traditional Chinese view of the human body. Although transplantation of solid organs and bone marrow "are now well-established and successful procedures in Hong Kong,"[27] cultural factors have jeopardized the potential health benefits by keeping the rate of organ donation at considerably lower levels than in Western countries. Surveys on organ donation have consistently confirmed traditional attitudes and beliefs about death and the need to preserve an unmutilated and complete body for afterlife.[28] These cultural perceptions are not shared by the medical community, which has tried to alleviate the shortfall of transplant organs through public education campaigns.

Secondly, the government's health care reform program based on the principles of "access to reasonable quality and affordable health care" for all citizens and the responsibility to share the costs provoked strong criticism from Hong Kong bioethicists of Confucian provenance.[29] The government's approach was charged with ignorance of the dominant values of Hong Kong and with a failure to truly understand Hong Kong society's basic moral vision by emphasising equity as the overriding value for its health care system. It was argued that Hong Kong society's approach to social policy is a care-based approach, which emphasises the value of care more than equality.[30]

Lastly, the evolution of the law has been a forceful influence on Hong Kong public morality. The enactment of Land Inheritance legislation, the establishment of the *Equal Opportunities Commission* and the

Human Reproductive Technology Bill are cases in point that prove both the complexity of the value system of contemporary Hong Kong society and its continuous, yet not uncontested development. The Land Inheritance legislation was initiated by liberal Chinese politicians and intended to outlaw the traditional practice of the original inhabitants of rural Hong Kong, which allowed only male descendants to inherit property. While this tradition was certainly in accordance with Confucian teachings and had been protected by the colonial administration for nearly one hundred years, the Chinese proponents of its abolition saw in this tradition nothing more than the remnants of feudalism and the violation of the principle of gender equality. Whereas their rural opponents put up strong resistance and mobilised noisy (and on occasion violent) street protests, they (cleverly) avoided to address the issue of equal rights and instead organised their campaign as a defense of the traditional way of life against urban intrusion under the slogan: "Protect the Native Homeland, Defend the Clans." It is perhaps not surprising that the attempt to preserve tradition for tradition's sake did not succeed, mainly because it was too obvious to the larger Hong Kong public that such a defense of tradition was merely a smoke screen for the protection of rather tangible economic interests.

The *Equal Opportunities Commission* (1996) is a further indication of the capability of the Hong Kong value system to integrate various moral traditions that have taken root in the hearts and minds of Hong Kong people. The Commission has been charged with administering laws that define specific individual rights, promote gender equality and provide for protection against discrimination on the grounds of sex, marital status or pregnancy in employment, education and services. The significance of this legislation can only be appreciated against the historical background of Chinese customary laws (with obvious Confucian moorings), which existed in Hong Kong side by side with the English Common Law. With very limited property rights and no entitlement to inherit family property women were financially and socially dependent on their husbands. Whereas they were allowed to be married only to one spouse during their entire lifetime, a man could have one principal wife and as many concubines as he could financially afford. The ease with which the centuries-old practice of polygamy was abolished (1971) is a further indication of the capability of contemporary Chinese society to integrate new moral values.

The bioethical discourse has been most strongly influenced by the long process towards the *Human Reproductive Technology Bill* (2000) and its three public consultations along the way.[31] The legislation is based on a broad consensus for reproductive technology to be regulated by a statutory supervisory body and a strict licensing regime for both medical practice and research employing reproductive technology. Although the government attempted to take "a value-free attitude" towards reproductive technology, the objectives of its regulatory body (*Council*) seek to ensure "the safe and informed practice of reproductive technology in a way that respects human life, the role of the family, the rights of service users and the welfare of children born through reproductive technology." Instead of "a value-neutral attitude," the Bill is in fact based on the principles of equal liberty, equal worth, and family integrity.[32] While the latter principle reflects an emphasis on traditional Chinese family values, the Bill is representative of the plurality of values that characterise contemporary Hong Kong society.

5. Ethical Perspectives on Human Cloning

In tackling the complex ethical assessment of human cloning, I restrict my exploration from the perspective of the bioethical discourse in Hong Kong to the following main areas of moral concern that have taken centre stage in the international debate ever since the successful application of somatic cell nuclear transfer (SCNT) brought human cloning a step closer to reality: personhood, integrity, reproduction, and human relationships. The resulting broad consensus holds that human cloning constitutes a moral wrong on the grounds that it:

(1) violates human dignity and instrumentalises human life,
(2) infringes the bodily integrity of the clone,
(3) interferes inappropriately with sexual reproduction, its nature and purpose, and
(4) distorts human relationships and forms of social bonding.

This view is almost unanimously shared across different moral traditions in Hong Kong, including all major religions, although the underlying reasons vary. The notable exceptions are, however, positions within the parameters of Confucian bioethics. Views also differ somewhat about the moral assessment of the extent, if any, of the harm the technique would potentially inflict on the clone.

A. Human Cloning Legislation

The near unanimity in the rejection of human cloning is also reflected in recent legislation. The *Human Reproductive Technology Bill* (2000) lists among the prohibited practices in section 13 (1) (e) and (f): to "replace the nucleus of a cell of an embryo with a nucleus taken from any other cell" and to "clone any embryo." Contraventions against these prohibitions are offences and liable to fines and imprisonment for six years (section 36 (1).

More controversial issues arise both internationally and in Hong Kong when the debate moves from the grand picture of human cloning in general to the small print by addressing issues such as the moral status of the human embryo and the moral implications of the distinction between reproductive and therapeutic cloning.

The moral position that stands in the background of the *Bill* has been called the intermediate position. It neither attributes to the embryo full moral status and personhood, nor does it see in it simply some human tissue that falls into exclusive ownership of the reproductive parties. Instead, the embryo is recognised as the earliest form of human life with the full genetic potential to develop into a human person. Although its moral status is weaker than that of a human person, it makes justified moral claims on us and deserves respect. The fact that the creation of human embryos for research purposes has been made an offence in law is an expression of such respect. This view accords the human embryo moral respect, which increases relative to its biological development, but is not absolute. Considerations of respect may be weighed against the significance of therapeutic benefits arising from research. On this basis, it has been considered ethically acceptable to use embryos up to the appearance of the primitive streak for clearly defined research purposes, if such use would result in specific and considerable therapeutic benefits.

It is obvious that the intermediate position is rather ambiguous, yet it is generally supported across the various sectors of Hong Kong society, with the exception of Christian groups.[33] The support may reflect a common attitude in the Chinese tradition towards prenatal life, which (according to Renzong Qiu) found one of its earliest expressions in Xun Kuang's (Xunzi) view that human life begins only at birth and ends in death.[34] Thus Confucianism is seen as tolerating not only abortion at any stage of fetal development but also infanticide, or, as Paul Unschuld put it, the practice of abortion "evoked little, if any, concern among Confucian thinkers."[35] This may also explain why neither abortion nor embryo research became a major issue in the public debate in Hong Kong.

While reproductive human cloning is expressly prohibited in the *Bill*, the question of therapeutic cloning is less clear. The issue has been raised recently in the context of the deliberations on human embryonic stem cell research. It was argued that as the resulting product of therapeutic cloning did not meet the requirements of the legal definition of the embryo as stipulated in the Bill (the product of fertilisation), the application of cloning in embryonic stem cell research could not be construed as a violation of the cloning prohibition. Although no final decision on the legality of human embryonic stem cell research has been reached, the Council on Human Reproductive Technology has extended its own definition of "embryo" so as to bring it under the ambit of its regulatory authority.

B. Major Moral Concerns

(1) Cloning and Personhood: Traditional Chinese ethics seems to have anticipated Locke's morally relevant distinction between a human being and a person without, however, defining the latter in terms of specific states of consciousness but rather through its relations with others. While not exclusively a "forensic term appropriating actions and their merit" (Locke), the Chinese concept of the person is similarly grounded in social interaction. With the exception of moral positions based on Christian and in particular Catholic doctrines and theological teachings, the dominant Hong Kong conception of human personhood is therefore not substantive but relational.

Several attempts have been made at reconstructing a contemporary version of the Confucian relational self, "whose individuality is constituted by a web of unique role relations," that could be utilised in the bioethical debate on human cloning.[36] While these reconstructions all draw on the same Confucian conception of personhood, their individual points of departure lead them to diverging conclusions. Edwin Hui's reconstruction of the Confucian concept of personhood is a direct response to what he takes as a "clear proclivity" in Western bioethics "to define human personhood in terms of higher-brain functions or some similar psychological criteria."[37] He argues that such criteria of personhood fail to account for accepted social practice grounded in "the dominant social ethos," which does not permit for "marginal life" to be "defined out of existence." As the lists of the criteria of personhood vary, such accounts reveal not only a certain extent of arbitrariness but also a remarkable alienation from public morality and law. Attempts to address this conflict by differentiating between "persons in the strict sense" and "social persons" (H. Tristram Engelhardt) reinforce the inherent ambiguity of such accounts.

Instead, Hui seeks to amend this conceptual basis by re-evaluating the Confucian relational self on the basis of the "perichoretic" (mutual interpenetration) conception of personhood as it was first developed in seventh-century Trinitarian theology. Hui claims that Christian Trinitarian theology has laid the foundations for a relational understanding of creation upon which a conception of personhood can be constructed that no longer shares the shortcomings of the Cartesian tradition. Against the backdrop of Christian onto-theology, the Confucian concept of *ren* can be seen as "a serious attempt by the Chinese people to probe the relational dimension of human nature as an instance of the manifestation of the perichoretic nature of reality."[38] As a relational term, *ren* constitutes personhood in a complex ontology of relationships that involves mutuality and commonality of the self and the other. "Instead of seeing a person as an independent, individualistic, and autonomous self, a Chinese person sees himself or herself as a member of the Heaven-Earth-man triad, inseparably connected to and mutually dependent on the transcendent heaven, the natural environment, and above all the community in which the person lives and from which the person derives his or her identity."[39] Thus Confucian personhood is a continuous process of "person-making," which is grounded in "the relation that exists between the self and the other both of whom attain personhood in that matrix." Hui concludes that on the basis of such relational understanding of personhood, human cloning would destroy natural relationships and therefore be morally wrong.

Similar conclusions have been drawn by Ruiping Fan, albeit from an exclusively Confucian viewpoint. Fan seeks to develop from his reconstruction of personhood a Confucian equivalent of the principle of human dignity, which would be violated by human cloning. While Fan concurs with scholars arguing that "a genuine concept of human dignity took form in the thought of Mencius," he extends the scope of the search for the most promising sources for a Chinese theory of human dignity beyond Mencius[40] and traces equivalents of the concept to a variety of ancient texts, including the *Book of Rites*, the *Book of Changes* (*Yijing*), particularly the *Great Commentary*, and the Doctrine of the Mean (*Zhongyong*). On his account, the Chinese version of human dignity is grounded in something "awesome" or "royal" that inculcates in human beings a sense of uniqueness and worth that must not be violated. As human beings have been assigned their position in the natural order by Heaven, they are placed in a system of natural relationships which make normative claims on them. Thus the natural order grounded in Heaven extends to human society and implies moral norms that define interpersonal relations. Fan concludes that as the relationship between parent and child is an integral part of the natural order established by Heaven and thus morally significant: "Any action or scientific innovation that jeopardises the relation of parent and child violates human dignity. From the Confucian view, the cloning of humans destroys the relation of parent and child. Accordingly, it is morally unacceptable for Confucians to practice human cloning."[41] The point Fan is apparently making is that human dignity is grounded foremost in filial relationships constituted by Heaven. Julia Po-wah Tao Lai has similarly argued that "the source of one's sanctity as a human being" that prohibits that either "self" or "other" can be a means to the other's end lies in "one's role relationships and role performance."[42] One implication for the issue of cloning is that human dignity can only be attributed to human persons in such relationships, but not to prenatal human life. Thus wasteful embryo research in the context of human cloning would not violate human dignity.

I believe Hui and Fan hold positions representative of the current bioethical discourse in Hong Kong. They both draw on classical sources of the Chinese ethical tradition in their attempt to develop alternative answers to contemporary bioethical problems that take a critical stance on the issue of Western bioethics. Interestingly though, the alternatives they offer are, to a considerable degree, defined by the model they wish to reject. This is a further illustration of the extent to which contemporary Chinese bioethics has in fact integrated components from alien traditions. It is therefore doubtful whether a purist reconstruction of Chinese bioethics could succeed. While Hui as a committed Christian seeks support for his version of relational personhood in both Confucianism and the Christian tradition, the latter's contribution towards the conceptualisation of autonomy as a presupposition for its role in contemporary bioethics remains obscure or at least not clarified. Fan assumes a strong position in Confucianism proper, yet his alternative version of Confucian bioethics is unable to break free from the libertarian model it is supposed to replace. What is problematic in both versions of the Confucian account of personhood is their essentialistic and naturalistic tendency to interpret relations within fixed onto(theo)logical parameters. Ultimately, the moral status of persons and their dignity are grounded in a metaphysical or cosmological order that defines all aspects of the web of natural relationships within which human life develops. In as far as human cloning interferes with this natural order, it is unnatural and morally wrong. It is, however, doubtful whether such a naturalistic reading of human relationships has faithfully encapsulated both the spirit of the classical Chinese tradition and the dominant notion of human nature. What the various accounts of human nature in classical Confucianism seem to have in common is not the idea of invariant and fixed relationships that leave no room for creative social interpretation.[43] Instead, human nature may be better understood in terms of process cosmology, which provides for continuous development and alteration "through changing patterns of growth and extension."[44] And it is far from clear whether the artificiality of human cloning alone would be a sufficient reason to reject it on the premise of a Chinese conception of processual nature whose continuous creativity is shared by human beings.

This alternative reading of the Chinese tradition seems to provide the main focus of Shui-chuen Lee's conceptualisation of relational personhood, which leads him, however, to a conclusion with regard to human cloning diametrically opposed to that of Hui and Fan.[45] Lee develops his version of Confucian bioethics by utilising Tsung-san Mou's Neo-Confucian ontology in the construction of an ethics that is able to integrate what Lee sees as "the two chief achievements of Western civilisation," i.e. science and democracy. Taking the guiding principle of his version of Confucian bioethics from the *Doctrine of the Mean (Zhongyong)* and its teachings about the Dao that "works sincerely and thus produces things in an unfathomable way," Lee offers an interpretation of nature in terms of a comprehensive, metaphysically grounded teleology of survival and sustainability. Linking it with the "mandate of Heaven," he arrives at what may be called the principle of self-realisation which enables "every being" to "manifest its mandate to the full[est]." Conceding that "the manifestation of the (D)ao could be somewhat incomplete or imperfect" as the occurrence of illness, birth defects and impairments amply illustrate, he points out that any remedy through scientific means would not only be justified but mandated by Heaven: "If biotechnology could help man to achieve this purpose, it is morally good." The mere fact then that human cloning employs novel techniques, which have no equivalent in natural reproduction, is not a sufficient reason to reject it on moral grounds. On the contrary, anything that serves the natural teleology by removing or preventing stumbling blocks in its path executes the heavenly mandate. "Confucianism basically supports the employment of biotechnology in the relieving of human defects whether they are inborn or man-made." To solve problems of infertility, Lee sees no morally relevant distinction between "the use of IVF test-tube baby biotechnology" and human cloning. Furthermore, it is an implication of the relational concept of personhood that neither wasteful embryo research in the wake of human cloning nor the artificiality of such forms of a-sexual reproduction would constitute a moral wrong. Taking the relational foundation of personhood literally, Lee argues that moral status can only be conferred by that other person who is most able to establish a direct relation to the developing human life, the mother. "As a member of the moral community, she can confer equal membership to the child that she is bearing, and this is final" and "nobody else has a right that overrides hers. If she does not want to confer this right to her future child, she has the right to abort." Since "no sensible communication" can be established between the mother and human life in the early stages of pregnancy, zygotes, blastocysts and early embryos are her exclusive property.

It may be worth noting that such a view of the moral status of prenatal life is not exclusively Confucian. On the premise that "it is persons [in the strict sense] who endow zygotes, embryos or fetuses with value," Engelhardt similarly states that their value must be primarily understood in terms of ownership and property:

> "Those who made or procreated the zygote, embryo, or fetus have first claim on making the definitive determination of its value. Privately produced embryos and fetuses are private property. (...) Unless their procreators have transferred their rights to others, they have a secular moral right to abort the fetus, even if others would gladly adopt the child it could become."[46]

(2) Cloning and Integrity: The ethical evaluation of the implications of cloning for bodily integrity (understood comprehensively as including both physiological and psychological factors) is usually based on a utilitarian cost-benefit calculation. Yet, in the absence of empirical data from and about the clone the question whether or not cloning will in fact cause harm cannot be answered conclusively. What remains is risk assessment, and that may vary considerably depending on the changing parameters and individual preferences for the evaluation of individual risk factors: "Whether a life is miserable or not is a matter of degree."[47] Risks most commonly listed include not only physiological conditions that may cause a life of illness, pain and premature aging, but also factors that seem to preclude the normal psychological development of the clone. At present, the case for the assessment of physiological risks seems stronger, since it can be based on numerous data from the cloning of animals, all of which

suggest that cloning is not safe and therefore cannot morally be used on humans. This view, which endorses earlier conclusions of the United States' *National Bioethics Advisory Commission* (NBAC) that current cloning technologies "involve unacceptable risks to the fetus and/or potential child," stands clearly behind the Hong Kong government's prohibition of cloning.[48] As the safety issue is largely an empirical question, a different reading is at least possible and has occasionally been put forward in Hong Kong.[49]

As the potential effects of cloning on personal identity, mental health and well-being are less straightforward, they have received more attention by Hong Kong bioethicists and been discussed more controversially. One of the major concerns is related to the implications of the knowledge about the genetic identity with another human being whose life lies open to scrutiny. It is assumed that such knowledge would exert undue pressure on the clone and thus seriously interfere with his or her own individual development by depriving him or her of the sense of human possibility in freely creating a unique and individual future. The negative effects would be subjective and psychological insofar as they are linked to the child's own knowledge about his/her origin and genetic make-up, but also objective and social since parental expectations would exert considerable influence on the child's life. The argument may be even reinforced in cases of "replacement" cloning when parents' expectations for a replacement of a lost child would be so great as to effectively deny the new child the open history of his/her own life and thus severely undermine his/her own personal identity. Such a child would then only have a "life lived in the shadow" with strongly diminished possibilities to live a life that is truly the child's own life.[50]

In Hong Kong as elsewhere there is a tendency to regard human cloning as an infringement of personal identity and uniqueness. In circles of Christian fundamentalists in particular, this objection has occasionally been taken even further when it is argued that the human genetic code is the equivalent of the soul of Christian theology and metaphysics. On this account, the parallel between the Christian soul and DNA is more than linguistic or metaphorical, since it is the genetic make-up that guarantees personal identity.[51] Needless to say such reading espouses both a crude theology and a crude theory of genetic determinism. What is more, it also confuses different issues of personal identity such as identity over time, identity in the sense of identification, identity as related to origins, identity as related to a sense of self, identity as related to uniqueness and individuality, and collective identity.[52]

The claim that human cloning would infringe upon uniqueness and personal identity is usually rejected with reference to the natural occurrence of identical twins and their possibility for different life experiences. Conversely, Jonathan Chan has pointed to the morally significant difference implied in being a genetic copy of a prior existing human being and being a twin, as well as to ambiguities in the presupposed theory of identity, which puts too great an emphasis on "the psychological make-up of a person" in explaining personal individuality or uniqueness. While on his account, the argument from monozygotic twins is flawed and of little help for deciding on the ethical issue of human cloning, a concept of identity and personhood deriving from a Confucian perspective can indeed show that cloning would not infringe on personal identity and uniqueness. Starting from the assumption that the "essential part of a person" can be found in individual moral consciousness or volition, Chan draws on Parfit's "teletransporter" example to show that it is counter-intuitive to punish the original for the clone's misbehavior and that such belief presupposes that both original and clone each have moral consciousness and will independent of each other and thus have different identities.[53]

Apart from the fact that such concerns are based on many unknown factors and at best informed conjectures, Hong Kong bioethicists are reluctant to subscribe to anything like a "right to an open future" (Joel Feinberg) or even a "right to ignorance" (Hans Jonas), since the basis and justification of such a right are unclear and at variance with the relational concept of personhood. Up to now nobody has endorsed a position similar to that of the European Parliament's *Resolution on Cloning* (13 March 1997) stating that "each individual has a right to his or her own genetic identity." Instead, it has been argued that objections based on the assumption that human cloning inevitably will cause physical and/or psychological harm to the clone and adversely affect the clone's personal identity are flawed and, consequently, do not justify a comprehensive ban.[54]

On the whole, one may conclude that based on a utilitarian cost-benefit calculation, there is agreement among Hong Kong scientists, ethicists, and the general public that at present no potential benefits can be identified that outweigh the risks to cloned children. Provided, however, that risks associated with human cloning technology could be reduced to an extent comparable to those resulting from other forms of accepted medical intervention, the ethical evaluation of human cloning would need to be revised. The moral wrong is therefore not associated with any intrinsic evil of the technology as such but a result of an application that on balance produces more harm than good. The fact, however, that unlike

NBAC the *Provisional Council* did not include a "sunshine clause" in its recommendation that would allow a revision of the ban on human cloning on the basis of new scientific evidence seems to suggest that the ban is not exclusively based on a utilitarian risk analysis but also supported by categorical arguments of human dignity and the relational concept of personhood.

(3) Cloning, Reproduction and Human Relationships: One of the most widely endorsed objections to human cloning arises from considerations about the moral significance of marital and familial relationships grounded in the natural order. Hong Kong bioethicists have argued that human cloning as asexual reproduction distorts the natural order, destroys parental relationships, objectifies the creation of human beings and transforms the sexual act from an expression of marital love into an impersonal laboratory procedure. "In nature's way, the inception of a new life occurs in the embrace and sexual union of man an woman, which is the culmination of the love between them."[55] Yet it should be noted that the argument about reproductive technology destroying marital love would not only apply to human cloning but to any other form of in-vitro fertilisation as well. While it is readily put forward in debates about human cloning, its comprehensive employment against all forms of artificial reproduction is largely restricted to religious positions. It is not supported by the general public, whose views are reflected in existing legislation regulating the use of reproductive technology or by current clinical practice.

The crux of the argument therefore cannot be found in the artificiality of human cloning, but in the fact that it employs asexual reproduction, which is regarded as a violation of the order of nature. The argument is usually extended to the disruptive implications of human cloning for parental and familial relationships. As the position of children resulting from asexual reproduction cannot be assigned in accordance with the natural system of genealogy and lineage, clones would be effectively denied their social and moral space both in the family and society at large; cloning is therefore seen as an assault on the traditional concept of parenthood and a threat to familial stability.

While the moral significance of parenthood and the family in the Chinese tradition is obvious, it is not so clear whether that would necessarily imply a ban on cloning. As the practice of adoption in traditional Chinese society illustrates, and as the donation of embryos, sperms or eggs to infertile couples in IVF-procedures confirms for contemporary Hong Kong society, genetic ties between parent and child are not a necessary requirement for the establishment of familial relationships. On Lee's account of personhood and moral status, it may be possible to extend the mother's act of moral recognition to the cloned embryo and thus accept the new child into the family. Indeed, Lee argues that as the clone bears a closer relationship to "the original" than one twin has to the other, "Confucians tend to regard them as parent and offspring, not as 'delayed' twins."[56]

The argument could even be further strengthened with reference to the concept of "filial piety." Hui reminds us of Mencius' statement that of the three unfilial acts "to have no posterity is the greatest of them all."[57] He suggests that the overriding concern for filial piety can justify "Chinese infertile couples to seek help from modern assisted reproductive technologies." While the use of donor eggs and the employment of surrogate mothers would "probably be less desirable," they would nevertheless be "not of primary importance."[58]

Especially, the issue of surrogacy has met strong objections from a Confucian perspective during one of the three public consultations that preceded the enactment of reproductive technology legislation. As the thrust of the objections concerned the distortion of familial relationships, it could be extended to human cloning as well. Opponents argued not only for a ban on commercial surrogacy (as recommended by the *Council*) but on surrogacy itself and in all its forms, since it was considered to be in conflict with the traditional Chinese values of parenthood, a violation of filial relations and detrimental to the well-being of the children concerned; above all, it was out of tune with the traditional Confucian concept of relational personhood.[59] This perception was corroborated by a telephone opinion poll conducted in March 1999 by researchers from City University on public attitudes towards surrogacy and related issues (such as "male pregnancies"). A majority of respondents rejected both commercial and non-commercial surrogacy as "immoral" and "unnatural" and neither in the interest of the children born through such methods nor in accordance with human social (parental) relationships. A similar view was expressed by the *Central Co-ordinating Committee on Obstetrics and Gynecology* of the Hospital Authority, which proposed actively to discourage surrogacy and to make commercial surrogacy and its arrangement or advertising a criminal offence.[60]

Nevertheless, it may be argued that from such a Confucian perspective and depending on the moral significance attributed to the filial obligation of procreation, scenarios could be envisaged for which human cloning would be the morally justified solution. One such scenario referred to in the NBAC report is the following: "The parents of a terminally ill child are told that only a bone marrow transplant can save the child's life. With no other donor available, the parents attempt to clone a human being from the cells of the dying child. If successful, the new child will be a perfect match for bone marrow transplant, and can be used as a donor without significant risk or discomfort. The net result: two healthy children, loved by their parents, who happen to be identical twins of different ages."

In conclusion: the question whether a definitive Chinese answer to the moral issue of human cloning can be found seems largely to depend on the possibility to construe a consistent version of an alternative Chinese (or Confucian) bioethics.

Notes

1. Beauchamp & Childress, 2001.
2. Gillon, 1994.
3. Clouser & Gert, 1990, 219-236. See also Gert et al., 1997.
4. Rendtorff & Kemp, 1999 and Beyleveld & Brownsword, 2001.
5. Herissone-Kelly, 2003, 65-77. Similarly Quante & Vieth, 2002, 621-649.
6. Beauchamp, 1999, 389-401.
7. Beauchamp & Childress, 2001, 4-5.
8. Beauchamp, 2003, 266.
9. Alora & Lumitao, 2001, 3.
10. ibid., 8-9.
11. de Castro, 2003, 301-319. See also de Castro, 1999, 227-235.
12. Fan, 1999.
13. Tu, 1989.
14. Angle & Svensson, 2001.
15. von Senger, 1998, 62-115.
16. Angle, 2002, 24.
17. Hansen, 1996, 109.
18. Ci, 1999, 334.
19. Fan, 1999, 22.
20. Shaw, 2000.
21. Chan (J.), 1998, 49-71, 191.
22. Nie, 2000, 250-251.
23. Hung, 1998.
24. Abbas, 1997.

25. Wong & Lui, 1993, 25. See also Ho & Leung, 1997, 331-358.
26. Cheung, 2002.
27. Hawkins, 2000, 17-18.
28. Chan (A.Y.) et al., 1990, 242-257 and Tao Li, 2000, 8-9.
29. Hsiao et al., 1999.
30. Tao Lai, 1999, 571-590.
31. For details see Becker, 2003, 261-284.
32. Gould & Chan (H.M.), 1995, 95-109.
33. Callahan, 1995, 39-40.
34. Qiu, 2000, 332. See also Qiu, 2004, 1697.
35. Unschuld, 1995, 465. Similarly Nie, 1999, 463-479.
36. Tao Lai, 1999, 576, 578.
37. Hui, 2000, 95-117.
38. Ibid., 104.
39. Hui, 1999a, 159.
40. Bloom, 1997, 94-116. Roetz, 1999, 236-261.
41. Fan, 1998, 73-93, 193-196.
42. Tao Lai, 1998, 606.
43. Kung, 2003, 554-558.
44. Ames & Hall, 2001, 83.
45. Lee (S.C.), 1999, 187-198.
46. Engelhardt Jr., 1996, 255-256.
47. Chan (J.), 2000, 199.
48. See *Cloning Human Beings. Report and Recommendations of the National Bioethics Advisory Commission* (Rockville, Maryland, June 1997). See also Becker, 2000, 1-31.
49. Chan (J.), 2000.
50. Kang, 1998, 95-123 and Holm, 1998, 160-163.
51. Nelkin & Lindee, 1995.
52. Chadwick, 2000, 183-193.
53. Chan (J.), 2000, 204-206.
54. Ibid.
55. Kang, 1998, 198.
56. Lee (S.C.), 1999, 196-197.
57. Chan (W.S.), 1973, 75.
58. Hui, 1999b, 133.
59. Tao Lai & Chan (H.M.), 1997, 137-155 (text in Chinese).
60. Lee (A.), 1997.

References

Abbas, Ackbar . *Hong Kong: Culture and the Politics of Disappearance.* Minneapolis: University of Minnesota Press, 1997.

Alora, Angeles Tan and Josephine M. Lumitao (eds.). *Beyond a Western Bioethics: Voices from the Developing World.* Washington, D.C.: Georgetown University Press, 2001.
Ames, Roger T. and David L. Hall. *Focusing on the Familiar.* Honolulu: University of Hawaii Press, 2001, 83.

Angle, Stephen C. and Marina Svensson (eds.). *The Chinese Human Rights Reader: Documents and Commentary 1990-2000.* Armonk, N.Y.: M.E.Sharpe, 2001.

Angle, Stephen C. *Human Rights and Chinese Thought: A Cross-Cultural Inquiry.* Cambridge: Cambridge University Press, 2002, 24.

Beauchamp, Tom. "The Mettle of Moral Fundamentalism: A Reply to Robert Baker." *Kennedy Institute of Ethics Journal*, Vol. 8, 1999, 389-401.

Beauchamp, Tom L. and James F. Childress. *Principles of Biomedical Ethics.* New York: Oxford University Press, 5th ed., 2001.

Beauchamp, Tom. "A Defense of the Common Morality." *Kennedy Institute of Ethics Journal.* Vol. 13, 2003, 259-274.

Becker, Gerhold K. "Reproductive Choice and Moral Responsibility: The Challenge of Human Cloning." *Prajna Vihara Journal of Philosophy and Religion,* Vol. 1/2, 2000, 1-31.

Becker, Gerhold K. "Bioethics with Chinese Characteristics: The Development of Bioethics in Hong Kong." In *The Annals of Bioethics: Regional Perspectives in Bioethics,* John F. Peppin and Mark J. Cherry (ed.). Lisse: Swets & Zeitlinger, 2003, 261-284.

Beyleveld, Deryck and Roger Brownsword. *Human Dignity in Bioethics and Biolaw.* Oxford: Oxford University Press, 2001.

Bloom, Irene. "Mencius and Human Rights." In *Confucianism and Human Rights*, Wm. Theodore DeBary and Tu Weiming (ed.). New York: Columbia University Press, 1997, 94-116.

Callahan, Daniel. "The Puzzle of Profound Respect." *Hastings Center Report*, 25/1, 1995, 39-40.

Chadwick, Ruth. "Gene Therapy and Personal Identity." In *The Moral Status of Persons: Perspectives on Bioethics*, Gerhold K. Becker (ed.). Amsterdam, Atlanta, G.A.: Rodopi, 2000, 183-193.

Chan, A.Y. et al. "Public Attitudes Toward Kidney Donation in Hong Kong." *Dialysis Transplant*, Vol. 19, 1990, 242-257

Chan, Jonathan. "From Chinese Bioethics to Human Cloning: A Methodological Reflection." *Chinese & International Philosophy of Medicine*, Vol.1/3, 1998, 49-71 and 189-192.

Chan, Jonathan K. L. "Human Cloning, Harm, and Personal Identity." In *The Moral Status of Persons: Perspectives on Bioethics*, Gerhold K. Becker (ed.). Amsterdam, Atlanta, G.A.: Rodopi, 2000, 195-207.

Chan, Wing-tsit. *A Cource Book in Chinese Philosophy*. Princeton: Princeton University Press, 1973, 75.

Cheung, Martha P.Y. (ed.). *Travelling With a Bitter Melon: Selected Poems (1973-1998) Leung Ping-kwan*. Hong Kong: Asia 2000, 2002.

Ci, Jiwei. "The Confucian Relational Concept of the Person and Its Modern Predicament." *Kennedy Institute of Ethics Journal*. Vol. 9/4, 1999, 325-346.

Cloning Human Beings. Report and Recommendations of the National Bioethics Advisory Commission. Rockville, Maryland, June 1997.

Clouser, K. Danner and Bernard Gert. "A Critique of Principlism." *Journal of Medicine and Philosophy,* 15, 1990, 219-236.

de Castro, Leonard. "Is There an Asian Bioethics?" *Bioethics*. Vol. 13, 1999, 227-235.

de Castro, Leonardo. "Bioethics in the Philippines: An Overview of Developments, Issues, and Controversies." In *The Annals of Bioethics: Regional Perspectives in Bioethics*, John F. Peppin and Mark J. Cherry (eds.). Lisse: Swets & Zeitlinger, 2003, 301-319.

Engelhardt Jr., H. Tristram. *The Foundations of Bioethics*. Oxford: Oxford University Press, 2nd Ed., 1996, 255-256.
Fan, Ruiping. "Human Cloning and Human Dignity: Pluralist Society and the Confucian Moral Community." *Chinese & International Philosophy of Medicine*, Vol.1/3, 1998, 73-93 and 193-196.

Fan, Ruiping (ed.). *Confucian Bioethics*. Dordrecht: Kluwer, 1999.

Gert, Bernard, Charles M. Culver and K. Danner Clouser. *Bioethics: A Return to Fundamentals*. Oxford: Oxford University Press, 1997.

Gillon, Raanan (ed.). *Principles of Health Care Ethics*. Chichester: John Wiley& Sons, 1994.

Gould, Derek B. and Ho-Mun Chan. "Organs and Embryos: Ethical Policymaking in a Moral Minefield." *Hong Kong Public Administration,* Vol. 4/1, 1995, 95-109.

Hansen, Chad. "Chinese Philosophy and Human Rights: An Application of Comparative Ethics." In *Ethics in Business and Society: Chinese and Western Perspectives*, Gerhold K. Becker (ed.). Berlin: Springer, 1996, 99-127.

Hawkins, Brian R. "The Role of Tissue Typing in Transplantation." *The Hong Kong Medial Diary*, Vol. 5/4, 2000, 17-18.

Herissone-Kelly, Peter . "The Principlist Approach to Bioethics, and its Stormy Journey Overseas." In *Scratching the Surface of Bioethics,* Matti Häyry and Tujia Takala (eds.). Amsterdam/New York: Rodopi, 2003, pp. 65-77.

Ho, Kwok-leung and Sai-wing Leung. "Postmaterialism Revisited." In *Indicators of Social Development: Hong Kong 1995*, Siu-kai Lau et al. (eds.). Hong Kong: Hong Kong Institute of Asia-Pacific Studies, 1997,

331-358.

Holm, Soren. "A Life in the Shadow: One Reason Why We Should Not Clone Humans." *Cambridge Quarterly of Health Ethics*, Vol. 7/2, 1998, 160-163.

Hsiao, W. et al. *Improving Hong Kong's Health Care System: Why and For Whom?* Hong Kong: Hong Kong Government Printer, 1999.
Hui, Edwin. "A Confucian Ethic of Medical Futility." In *Confucian Bioethics,* Ruiping Fan (ed.). Dordrecht: Kluwer, 1999a, 127-163.

Hui, Edwin. "Chinese Health Care Ethics." In *A Cross-cultural Dialogue on Health Care Ethics*, Harold Coward and Pinit Ratanakul (eds.). Waterloo: Wilfried Laurier University Press, 1999b, 128-138.

Hui, Edwin. "Jen and Perichoresis: The Confucian and Christian Bases of the Relational Person." In *The Moral Status of Persons: Perspectives on Bioethics*, Gerhold K. Becker (ed.). Amsterdam/Atlanta, G.A.: Rodopi, 2000, 95-117.

Hung, Ho-fung. *Rethinking the Hong Kong Cultural Identity: The Case of Rural Ethnicities*. Hong Kong: Hong Kong Institute of Asia-Pacific Studies, 1998.

Kang, Phee Seng. "To Clone of Not to Clone: The Moral Challenge of Human Cloning." *Chinese & International Philosophy of Medicine,* Vol. 1/3, 1998, 95-123.

Kung, Shun-loi. "Philosophy of Human Nature." In *Encyclopedia of Chinese Philosophy*, Antonio S. Cua (ed.). New York, London: Routledge, 2003, 554-558.

Lee, Anita. "Opposed to Surrogacy." *South China Morning Post.* 13 May 1997.

Lee, Shui-chuen. "A Confucian Perspective on Human Genetics." In *Chinese Scientist and Responsibility: Ethical Issues of Human Genetic in Chinese and International Context*, Ole Döring (ed.). Hamburg: Mitteilungen des Instituts für Asienkunde, 1999, vol. 34, 187-198.

Nelkin, D. and M. S. Lindee. *The DNA Mystique: The Gene as a Cultural Icon*. New York: W. H. Freeman, 1995.

Nie, Jing-bao. “The Problem of Coerced Abortion in China and Related Ethical Issues.” *Cambridge Quarterly of Healthcare Ethics*, Vol. 8, 1999, 463-479.

Nie, Jing-bao. “The Plurality of Chinese and American Medical Moralities: Toward an Interpretive Cross-Cultural Bioethics.” *Kennedy Institute of Ethics Journal*. Vol. 10, 2000 , 239-260.

Qiu, Renzong. “Medical Ethics and Chinese Culture.” In *Cross-cultural Perspectives in Medical Ethics*, Robert M. Veatch (ed.). Boston: Jones & Bartlett Pub., 2nd ed., 2000, 318-337.

Qiu, Renzong. “Contemporary China.” In *Encyclopedia of Bioethics*, Stephen G. Post (ed.). New York: Macmillan, Thomson, 3rd ed., 2004, Vol. III, 1693-1700.

Quante, Michael and Andreas Vieth. “Defending Principlism Well Understood.” *Journal of Medicine and Philosophy*. Vol. 27, 2002, 621-649.

Rendtorff, Jacob and Peter Kemp. *Basic Ethical Principles in European Bioethics and Biolaw, vol. 1: Autonomy, Dignity, Integrity and Vulnerability*. Copenhagen: Centre for Ethics and Law, 1999.

Roetz, Heiner. “The ‘Dignity within Oneself’: Chinese Tradition and Human Rights.” In *Chinese Thought in a Global Context: A Dialogue Between Chinese and Western Philosophical Approaches*, Karl-Heinz Pohl (ed.). Leiden: Brill, 1999, 236-261.

Shaw, Sin-ming. “Back to a Culture of Subservience.” *The South China Morning Post*, 27 August, 2000.

Tao Lai, Julia Po-wah. “Confucianism.” In *Encyclopedia of Applied Ethics*, Ruth Chadwick (ed.). San Diego: Academic Press, 1998, vol. I, 597-608.

Tao Lai, Julia Po-wah. “Does it Really Care? The Harvard Report on Health Care Reform for Hong Kong.” *Journal of Medicine and Philoso-*

phy, Vol. 24/6, 1999, 571-590.

Tao Lai, Julia Po-wah and Ho-mun Chan. "Should Hong Kong Government Ban All Forms of Surrogate Arrangements - Some Ethical Considerations." *Values and Society*. Beijing: China Social Sciences Publishing House, 1997, Vol. 1, 137-155 (in Chinese).

Tao Li, Philip Kam. "Public Attitudes Towards Organ Donation in Hong Kong." *The Hong Kong Medial Diary*, Vol.5/4, 2000, 8-9.

Tu, Wei-ming.*Confucianism in an Historical Perspective*. Singapore: The Institute of East Asian Philosophies, 1989.

Unschuld, Paul. "Confucianism." In *Encyclopedia of Bioethics,* Warren T. Reich (ed.). New York: Simon & Schuster MacMillan, 1995, vol. I, 465-469.

von Senger, Harro. "Die UNO-Konzeption der Menschenrechte und die offizielle Menschenrechts-Position der Volksrepublik China." In *Die Menschenrechtsfrage: Diskussion über China - Dialog mit China,* Gregor Paul (ed.). Göttingen: Cuvillier Verlag, 1998, 62-115.

Wong, Thomas W. P. and Tai-lok Lui. *Morality, Class and the Hong Kong Way of Life.* Hong Kong: Hong Kong Institute of Asia-Pacific Studies, 1993, 25.

Current Debates on 'Human Cloning' in Korea

Chin Kyo-hun

Abstract: The first part of this essay gives an introduction to the circumstances of recent developments in biomedical research and outlines important Korean discourses on bioethical regulations (e.g., the draft bill on the Basic Safety Law on Bioethics in Korea etc.). The second part analyses the inherent concepts of identity and integrity of the human being in the light of Korean traditional thought in the setting of the DFG Project on "Culture-Transcending Bioethics." Consequently, all forms of human cloning should be prohibited, because an embryo is a human being and should not be destroyed in the interests of research cloning. The most effective way of banning reproductive cloning is to inhibit the process at the very beginning, by prohibiting the creation of cloned embryos.

Key Words: All forms of human cloning, Parthenogenesis, Ethics, Philosophical anthropology, Korean traditional thought, Natural law, Human dignity, Filial piety, Respect for life.

1. Introduction

The first part of this essay gives an introduction to the circumstances of recent developments in biomedical research and outlines important Korean discourses on bioethical regulations (e.g. the draft bill on the Basic-Safety Law on Bioethics and the bill concerning Bioethics and Security of Life in Korea). The second part analyses the inherent concepts of identity and integrity of the human being in the light of Korean traditional thoughts in the setting of the DFG Project on "Culture-Transcending Bioethics."

2. Recent Developments in Biomedical Research in Korea and Its Regulation

Six years after the publication of Ian Wilmut's experiments,[1] which led to the birth of the first cloned mammal via somatic cell nuclear transfer, human cloning is still the object of ongoing biological research, accompanied by anthropological, ethical and juridical suppositions and passionate public debates, also in Korea

Korean bio-scientists have reached a high level of progress in biotechnological research, including animal cloning and embryonic stem cell research. In 1998, a medical team from an infertility clinic in Seoul carried out a human cloning experiment by using stem cells taken from a human embryo. At almost the same time a team of veterinary surgeons experimented to combine different species, but the experiments failed. The first of these experiments met with strong criticism from religious circles and many civil organisations (NGOs) on the grounds that a human embryo is human life and not less.

But the Korean government was not so much interested in the ethical implications of biotechnology. It was more concerned with biotechnology as an economic challenge for the development of the Korean economy. By 2010, the government will have expanded the budget to $1.8 billion to support research on genetic engineering, proteomics, bioinformatics, and treatments for incurable diseases. Already, there are over 700 bio-venture companies, 137 infertility clinics (utilising in-vitro fertilisation), over 50,000 frozen surplus embryos, and 8 stem cell lines (confirmed by NIH, USA) in Korea. But there is no provision for penalising violation of the moral standards which should govern biotechnology. This laissez-faire attitude of the government, of many research groups and business circles without rules and standards based on humility and the assumption of dignity of the human being could lead not only the Koreans but - if practised everywhere - all of mankind to a global disaster. Therefore, I am convinced that we need one unified and global bioethics. Since 1998, the Korean Bioethics Association, religious circles and NGOs have pressed for a tightening of the basic laws concerning bioethics in order to regulate various issues, including the banning of human embryo cloning.

To begin with I would like to mention the results of The Korean Consensus Conference on Cloning, which was held by the Korean National Commission for UNESCO, in 1999. Modelled on experience gained in Denmark, the conference was advertised nationwide, and the public was invited to participate. A sixteen-member citizens' panel was selected from volunteers according to gender, profession and other criteria. The citizens' panel formulated the following ten key questions: 1.What is cloning? 2. What are the potential benefits of both human and animal cloning? 3. What are the immediate and long-term risks of cloning? 4. When does human life begin? 5. Up to what point should we allow cloning? 6. What are the vested interests in cloning research? 7. What are the current domestic and international regulations on cloning?

8. To what extent do citizens need to participate in cloning policy formation and how can we accomplish that participation? 9. What are the social and ethical responsibilities incumbent on scientists; and how can we raise awareness of these among scientists and lay people? 10. What role can the religious communities play in the cloning debates?

To begin to discuss these key questions, a group of experts was chosen by the citizens' panel with the help of a steering committee on the final day of the Conference. The expert panel comprised senior scientists working in the field of biomedical sciences as well as in ethics, theology, law, science policy and science studies, specialists as well as NGO representatives. The selection of expert panellists aimed at a balanced and well-informed view on various issues and tried to include a wide range of diverse viewpoints, from those avidly in favour to those who strictly oppose cloning research on ethical grounds. The expert panel proposed statements covering each of the key questions and answered supplementary questions from the citizens' panel. The citizens' panel continued the discussions and drafted the final statements, which were then presented at a press conference.

Finally the citizens' panel called for a ban on all attempts to clone a form of human life, including pre-embryos (embryos before the 14th day after fertilisation). Only two panellists maintained that the cloning of pre-embryos strictly for therapeutic purposes could be potentially beneficial to human health. The majority of the panellists adopted the view that a new human life begins at conception and concluded that experiments with pre-embryos are ethically unacceptable.

The citizens' panel recommended that both the government and the scientific and medical communities should act immediately to examine the current state of cloning research in Korea and to establish an appropriate regulatory system, since many projects of cloning research are promoted for commercial interests. The panel recommended that animal cloning should be carried out exclusively under strict control and that regulation should be inaugurated by the government. It emphasised the urgent need to educate scientists in bioethics and to inform the public about various aspects of biotechnological research problems.

Consequently, the Korean government established the "Korean Bioethics Advisory Commission" in the year 2000. The main objective of the Commission was to establish a consensus and to propose basic regulations of bioethics.

The Commission was made up of 20 professionals from the fields of ethics (1), religion (3), genetic engineering (5), medicine (5), law (2), and sociology (1) as well as from NGOs (3). The Commission elaborated the basic framework of a bill regulating the ethical issues of biotechnology. This framework included the constitution and procedures of the National Bioethics Committee under the direct control of the President of Korea, regulations on cloning research, human embryo research, genetic treatment, genetic mutation of animals, utilisation of information from the human genome, the patenting of genetic research results, and the provision of bioethical education for research workers etc.

The following passages present only the central lines of argument in the Basic Law on Bioethics concerning human cloning by the Commission in November 2001. The most important ethical points of view concerning human embryonic research are the following:

Cloning and combining the genetic code of human beings or of animals

1. It is prohibited to clone human beings by using somatic cell nuclear transfer into oocytes, likewise to support or instigate cloning.
2. Animal cloning by using somatic cell nuclear transfer is permitted. However, the Korean National Bioethics Committee should examine whether or not it is to be permitted in cases where the ecological balance or the diversity of species could be endangered.
3. All experiments on the genetic combination of human beings and animals are prohibited.

Research on human embryos and its application

1. The procreation of embryos by in-vitro fertilisation (IVF) intended for infertility treatment should be allowed.
2. Research utilising the surplus human embryos procreated by the IVF method employed for infertility treatment should be allowed for a certain period of time until the research on adult stem cells proves more effective than the research on embryonic stem cells. Nonetheless, research workers should obtain informed consent from both the egg donor and the sperm donor.
3. Research on stem cells utilising aborted foetal tissues should be permitted only on the basis of a written consent from the donor of the sperm and egg cells used in the creation of the foetus. Experiments using tissues acquired from illegally aborted human foetuses should be prohibited.

4. Any attempt to create a human embryo by using nuclear transfer from a somatic cell into the oocyte or by creating human embryos solely for research purposes should be prohibited. The research on human embryos or on stem cells created by such a method should also be prohibited.
5. The government should support research on stem cells derived from adult somatic tissues and encourage a change in the direction of stem cell research from embryonic tissues to somatic tissues.
6. To implement the above guidelines and regulations, the Korean National Bioethics Committee (KNBC) will form a special subcommittee called the Human Embryo Research Advisory Committee with the following functions:

- Periodic reviews of the status and subject matter of human embryo research by the Institutional Review Board (IRB)
- Provision of education in bioethics for the researchers and research assistants involved in human embryo research etc.

If these regulations are violated, the institute or the individual responsible and the collaborator will face criminal or civil charges.

Yet there are some scientists who support human cloning for medical research purposes. They wish to harvest stem cells from cloned human embryos for the purpose of therapeutic research; and they insist that it should be permissible to make use of the surplus of frozen human embryos for medical research without implantation of the embryo into the womb. The government plans to introduce further bills that are designed to protect the biotech industry's experimental and business opportunities by criminalising the implantation of cloned embryos and permitting embryo banks (farms), in which millions of embryos would be manufactured and destroyed to harvest stem cells under the strong pressure of the politican who has supported the bio-scientists and the bio-ventures.

Anyway, the Parliament passed the Legislative Bill Concerning Bioethics and Security of Life on December 29, 2003 at 11 p.m., without further discussion suddenly and together with another bill, and it is put into force from January 1, 2005. The Bill, on the one hand, intends to prohibit the cloning of human beings by somatic cell nuclear transfer, the implantation of embryos which are produced by SCNT, and the production of embryos by heterogeneous nuclear transfer. On the other hand, it permits embryonic cloning by SCNT in exceptional cases where the goal is to explore treatments for hitherto untreatable diseases which are defined in the Presidential decree, for which effective regulation was established in December 2004. But christian circles are preparing for the revise of the Bill and also are going to sue for the unconstitutionality of the Bill to the Constitutional Court.

However, many people are convinced that this particular type of embryonic cloning contradicts ethical standards for the following reasons: It is never morally acceptable to sacrifice one human life for the real or potential benefit of many. Furthermore, to treat a human being as a means to an end (especially in research cloning) will undoubtedly lead to the exploitation of women because researchers need to obtain large quantities of ova in order to produce a certain quantity of cloned embryos from which a sufficient number of viable stem cell lines will develop. In doing so, women must be injected with superovulatory drugs and undergo an invasive procedure, the side-effects of these injections being abdominal pain and nausea.

In addition to the moral considerations above, the cloning of embryos should be prohibited because it increases the likelihood of human reproductive cloning. It will be impossible to prevent the implantation and subsequent birth of cloned embryos once they are available in the laboratory. Consequently, all human cloning should be prohibited. An embryo is a human being, and we should not destroy embryos in order to pursue cloning research. To convert embryos into stem cells is somehow analogous to killing a human being. According to an old tradition it is 'unnatural,' it is against the natural law (*Naturrecht, lex naturalis*; i.e, "Thou shalt not kill"), against a self-evident primordial moral principle not only in Christianity, but also in Buddhism and Confucianism. The most effective way of banning reproductive cloning is to inhibit the process at the very beginning, with the creation of cloned embryos.

3. Human Cloning From a Traditional Korean Perspective

Rudolph Jaenisch, the embryologist at MIT (Cambridge/USA) has repeatedly stated that it would be virtually impossible to clone a healthy baby. The efforts of Michael West's Institute of Advanced Cell Technology (based in Massachusetts) revealed that highly qualified research workers were not able clone an embryo.

But we cannot ignore that the majority of scientists once believed that the cloning of a sheep was impossible (but they had finally to admit that it happened). With Dolly even the possibility of human cloning became more probable. No one has yet managed to clone a human being, but several groups have announced that they plan to do so.

The team led by Woo Suk Hwang and Shin Yong Moon of Seoul National University in February 2004 revealed that they had grown the first embryonic stem cell line derived from a cloned human embryo by using nuclear transfer from a somatic cell.[2] Their embryonic stem research was done before the effectuation of the bill.

Hwang's experiments pose a serious ethical problem to the world and undermine the dignity and sanctity of human beings, because the embryo is the moral equivalent of an adult human life. We do not sacrifice one human life in order to possibly help another human life. Therefore, a national and worldwide effort to ban human cloning is needed more urgently than ever. We would ask the United Nations to proceed with cloning bans of all forms of human being (inclusive embryos). Fortunately the declaration to "prohibit all forms human cloning inasmuch as they are incompatible with human dignity and the protection of human life" was adopted and approved by the UN's General Assembly on the 8th of march 2005, but not as legally binding.

Among the main reasons for experimenting with embryos is the desire for a supply of donor organs, the eradication of genetic diseases, and the possibility of enabling infertile couples to reproduce. A more curious reason is the desire to seek immortality by reproductive cloning. Human cloning is a test for public policy, and legislative decisions are often made with regard to the future of certain scientific research. Research in cloning will continue to be defended by those who wish to justify their objective in the name of scientific freedom. But scientific freedom without regard to morals is not a fundamental human right.

The debates on human cloning have so far resulted in a conflict between two forces in Korea: deeply held public beliefs that cloning should be opposed on any terms and the intentions of some bio-engineers who claim that they have to experiment with embryo cloning especially to succeed in curing the sick.

Before some details of Korean traditional thought are displayed to evaluate the idea and the practice of cloning, a glance at the history of Korean culture could help to understand Korean thought.

A. A Glance at Korean Traditional Thought

Korea has a very distinct culture from China, although the Korean people were long influenced by Confucianism, Taoism and Buddhism through China. In the course of time, the three religions were acculturated into Korean traditional thought and thereby koreanised. In recent times, Christianity has also been assimilated into Korean culture. The Korean traditional reception of religious thought knows four consecutive periods: the periods of shamanism, Buddhism, Confucianism and Christianity.

Since the beginning of Korean history shamanism has influenced Korean tradition, particularly the Korean view of life. In the 4th century Shamanism was marginalised by Buddhism, which came from China. Buddhism was the national religion and dominated Korean moral thought until the 14th century. Later, in the Cho-sun Dynasty, Confucianism took its place and became the general moral value system and the single dominant principle in all areas of public and private life for approximately 500 years. In the 18th century some Confucianist scholars were drawn into a conflict with the regime of the Cho-sun Dynasty and its doctrine and began to study Catholic sources in a Buddhist temple with the help of the books of Matteo Ricci (1552-1610) which were imported from Peking. In 1784 the Korean Catholic church was established, but for more than 100 years Catholics were persecuted mainly because they disapproved of ancestor worship and sacrificial rites. This was mainly inspired by the Vatican's misunderstanding of ancestral respect and filial piety. Nevertheless, filial piety has shaped all ideas of respect for life until now. When the isolationist politics of the Kingdom of the Cho-sun Dynasty ended, Korea opened its borders; the Protestant Mission began with the arrival of American missionaries in 1884. The Protestant Church has since become the country's largest religious community.

Of Korea's total 47 million population, Protestants account for over 8 million, Catholics 4.8 million, Buddhists ca. 4.5 million Buddhists, and adherents of other religions over 2 million. Korea has seen no wars of religion in its history, which reflects the Korean people's religious tolerance. There are many cooperative movements; and cultural exchange among religions is growing.

B. The Effect of Filial Piety on Respect for Life

The central motive behind the idea of respect for life is the principle of filial piety solidly implanted in Korean traditional thought. Ethics considers as a main problem the field of human relations, and its special characteristic is that it reflects these relationships from the point of view of the whole human being according to the basic ideas of philosophical anthropology, particularly in the 20th century. Western philosophy has been inclined to interpret the human being dualistically, but in ethics the dualism, especially in the sense of a dichotomy (separation of nature and reason) contradicts the idea of responsibility for one's own life. 20th century reflection on the human being shows a trend towards holistic concepts. When Karl Barth speaks of the human being as an 'embodied soul' (leibhafte Seele) or a 'besouled body' (beseelter Leib), he is trying to frame concepts which include the person in its totality. Max Scheler's concept of 'spirit' (Geist) has a similar intention as Karl Jasper's emphasis on 'existence.' Life is all-inclusive, comprehending the whole person: Life's unique dignity could be answered by a 'reverence for life' (Albert Schweitzer).

We find a similar reverence for life in the conceptions of filial piety that permeate Korean traditional thought. The most fundamental aspects of filial piety are set out at the beginning of the Chinese Classic of Filial Piety: "Since we receive our bodies, flesh, hair, and bones from our parents, allowing no harm to come to our persons is the beginning of filial piety." (Classic of Filial Piety, chap.1). Our bodies are not our own, for they have been passed on to us from our parents. Karl Barth expresses a similar idea by calling our bodies a 'loan' (Leihgabe).[3] If we embrace the opposite idea that our bodies are simply our own, we could on the one hand develop an inclination to maintain them, and on the other succumb to the temptation to mistreat them excessively. If we recognise our bodies as loans, it becomes a self-evident principle and a matter of responsibility that we take good care of them. We are obliged to preserve this gift as far as possible from harm in order to be able to pass bodily life to our descendants; this is the beginning of filial piety. What is borrowed must be returned into the stream of life; passing on bodily life in a wretched condition to the next generation or causing our children to grow up unhealthy because they are born of a mistreated body cannot be considered filial. The acculturated Buddhism in Korea has also emphasised the importance of filial piety.

Korean traditional thought concerning filial piety proceeds from the concept of thankfulness for and the grateful response to the fact that one exists today. To live gratefully means to have received this gracious favour without obligation to recompense, but only to be responsible for it. I repay Heaven (in the Confucianist sense) or God (in the monotheistic sense) and my parents for it by passing it on to my descendants without anticipating a reward. Therefore, working with a sincere heart to accomplish something for our community is a response but not a recompense vis-à-vis the religious authority (God or Heaven) or my parents for their altruistic gift of life; the aim is not to oblige God, Heaven or a god to grant transcending beatitude. Christians cannot but entrust all that to the will of the Heavenly Father. Traditional Korean ethics transports analogous attitudes: Since human beings have received their bodies from Heaven or divine authority and from their parents, they respond to this gift by taking care of it, and see it as a duty to the community to pass it on in good condition to their descendants. Thus 'reverence for life' does not mean to affirm oneself as a solitary being or to concentrate exclusively on one's own personal life, but to promote 'our' life. This kind of respect for life is characteristic of the ethics of filial piety. We find it not only in Korean Confucian ethics, but also in Korean Buddhist and Korean Christian ethics.

In terms of their deep respect for life, Taoists, Confucians, Buddhists and Christians are fully congruous in spirit. Whenever they think of life, they always trace it back to what originated in or what is conferred by Heaven, Tao, Buddha, and God. Though they use a different set of terms, there is no real divergence in the practical consequences. Therefore, I think, Heaven, Tao or God all designate the same primeval source of life.

We should respect all human life because it has absolute value. Although Buddhism, Taoism, Confucianism and Christianity differ theoretically, they all share a deep respect for life. The meaning of life is self-evident and could be understood by intuition (*Wesensanschauung*). We cannot derive but can only state that all life is meaningful and invaluable. We believe that all human life, and this includes the embryo, is dignified and must be respected.

The profound respect for the newly conceived child cannot be based on any explicit teaching of Buddha, Laotzu, Confucius, and of

Jesus and the apostles, as it is transmitted in tradition. We find hardly any direct reference in the classical works to practices or customs that constitute a threat to the unborn child (foetus). Although Buddha, Confucius, Laotzu, and Jesus did not pay particular attention to the unborn child, we can develop a certain analogy between Korean traditional ideas and respect for life of the foetus or embryo. Because we can relate the moral principle of respect for life to Korean traditional thought, we firmly hold that it is always wrong to opt against any form of life. Equipped with these principles, we can focus on intentions that undermine respect for life, and we can apply this principle to a comprehensive list of contemporary bioethical issues such as cloning, embryo research, reproductive technology, and genetic manipulation.

Korean traditions usually consider the foetus to be a human being. They consider the age of a newborn baby to be one year, because they regard the act of fertilisation as the beginning of human life. The Koreans count life in the mother's womb - although only nine months - as the first year of life.

In the event of miscarriage, a noble family including the royals performs a funeral service for the dead foetus, because the unborn baby is considered to be a genuine human being. A further example: It is customary in Korea for a pregnant woman to observe special behavioural rules for the prenatal care of her unborn baby (*Tae-kyo*), because the foetus is believed to be equally capable of learning as a person already born.

4. Conclusion

The Korean tradition concerning respect for life has begun to be neglected in the wake of modernisation, i.e. technocracy and industrialisation under the influence of Westernisation. A certain spirit of commercialism is emerging and - according to religious critics - is being exacerbated by mammonism and indifference. A reductive rationalism ignores the emotional aspects of the human being and the intuition of the reverence for life.

Koreans have enjoyed the benefits that advances in science and technology have brought. As a result, people are occasionally tempted to believe blindly that science and technology can solve all our mortal problems. There is so much euphoria about the pure functionality of scientific and technological advancement that many are inclined to forget the side-effects and possible backlashes. As a result, science is losing touch with morality and drifting towards ethical insensibility. The Korean biotechnology industry has not seriously discussed the urgent questions concerning the identity of the human being and respect for life. Other problems confronting it are the unanticipated side-effects of industrialisation; the ecological crisis, the endangering of the dignity of human being, and the undermining of human rights.

The consequences concerning human cloning can be summarised as follows: Human cloning means that human life can be born without the union of sperm and ovum. So far, we have maintained that the origin of a human being stems from the union of a man and woman, and this intuition was seen as a universal value and a standard criterion that informs the dignity of a human person. With the possibility of human cloning, we are faced with the relativity of these concepts, with the possibility of creating an extraordinary human being through human intervention or human manipulation.

I am opposed to both the reproductive cloning of human beings and human therapeutic cloning. Hence, I have argued against human cloning in a great number of publications in Korea, but my arguments are hidden beneath formulae such as "trampling on living beings," "a dangerous act that injures the dignity of a person," "destruction of the sublimity of marriage," "an act causing the destruction of the fundamental order of the family, i.e. the foundation of our human society," "an evil act which will bring about calamity upon all humanity," "unethical conduct of certain of groups to attain their goal, e.g., to make monetary profit and to fulfil secular (i.e. contradicting the religious and ethical tradition) ambitions," "disclosure of an insensitive mind without moral principles," "an act causing an imbalance of the human integrity that the Heaven (God) created," "an act against the natural law" (*lex naturalis*, Naturrecht).[4] The natural law is interpreted as the primordial moral principle of human acts, as Tao and The Heavenly order. An almost universal principle is the commandment "Thou shalt not kill." Moreover, "parthenogenesis (monogenesis) is an act causing the destruction of the harmony of Yin and Yang, and there is the serious danger that a woman's body will be reduced to a provider of oocyte cells and an abused object."

As a consequence of these intuitions, we should ban human reproductive and therapeutic cloning. In addition to demanding such a ban, I would like to appeal first to the conscience of scientists. Beyond the proclamation of a basic law of bioethics, which is urgent and important, I urge scientists to stop all human embryonic cloning experiments, no matter what their pursuits or objectives may be. The reason for this is that such experiments constitute a dubious act that infringes on the dignity of the human being by exploiting a human life in order to benefit another. We should not utilise a human being as a means of reducing other people's suffering. It is tantamount to destroying the life of one human being in order to save the life of another.

It is not my intention to damage the freedom and autonomy of science. However, scientists should not neglect the fact that the freedom and autonomy of science can be guaranteed only insofar as they respect the dignity of the human person.

We must recognise that biotechnology will be a moral good only if we are able to use it in ways that protect the very foundation of human life and enhance human dignity. We should not abuse our autonomy simply because we are able to do things, since any human act requires moral reasoning and responsibility. We should never support scientific technology that undermines the dignity of the human person. I sincerely hope that we will always advance in science in accordance with the imperative to protect the dignity of the human person, which is a universal and permanent value.

Notes

1. Nature, 1997, 385: 810-813.
2. W.S. Hwang, published online Feb 12, 2004; 10.1126/science.1094515 (Science Express Reports)
3. Barth, 1957, 368.
4. Rommen, 1936, 224-255.

References

Barth, Karl. *Kirchliche Dogmatik, Bd. III/4 Die Lehre von der Schöpfung*. Zürich, 1957.

Hwang, W. S. et al., published online Feb 12, 2004; 10.1126/science.1094515 (Science Express Reports)

Rommen, Heinrich. *Die ewige Wiederkehr des Naturrechts*. Leipzig, 1936.

The Ongoing Debate in South Korea

The Cloning Debate in South Korea

Phillan Joung and Marion Eggert

Abstract: In early 2004, the announcement by a Korean research team that they had successfully cloned human embryos and that, for the first time, they had derived a stem-cell line from a cloned embryo sparked an ethical debate world wide. The controversy reflects the still unresolved moral problems relating to this scientific breakthrough and the prospect of biomedical innovations which promise remedies for incurable diseases as well as competitive advantages in the global economy. This is why South Korea regards stem-cell research and biotechnology as epochal challenges and pursues ambitious bio-political aims. However, the young democracy has embarked upon a highly controversial public debate on the dangers and potentials of new genetics and biosciences which, eventually, led to the national „Bioethics and Safety Law." In this context and with regard to the various international reactions to the recent groundbreaking Korean achievement, the present article attempts to address the ethical problems of cloning technology by analyzing the South Korean discourse on the basis of the nation's peculiar historical and cultural preconditions. It is shown that due to Korea's complex intellectual and political history, the heated inner-Korean debate on cloning can be regarded as representative of cross-cultural bioethics.

Key Words: Korea, Cloning, Stem cell research, Bioethics, Biopolicy, Culture, Cloning debate, Egg cell donation, Collective identity, Woo-suk Hwang.

1. Introduction: Toward the Seoul Cloning Experiment

At the annual meeting of the American Association for the Advancement of Science (AAAS) on February 13, 2004, scientists of Seoul National University caused a sensation by presenting their report *Evidence of a Pluripotent Human Embryonic Stem Cell Line Derived from a Cloned Blastocyst.*[1]

According to this report, the research team under Woo-suk Hwang and Shin-yong Moon extracted 242 oocytes from 16 donors and denucleated these oocytes by using a method which they had developed themselves. Into the denucleated oocytes they implanted the so-called cumulus cells of the respective donor. These cumulus cells cling to the surface of the donated oocytes. Cell division was accelerated through the use of special chemicals. Through this procedure, the scientists produced 30 embryos, which they managed to grow to the blastocyst stage. The embryos have the same DNA as the egg donors. Finally, the researchers managed to produce a stem cell line out of 20 suitable blastocytes. This is commonly regarded as a breakthrough in bioscience. The pluripotent embryonic stem cells are much desired because of their capacity to develop into all the various cell types of the human organism. According to the researchers, cell therapy and development of organs for transplantation without any risk of rejection will be possible in ten years. From an ethical perspective, however, this biomedical innovation, which requires so-called therapeutic cloning and embryonic research, is highly controversial. The following takes the 'case Hwang' as paradigmatic, for it highlights the peculiarities of Korean bioethics as well as its interplay with the international debate.

Experimental bioscience has now proved that it is possible to produce a human embryo capable of development through somatic cell nuclear transfer (SCNT), and also that such an artificial embryo is then a potential resource for producing stem cells. This has caused a variety of reactions worldwide: on the one hand, the successful South Korean cloning experiments were celebrated by experts and by biotechnological enterprises as a milestone in the field of therapeutic cloning. For, since the birth of Dolly, SCNT has been successful only with animals (but not yet with primates) so that the cloning of humans had hitherto been regarded as unrealistic.[2] On the other hand, the cloning of human embryos, now conducted by the South Korean researchers, is viewed as the violation of a taboo because, in many respects, it shakes the very foundations of the *conditio humana*.

The most important feature of this blow is the fact that it is now possible to produce human life artificially and asexually. The age-old question of the Sphinx - "What is man" - has thus been reduced to a merely technical problem. Consequently, anthropological, theological and ethical reflections have become groundless. That is why, compared with previous transgressions in the name of modern scientific research, this violation of a taboo is by far the most serious one. The success of the Seoul research team is an irreversible leap in the field of reproductive medicine and its history of gradual transgressions since the introduction of in-vitro fertilisation: through cloning, the production of human life without fertilisation of a human egg cell with a sperm cell has become possible - in the near future, even the human egg cell will be dispensable. Yet, it has been shown that it is possible to produce and cultivate egg cells as well as sperm cells;[3] and, in order to allay the reservation of his critics, Woo-suk Hwang has declared his intention to use such artificial oocytes exclusively for his future cloning experiments. However, this does not solve all of the fundamental problems by any means. Rather, such a project raises the following questions:

Do we have to consider embryos produced in this way as human beings? Can we treat these artificial embryos differently, as many researchers suggest? What are the concrete ethical and juridical problems which result from cloning? In sum, it seems that the traditional boundaries between nature and the artificial, between human being and mere biomaterial are no longer tenable. Notwithstanding, the present article makes an attempt at addressing the ethical problems of cloning experimentation by analysing the controversial cloning debate in South Korea as an example.

The starting point of this analysis are the many different reactions which the report on the successful experiments has elicited both in South Korea, a newly industrialising country, and in classic industrialised countries, such as Germany and various parts of the USA. In these countries, the attitude toward the South Korean cloning experiments is dominated by ethical reservations toward cloning in general, which are largely based on the danger of abuse. In South Korea, on the other hand, the fact that the scientific innovation has earned the country international recognition has led to optimism and enthusiasm, which have superseded the harsh critique of Woo-suk Hwang's research - at least for the time being. The following section attempts to explain this evident change in attitudes toward scientific and technological development in the light of Korea's historical and political background.

2. The Post-colonial Condition of a Newly Industrialising Country

A. Problems of Altruistic Egg Cell Donation and Informed Consent

It is generally assumed that the main reason why South Korea has succeeded where other countries with similarly ambitious cloning experiments have failed lies in conditions favourable to scientific research. South Korea shows four characteristics which are typical of a newly industrialising country:[4] scientists with Western training, an optimistic attitude toward progress, liberal laws designed to improve national competitiveness, and strong patriotism as a connecting link between various social interest groups. It is thus an oversimplification to interpret as mere rhetoric Woo-Suk Hwang's appellation of the 16 egg cell donors as "national heroines," who have sacrificed themselves for the future of Korean biotechnology voluntarily and without any financial interest.[5] At the press conference in Seattle, Hwang himself declared that without the donors' sacrifice the spectacular results would not have been possible, and duly expressed his gratitude to them. In the supplement to their report, the Seoul research team documented the legality of their experiments.[6]

This manifestation of ostentatious patriotism and altruism has raised doubts, especially in Germany, whether the team had indeed observed the principle of informed consent, which presupposes individual autonomy and freedom of will, before extracting the egg cells. Considering the socio-cultural Korean background, i.e. a patriarchal social structure deeply rooted in Confucianism, these doubts seem justifiable. In this traditional social order, female willingness to make sacrifices for the general good is regarded as a central virtue. However, traditional gender roles are changing, especially among the younger generation.

Another criterion that sheds doubt on the claim that the egg cell donors had given their informed consent is the establishment of an egg cell market in South Korea. It is a well-known fact today that there are companies offering oocytes to infertile couples, which have been extracted - in exchange for 'compensation' - from young students often attending reputable universities. In South Korea, the trade in human egg cells and sperm cells will not be banned by law until January 1, 2005. This regulation will then put an end to a practice which is deeply rooted in the traditional Korean acceptance of surrogate motherhood. This fact highlights the contradictions in a country torn between developments toward modernity on a socio-political level on the one hand and the perpetuation of traditional everyday realities on the other.

The declared motivation of the egg cell donors - making a sacrifice for a collective goal - raises a new ethical question concerning egg cell donation for research purposes. This individual motivation differs fundamentally from egg cell donation to infertile couples because donation for research purposes is not linked to a concrete and immediate goal. The more abstract aim of developing therapies in a distant future or for patriotic reasons, however, sheds a new light on egg cell donation. From this perspective, especially considering the disproportionately high number of egg cells needed, therapeutic cloning becomes an ethically explosive topic.

B. Research and Development, and the Destiny of the Nation

The emphatic form of Korean nationalism and altruism - which causes a certain degree of bewilderment in the West - is characteristic of a newly industrialising country: a faith in technological and scientific progress.[7] After its liberation from Japanese colonial hegemony in 1945, South Korea, a 'frontier state' in the conflict between East and West, has rapidly advanced along the path toward industrialisation and modernity. Today, with the age of biotechnology dawning, the country perceives itself as a democratic state. Yet, in South Korea as well as in the classic industrialised countries, the speedy economic growth of recent decades has cast a shadow on modernisation, and the unwavering faith in technological and scientific progress is being increasingly challenged. This is the background of a tough biopolitical fight between modernists and conservatives in South Korea as it has been epitomised by the struggle for a national law on bioethics over the past years. Woo-suk Hwang has played a prominent role in this controversy; and his research is highly charged with symbolic meaning.

The most important argument of the modernists is the question of national competitiveness. For South Korea, this argument has gained a special significance because obstacles to bio-scientific research and technology have been represented as a threat to national sovereignty. The fear that failure to respond to the epoch-making biopolitical challenge will lead to a renewed dependence on more powerful states thus adds to the gnawing national concern. In this context, Hwang's announcement after his return from Seattle that he will suspend his cloning experiments with human egg cells until the Korean government and society have confirmed their moral acceptability, has become a decisive signal. This is shown by an internet opinion poll carried out the following day: 65.7% of participants voted for a continuation of his research, 29.5% voted against it.[8] Although the results of this survey might be rejected as hardly sustainable and as an emotionalised and rash response, they nevertheless confirm the hypothesis that the climate in South Korea is quite favourable toward scientific research and, moreover, that the public discourse on bioethics - at least the major issues - develops quite differently from what is feared in Germany.

Considering the international recognition of the Seoul cloning experiments, it appears that, for the moment, the modernists have the upper hand. The decision for clinical research seems to confirm this assessment.[9] And while Hwang has announced the suspension of his cloning experiments with human egg cells, there are plans to set up a new institute for embryonic stem cell research. This institute, scheduled to commence operation next year, will coordinate interdisciplinary stem cell research, physics and neurobiology; and prominent Korean scientists currently employed in the United States have already been recruited. The integration of the research activities in these fields within one institute, so the Ministry of Health argues in response to its critics, will ensure a better control of research in the future - including Woo-suk Hwang's cloning project. Moreover, the Ministry of Health has predicted that his research project will be approved by the National Bioethics Advisory Commission once the "Law on Bioethics and Security" comes into force on January 1, 2005.

C. The Ethics of Curing versus the Natural Order

Woo-suk Hwang himself is convinced of his own mission, and he defends his research project by referring to the ethics of curing and relieving from pain. As, in any newly industrialising country, caring for the ill is still largely a family affair and thus a heavy burden for those who are affected, he touches a sore spot. The immediate reactions to the Seoul cloning experiments show that it is these people in particular who support Hwang. They, in turn, challenge Hwang's critics and accuse them of being without compassion. The illusion of a speedy cure alone, however, is not the most important topic. Rather, the anticipation of a utopia becomes a redemptive ritual. By professing Buddhism and by representing his project as missionary work, Hwang plays on the quasi-religious attitude of the Koreans, whose collective consciousness has been shaped by a history full of deprivation and suffering and which, consequently, is easily susceptible to promises of redemption. Accused of violating both 'divine order' and human dignity, Hwang has chosen to go on the attack by emphasising that an achievement destined to cure human beings of their sufferings cannot be immoral. Dong-ho Lee, a catholic priest and one of Hwang's many critics, accuses him of meddling with Creation by producing human life asexually and, in addition, of paving the way for reproductive cloning.[10] Hwang, however, rejects reproductive cloning altogether and has declared his support for an international ban.[11] In addition, Hwang circumvents the argumentation of his critics by grounding his personal ethos not on the atheist or materialist stance typical of a scientist but rather on the principles of the enlightened humanism that has, from the beginning, characterised modern experimental sciences. Due to their international training,

South Korean scientists are familiar with these principles, too. In this view, the world is no longer regarded as a perfect creation. Rather, it is the task of man to improve and perfect it in order to enhance the quality of life in this world.[12] By subscribing to such a utilitarian sense of mission, Hwang distinguishes himself from religious and conservative bioethicists, who consider technological and scientific progress as an illegitimate intervention in the natural order. Such references to natural right conceptions usually lack any further reflection, but invoke the assumed deeper attachment to nature which is characteristic of the 'Korean soul.' These various religious and ideological worldviews imply different ranges of options for man and thus, in effect, different ethical conceptions.

In order to understand the intensity of the bioethical controversy - and in South Korea, this controversy is especially sharp -, it is necessary to consider two crucial factors in particular: firstly, the post-colonial condition of a newly industrialising country described above and, secondly, the condition of a young democracy, in which the bioethical debate has become a public issue. Eventually, the controversy was sparked off by the unsolvable tensions between the various rights and interests to be guaranteed by the modern secular constitution: above all, that between freedom of research and protection of human dignity. It seems that in the debate on bioethics the various positions, especially those for and against therapeutic cloning, are irreconcilable not because the disputants assume the priority of one these two concepts over the other, but, rather, because the concepts and terms are in fact interpreted differently. Rejecting outright the dichotomy of scientific freedom on the one hand and human dignity on the other, those who subscribe to the ethos of curing regard innovation in the field of medicine as the most important pledge of human dignity. Considering its intensity as well as its publicity, it seems that the South Korean bioethical debate is an exception among the post-colonial pacific countries, especially when compared with China or Japan. This intensity is explained, at least partly, by South Korea's outstanding achievements in the field of bioscience. An analysis of the inner-Korean debate on cloning will show how far the peculiar South Korean situation is representative of cross-cultural bioethics.

3. Bioethics Between Tradition and Modernity

A. NGOs and Religious Groups as a Democratic Corrective

If we consider the recent developments in the field of bioethics as an arena for reflecting upon the dangers and potentials created by new biotechnologies and genetic engineering in South Korea, we might describe the year 1998 - roughly 22 months after the birth of Dolly - as a historical turning point. Until then, bioethics, like traditional medical ethics, had widely been regarded as an academic discipline; thereafter, however, bioethical discourse in general and the debate on cloning in particular became a controversial public issue. It was thus the starting point for a lively controversy between NGOs, religious groups, and the Korean Bioethics Association on the protection of embryos and human dignity in the under-regulated field of genetic engineering and reproductive medicine. These groups criticised the biopolitical ambitions of the South Korean administration, which, since the passing of the law for the promotion of biotechnology in 1983, had steadily increased its investment in this future key technology. These promotional programs still breathed the spirit of progressive modernism and global economic competition. Negative consequences of biotechnological achievements, on the other hand, were considered only in the late 1990s in the course of the democratising process. The goal of the biotechnology critics was the revision of the abovementioned law. At the same time, the opposing groups negotiated concepts designed to achieve a consensus; and, eventually, these negotiations led to the national "Law on Bioethics and Security," which was passed by Parliament on December 29, 2003. Academics from a wide range of disciplines as well as representatives of various interest groups were involved in the making of this law. Considering the public and interdisciplinary character of the present discursive culture in the field of bioethics, South Korean bioethics is neatly described as a "science of survival," a term that was coined by Rensselaer Potter.[13] The heated and very emotionalised debate indicates that bioethical problems have indeed become a vital question of global significance.

The immediate occasion for this turning point was the announcement on December 14, 1998 that a team of Korean researchers had successfully cloned a human embryo through SCNT.[14] This experiment was perceived as the first violation of a taboo in the field of cloning and thus caused a great stir on a national and international level. According to the Korean researchers, this experiment had been stopped, allegedly because the transplantation of genetically manipulated embryos into the uterus had been prohibited through the directives of the Korean Medical Association since 1993. The responsible researcher, Bo-yong Lee, pointed out to the investigating committee, which was set up immediately, that the goal of the cloning experiment had been to find a remedy for infertility.

Although infertility is a particularly delicate issue in Korea and although, consequently, modern reproductive technologies are widely accepted, the public reacted with a surprisingly harsh critique of the opaque research practices and demanded immediate legal action, which should take into account and be based on a general consensus. This public agitation was caused by the notion of the swift advancement of scientific research and practices and, moreover, by the realisation that these rapidly advancing developments were taking place in a legal 'gray area' - an awareness that mobilised unconscious fears. The repeated announcements by the Raelian sect that several Korean surrogate mothers were involved in their cloning project further intensified the already emotional public debate which has been ongoing since 2001. Although these claims have remained unconfirmed, the South Korean administration has taken juridical measures against the sect.[15]

These developments form the background for the emergence of a dense bioethical network in South Korea, which is closely connected with the national democratising process since the 1980s. Interdisciplinary competence, the integration of various social interests, and orientation towards international developments are the main characteristics of this network. The foundation for the bioethical exchange between experts and laypersons was laid at the Consensus Conference, held by the Korean National Commission for UNESCO in 1999. Mass media and internet have gained eminence by providing transparency and exchange of information on recent developments in biotechnology and biopolitics. The *People's Solidarity Participatory Democracy* (PSPD), which regards itself as an extra-parliamentary opposition designed to enhance civil rights, to prevent the abuse of power, and to develop political alternatives,[16] coordinates and documents the activities of the various movements. Membership includes experts from various disciplines, who comment critically on national and international developments.

Two central and integrating aims of the heterogeneous NGOs and religious groups are the establishment of an opposition to the power politics in South Korea and the democratisation of sciences. The latter implies the demand for more transparency of those research projects which are debated in public. Moreover, these critics search for possible alternatives themselves, too. Due to their ethical pretensions, the NGOs possess a normative character in South Korea and thus operate as a democratic corrective, aiming to increase public awareness of the rapid biotechnological developments. The representatives of the critical civil society perceive the passing of the above-mentioned "Law on Bioethics and Security" as a bitter disappointment in two ways: first, it renders the decision of the 1999 Consensus Conference (general ban on any kind of cloning) null and void. And, second, parts of that public they themselves had helped establish are now opting for further research and therapeutic cloning, especially self-help groups of disabled or the seriously ill or their families, which account for a significant part of the NGOs. Taking into account the immense difficulties of harmonising the divergent religious and ideological positions through long-lasting legal processes, we should assess such a re-orientation as confirmation of the corrective function of NGOs and religious groups. Without them, any attempt at setting up the National Bioethics Advisory Commission (NBAC) would be doomed to fail.

B. Human Dignity and the Internationalisation of Bioethics

The first article of the South Korean "Law on Bioethics and Security" contains the following declaration:

> It is the aim of this law to provide protection from the dangers connected with biotechnology, which jeopardise the dignity and value of human beings and their bodies, by laying the foundations of bioethics and security. The aim of this law is also to create such conditions for biotechnology as will ensure that biotechnological developments will serve the purpose of the prevention and cure of human diseases, and increase the health and quality of the life of the people.

This formulation documents that it is the goal of the legislator to guarantee, in accordance with the South Korean constitution, both the protection of human dignity and the freedom of research in the field of biomedicine. It remains unclear, however, whether this pledge to protect human dignity also applies to embryos, which are defined by the same law as fertilised oocytes, i.e. egg cells from the moment of fertilisation up to the development of that cell mass out of which all organs develop." (§2.2). The uncertainty concerning the juridical and legal status of the embryo - the question of whether or not it is a human being - has left room for different interpretations of the ethical status of embryos and has thus ignited a highly controversial cloning debate in South Korea.

The majority of Korean jurists argue that the moral status of the embryo depends upon social consent and they refer to the diversity of regulations abroad. For them, the core issue is whether South Korea respects international agreements concerning the protection of human dignity and human rights.[17] The crucial question is whether embryos are worthy of legal and juridical protection: if the embryo is regarded as a human from the very beginning, it has a right to human dignity; and, according to the modern South Korean constitution, it must not then be used as a means to an end. It follows that any kind of cloning and any research using embryos is unethical.

On the other hand, some researchers and medical experts reject this interpretation and maintain that human existence commences only with the development of an individual personality, roughly two weeks after the fertilisation of the egg cell. According to this view, frequently expressed in the Anglo-Saxon world and in South Korea too, humans do not possess personality from the very beginning but evolve it in the course of their development. Supporters of stem cell research and therapeutic cloning categorise the first stage in the development of an embryo, i.e. up to the 14th day after fertilisation, as *pre-embryonic*.[18] This hypothesis is based on the so-called 'biogenetic basic law.'[19] According to this basic law, the individual development (ontogenesis) recapitulates the entire phylogeny so that it is possible to determine by analogy the boundary between the human and the 'pre-human.' As the development of twins shows, so the argumentation runs, the individual development commences only on the 14th day after the fertilisation of the egg cell. Yet this hypothesis conflicts with the fact that the human genome of the individual is determined at the moment of fertilisation. If the *pre-embryo* is denied the status of a person, the defense of its (not her or his) dignity is hardly justifiable.[20] The fact that these divergent positions are central to the debate on cloning in South Korea indicates that there is no general consensus concerning the ethical status of the embryo. By endeavouring to reach such a consensus, however, South Korea distinguishes itself from those countries which, faced with the crucial question of determining the beginning of life, resort to religious codices, like e.g. Islamic countries.

In the aftermath of the successful cloning conducted by the Seoul research team, the critique of the recent legal regulation of cloning has intensified in South Korea. There is a growing number of people who demand a more precise formulation concerning the protection of embryos, and, presumably, this tendency will have an impact on the future NBAC's decision on therapeutic cloning. The Korean discourse thus takes a direction which is opposite to that, e.g., in Germany. While in South Korea under-regulation in the field of biotechnology has until recently facilitated research, scientific success has led to more comprehensive and stricter legal regulation. In Germany, on the other hand, research has been constrained by the strictest legal protection of embryos from the very beginning. Yet, permission for research with imported stem cell lines has led to a certain relaxation, and further liberalisation is demanded. This development has not gone unnoticed in South Korea.

Well before the passing of the law on bioethics and Hwang's breakthrough, the ethical status of the embryo had been at the centre of the South Korean controversy on cloning. As pointed out above, this controversy is shaped by basically the same arguments as in Western debates and range from the theory of potentiality and continuity to the biomedical 14-days argumentation. One reason for this similarity is that since the establishment of a modern system of public education in 1948, the academic elite has been well-informed about and much interested in Western research. More importantly, however, the ethical problems resulting from bioscience have a universal dimension. Korean philosophers, on the other hand, tend to maintain the deontological position, which regards embryonic research as utilising and exploiting human beings. They reject the division of embryonic and prenatal development into different phases because, according to their critique, it serves the utilitarian motif of research into and experimentation with human beings.[21] This scepticism is based on the slippery-slope hypothesis, according to which moral indifference to operations and intervention during the first phase of human existence will unavoidably lead to a further erosion of moral standards.

In sum, the debate on cloning reflects a fundamental dilemma of ethical considerations in South Korea as well as in other democratic countries. It is extremely difficult to reach a general consensus concerning the protection of human life at the earliest stage of its development. Yet it seems that the problem lies not in the recognition of human dignity as an ethical category but, rather, in the divergent definitions of the embryo. As the ethical legitimacy of cloning experiments and embryonic research depends on this definition, the present Korean controversy renders the status of the embryo an *experimentum crucis*.

Some, however, have not yet accepted the challenge. Either the embryo is redefined as an object for research on the grounds that it has not sprung from the fusion of an egg cell and a sperm cell and thus does not enjoy the full protection of its dignity - scientific facts weaken and even supplant ethics. Or it is maintained that, from the moment of fertilisation, the embryo's human dignity and human right must be considered worthy of protection, even if the embryo has been produced through therapeutic cloning (SCNT). For an adequate and sustainable solution of the conflict, it is vital that this central question is discussed openly.

4. Changing Perspectives on Cloning

Until the end of 1998, the South Korean cloning debate had to a large extent remained a passive reaction to developments abroad. Yet due to successful South Korean cloning experiments this changed abruptly.[22] Since then, the public has been kept in suspense by repeated reports of ground- and taboo-breaking achievements. The publication of the above-mentioned report on the first South Korean cloning experiment on December 14, 1998, was followed by further announcements of success, which sparked off a heated bioethical debate. As early as August 2000, Woo-suk Hwang, a veterinary, and, in March 2002, Se-pil Pak, an expert in reproductive medicine, claimed to have cloned human embryos through cross-species SCNT - these announcements caused much public agitation.[23] Therefore, the 'sensational' success announced at the AAAS conference in Seattle was simply a logical consequence of their research work. As these two researchers represented their work as therapeutic cloning designed to produce embryonic stem cells from the beginning, the national controversy about cloning focused on the ethical problems connected with therapeutic cloning and embryonic research.

Yet the starting point of this debate in South Korea are the responses to a report by Jerry Hall of George Washington University, who had cloned a large number of human embryos through IVF-embryo-splitting in 1993.[24] His results proved that it is possible to clone human beings artificially. The tenor of the reception was that this procedure had previously been carried out with animals before so that Hall's results were technically not an innovation. The news of the birth of 'Dolly' through SCNT in early 1997 generated much stronger reactions among South Koreans, especially among South Korean Christians. The latter responded immediately by petitioning government for a ban on cloning. Their apocalyptic argument was that cloning was an "attack on Creation" and the 'ultimate sin.'[25] This reaction triggered a lively public debate in South Korea, largely dominated by the Christian community. At the same time, however, there have been moderating voices warning of the danger of reacting too emotionally and spreading fear. Instead, so they demanded, the new challenge should be tackled soberly.[26] Following and referring to these public controversies, a critical philosophical discourse developed in the academic world - the institutionalisation and internationalisation of

Korean bioethics set in. In addition to the above-mentioned PSDP, the Korean Bioethics Association, the Korean Society for Medical Ethics Education, the Catholic Institute of Bioethics, and the ELSI Korea have gained prominence. Through their interdisciplinary studies, which also take into account international developments, these institutes contribute a sound theoretical frame to the public discourse.

Apart from the highly emotionalised and biopolitically motivated part of the controversy, it has become clear that the debate on cloning is shaped by two irreconcilable positions, which are neither new nor typically Korean. On the one hand, there is the empirical and materialist position, which has a progressive conception of history and assumes the changeability of the conception of the human through the change in living conditions. According to this view, it is the task of science and technology to constantly improve the quality of life by controlling nature. A crucial tenet of this position is the assumption that problems can be resolved only through scientific innovation. Consequently, according to Chông-sôn Sô, a prominent medical expert at Seoul National University, the biotechnological age and the concrete vision of a biomedical revolution make a new attitude towards life imperative.[27] On the other hand, a harsh critique of such fantasies of 'scientific omnipotence' has been expressed by those who argue in favour of holistic ethics. Hoe-ik Chang, a professor of physics at Seoul National University who supports this position,[28] has won much approval among the critics of biotechnology on various sides, i.e. not merely traditional and religious bioethicists. It is clear that, in South Korea, the bioethical conflict runs through all disciplines and religious creeds although it is represented as a simple opposition of pro and contra in the public debate. The following is therefore dedicated to elucidating the critical ethical arguments.

A. Ethical Problems of Human 'Reproductive Cloning'

The central issue in the critique of the successful South Korean cloning experiments is the danger that this technique might be abused for reproductive cloning. As these two forms of cloning do not differ technically, the results produced by the Seoul research team are condemned as a 'recipe' for reproductive cloning. According to experts, the goals of reproductive and therapeutic cloning can be distinguished morally. While they defend therapeutic cloning as scientists and medical practitioners consistently and with much verve, they reject reproductive cloning categorically, mainly because of its unreliability, and remind their audience of the difficult birth and early death of Dolly. They predict high rates of deformity as well as incalculable physiological and psychological risks. In addition, the small probability of successful reproductive cloning exacerbates the ethical difficulties of egg cell donation and surrogate motherhood.

Such difficulties result mainly from technical problems; yet social ethics and anthropology supply the South Korean critics with further arguments against reproductive cloning, which deal with asexual reproduction and eugenics. While in the aftermath of the 1962 London Ciba conference the critique of "collective fantasies of eugenics" and the violation of the principle of contingency of individual human life emotionalised the debate in Western countries, the arguments are surprisingly concrete in South Korea:[29] disturbance of the social equilibrium through a genetically manipulated elite;[30] disturbance of traditional family structures and marital habits; disturbance of the most elementary human relationships. In addition, the critique of the exploitation of human life through experimentation has gained prominence, for such abuse implies a fundamental disrespect of human life.

As early as 1999, the citizen panel of the Consensus Conference described the risks of cloning:

The most problematic issue with regard to cloning technology is the possibility of human individual cloning using nuclear transfer techniques. Human beings are considered to have a unique genetic constituency. Thus, cloning human beings would pose grave threats to human identity and dignity. Human embryonic cloning, as well as the abuse and careless disposal of human embryos may cause a negative impact on the dignity of human life. The attempt to clone babies with only favourable genetic traits while eliminating those with inferior heredity will lead to serious social problems, raising the question of eugenic discrimination and creating something called a 'genetic class.' A cloned human would feel alienated in terms of familial status, resulting in, among others, an inheritance problem. In sum, a birth independent of a union of man and woman would mean great turmoil for the current marriage system.[31]

For these reasons, the citizen panel voted unanimously against reproductive cloning; yet two out of sixteen participants voted for therapeutic cloning. In this context, the formulations used for describing the differences between "individual cloning" and "embryonic cloning" are noteworthy. In South Korea, the two expressions frequently replace the terms "reproductive" and "therapeutic" cloning, which are preferred by Western experts. However, the distinction between individual and embryo is - perhaps strategically - quite misleading, for personality is a topic only in the context of reproductive cloning. It is therefore surprising that, in addition to critical NGOs and religious groups, critical philosophers too use and accept these terms.

Young-mo Koo and Sang-ik Hwang have argued in favour of a general ban on cloning by referring to the slippery-slope argument: the legalisation of "human individual cloning" - though possibly curing infertility - might lead to abuse on a massive scale. They, too, anticipate major ethical problems resulting from the commercialisation of cloning technology and that, eventually, biopower will get irreversibly out of control.[32]

B. Ethical Problems of Therapeutic Cloning

While there is a broad consensus that "reproductive" cloning should be banned, the limited permission for so-called "therapeutic" cloning for research purposes, allowed by the "Law on Bioethics and Security," which was passed by Parliament on December 29, 2003, is highly controversial. The extent and mode of such research will be controlled by the newly created National Bioethics Advisory Committee. Research projects using so-called surplus embryos, which are left over from in-vitro fertilisation and have been frozen for at least five years, will be permitted. On the other hand, there will be a strict ban on the trade and sale of human ova and sperm, on the fertilisation of ova and sperm after a sex selection, as well as on the fertilisation of ova and sperm cells of defunct persons or minors. Genetic tests of embryos and foetuses will be allowed for the diagnosis of severe hereditary diseases. It will be in the Korean president's authority to define such exceptional cases. Moreover, the law includes measures that are designed to secure the protection of genetic information. While the prohibition of both reproductive cloning and the production of hybrids came into force immediately, other measures will do so only on January 1, 2005. Although in comparison with other countries these regulations are rather restrictive, critical bioethicists, who demand a complete ban on any kind of cloning, hope to achieve a revision of the law before it comes into force. Jurists like Jung, however, point to the fact that a rigorous prohibition of cloning would contravene the South Korean constitution, which guarantees freedom of research independent of moral judgements.[33]

Ethical problems result from the fact that through cloning procedures (SCNT), embryos are produced with the objective of gaining pluripotent stem cells, which are then used for both research and therapeutic purposes. Out of these stem cells it is possible to produce cells of the desired type in order to create an 'individual' transplant which will not be rejected. Supporters of therapeutic cloning assume that by using cells which have the patients' own DNA it will be possible to cure serious diseases, such as, e.g., Parkinson's and Alzheimer's, and, moreover, that it will be possible to solve the problem of the scarcity of organs for transplantation. In the aftermath of the successful cloning in Seoul, expectations based on biomedical innovations have increased considerably. Woosuk Hwang has predicted that within ten years the application of the technique in clinical medicine will be possible. As his preparatory experiments in the field of hybrid and chimera research have been rejected across the board because they are seen to violate the natural order and transgress the boundaries of the human species, it seems that the critical perspective is narrowing. For, in South Korea, too, the controversy on cloning concentrates upon either the status of the human embryo or vague ideas of future application. The concrete ethical problems, which result from the application of the cloning technology, however, are neglected. Yet, it can be assumed that they will soon become the focus of the debate. In addition to the above-mentioned moral implications of egg cell donation and surrogate motherhood, the crucial ethical problem of the production of embryonic stem cells for therapeutic purposes is the exploitation and fragmentation of the human body.

One of the reasons for the complexity of this problem which, due to a "scientific and technological fragmentation,"[34] aggravates and even impedes a differentiated and more open ethical discussion, is that prospects and hopes of cures are inextricably interwoven with scenarios and fears of abuse. For instance, it would be a step towards a more individual and gentle therapy if remedies could be tested with the patients' own tissue produced through cloning procedures. At the same time, however, the patient's body would then be extended by that tissue; and that tissue would eventually become a test laboratory. It is thus almost impossible to gauge the impact that such options will have on human self-definition and on our understanding of the unity or duality of soul and body. If many South Koreans today view present developments as the dawning of a new age of biotechnology and demand the readjustment of the view on life, we should expect that such far-reaching changes will indeed be realised; and we might expect, too, that after the controversial success of the cloning experiments conducted in Seoul and with the national biopolitical ambitions, a new chapter of the cloning debate will be opened in South Korea.

5. In Search of a Collective Identity in Bioethics: A Critical Note

Bioethical debates in Korea seem to centre around four basic fears which Korean citizens share with people all over the world, although their intensity and forms of expression are shaped by the specific Korean situation.

First, there is the fear of falling behind in the global competition for money and power, a fear that still finds nourishment in the historical experience of having fallen prey to a stronger neighbour due to economic and technological backwardness. Soterial expectations towards "Mr Science" that had evolved in this context are still influential today, thus biotechnology is viewed as a key to future prosperity. Ethical arguments against stem cell research may, from a perspective which is coloured by social Darwinist thought,[35] be seen as a foreign import intended to impede developing or newly industrialising countries like Korea or China in their efforts to fully realise their obvious potentials in these technologies.

Second, and in direct contrast, there are the ubiquitous anxieties produced by the industrialisation process with its social and environmental side-effects. Deep-seated fears of social, family and natural order being disrupted and of technological processes taking precedence over human agency help to demonise biotechnological research as the door-opener to a process that will lead to final and total de-humanisation. Such anxieties accompany social and technological change everywhere. Yet in Korea, with its history of an extremely swift and heteronomous modernisation process under colonial and dictatorial regimes, which did little to soften its impact on social fabric and individual fates, they may easily combine with ideas about biotechnology as a "foreign influence" running counter to a "natural Korean" state of being.

While these two clusters of ideas are founded on fears concerning personal well-being (reaping profit or suffering damage from the modernisation process), parallel fears arise from the potential gains or losses of belonging to the Korean nation. Efforts to gain respectability for Korea as a member of the international community of nations are accompanied, or balanced, by fears of losing collective identity. Both sentiments can be directed for or against the limitation of stemcell research.

As Woo-suk Hwang's example has shown, success in this field brings rich rewards in the "currency of the soul," attention and respect: a Korean scientist as a world star is worth even more than playing host to the Football World Cup in terms of "feeling good to be Korean." Cutting-edge scientific results are better evidence than volatile statistics of GNP and the like that the country has caught up with developed nations. But respectability as a member of civilised nations can also be seen to hinge on the existence of "up-to-standard" bioethical discourse and legislation, especially in the face of an international (i.e. Western) journalistic discourse that still subscribes, more or less unconsciously, to the stereotypes of "Asian despotism" and assumes a deficiency in the valuation of human life in Asian societies. In a cultural community that tends to equate modern civilisation with Christianity, the need for ethical standards compatible with "international" conventions is acutely felt.

The widely shared value of maintaining collective identity can be applied in even more contradictory ways. On the one hand, it can serve as a counterforce to both uses of the quest for international status outlined above when this quest, in its respective forms, is denounced as disloyal to the intrinsic values of Koreanness. On the other hand, justifications for both kinds of ethical decisions - pro and contra embryo research - can be drawn from references to a construed national identity. When Woo-suk Hwang points to Buddhism as the philosophical source of encouragement for his research, he clearly wishes to imply that his actions are not ethically unreflected, but tied to a traditional "Korean" ethics. In the same vein, but with the opposite effect, Korean Christian bioethicists put much effort into the search for more indigenous foundations for their verdict against the destruction of embryos for research. Kyo-hun Chin's contribution in this volume is a prime example.

As these contrasting examples clearly show, simplistic culturalistic explanations of Korean bioethical behaviour are doomed to fail. Not only is "tradition" necessarily at a loss for answers to these contemporary problems; in Korea, with its complex cultural heritage, "tradition" itself has long been a field of contestation. None of the teachings that are referred to in each and every discussion of a Korean ethics - Shamanism, Buddhism, Confucianism, Christianity - has ever held authority over the whole of the Korean population as far back as history can be traced.

Little is known about ancient Shamanism; but, certainly, what is meant by this not very concise appellation comprised local cults: Buddhism was brought to the peninsula during the state-building processes from the 3rd century AD onward, precisely to counter, with a unifying force, the prevalent spiritual attachment to locality. In this early stage, it was received together with Chinese administrative knowledge ("Confucianism") as a tool of elite empowerment. By the time Buddhism had gained appreciable influence on the lower echelons of society, it was more or less discarded by the literati elite in favour of Neo-Confucianism, which in turn entered a process of downward diffusion. During these centuries, the possibility of a Buddhist literati identity never slipped completely out of sight. With the advent of modernity and a contemporary ethnic consciousness, there was no lack of advocates for building a nation on the foundations of Buddhism, Confucianism and Christianity, respectively; at the same time, none of these creeds could escape criticism as fundamentally un-Korean. While Shamanism is rather immune to this kind of criticism, it has, on the other hand, been identified with cultural backwardness for too long - probably since the advent of Buddhism, but certainly since the Neo-Confucian renaissance in the 15th century, and again, very plainly, under the Chung-hee Park modernisation regime of the 1960's and 70's - to be universally accepted by Koreans as common cultural ground.

Having co-existed for long periods in different parts and layers of society, obviously no one of these strains of ideological traditions in Korea can be isolated as the source of a collective ethical identity. At the same time, the history of ideological strife surrounding the establishment of each of the religions suffices to bar any expectation that such a collective identity could be found in the common ground they share: if such a ground is to be found, it can be no more than ethical commonplace, not sufficient to build a distinctive identity on it. Syncretistic arguments of this kind should, therefore, not be taken at face value. While they must be made in order not to violate the shared value of maintaining cultural identity, they are not so much a *reflection* of such an identity than an attempt at *creating* an ethical identity in line with the author's agenda. The stance taken by individual Koreans on present scientific developments will be informed less by a cultural matrix than by personal convictions derived from multiple sources, including family traditions; the stance of the Korean populace at large will probably be shaped primarily by the balance of their hopes and fears concerning the role of science in social processes.

Notes

1. Science Express Reports, 2004.
2. Simerly, 2003.
3. Denker, 2003.
4. cf. Song, 2002, 41-43.
5. FAS, 2004.
6. Hwang, 2004.

7. Song, 2002, 41-43.
8. http://www.Mediadaum.net.
9. Soh and Kim, 2004; Kim, 2004.
10. Kim, 2004.
11. cf. Koehler, 2004.
12. cf. Siep, 1996, 310.
13. Potter, 1971; Engels, 1999, 11-13.
14. Kim, 1998.
15. BBC News, 2002.
16. http://eng. Peoplepower21. org.
17. Park, 2000, 113-118.
18. cf. Lee, 2002.
19. Rager, 1996, 264.
20. Ku, 2002, 181-187.
21. Lee, 2002.
22. Park, 2002, 47.
23. Ahn, 2002.
24. Hall, 1993.
25. Park, 2002, 49-50.
26. Sang-ik Hwang, 1997.
27. Sô, 1997.
28. Park, 2002, 51-52.
29. Schneider, 2003, 267.
30. Consensus Conference, 1999.
31. Citizen Panel Report, 1999.
32. Koo and Hwang, 2000, 199-201.
33. Jung, 2000, 27-28.
34. Kollek, 1998, 22.
35. For the influence of Social Darwinism on Korean intellectuals of the early modern period, see Tikhonov, 2003.

References

Ahn, Yong-hyeon (An, Yong-hyôn). "Human Stem Cells Created in Cow Ova." In *Digital Chosun*. March 8, 2002.

BBC News. *S. Korea Launches Cloning Inquiry*, July 26, 2002.

Citizen Panel Report. *Progression of the Korean Consensus Conference.* Organised by the Korean National Commission for UNESCO. March-September, 1999. Seoul, Korea.

Denker, Hans-Werner. "Totipotenz oder Pluripotenz? Embryonale Stammzellen, die Alleskönner." *Deutsches Ärzteblatt.* Jg. 100. Heft 42, Oct 7, 2003.

Engels, Eve-Marie. "Natur- und Menschenbilder in der Bioethik des 20. Jahrhunderts. Zur Einführung." In *Biologie und Ethik*, Eve-Marie Engels (ed.). Stuttgart, 1999.

Frankfurter Allgemeine Sonntagszeitung (FAS). "Das Ei des Dr. Hwang Woo-suk." (Feb 15, 2004).

Hall, J. D., D. Engel, P. R. Gindoff et al. "Experimental cloning of human polyploid embryos using an artificial zona pellucida." In *Fertility and Sterility* 60 (Suppl.), S1. (The American Fertility Society conjointly with the Canadian Fertility and Andrology Society, Program Supplement, 1993 Abstracts of the Scientific Oral and Poster Sessions, Abstracts 0-001).

Hwang, Sang-ik. *How to Understand the Cloning of Humans: A New Spelling of Life* (Korean), *Sindonga*, 4, 1997, 560.

Hwang, Woo Suk, Young June Ryu, Jong Hyuk Park, Eul Soon Park, Eu Gene Lee, Ja Min Koo, Hyun Yong Chun, Byeong Chun Lee, Sung Keun Kang, Sun Jong Kim, Curie Ahn, Jung Hye Hwang, Ky Young Park, Jose B. Cibelli, and Shin Yong Moon. "Evidence of a Pluripotent Human Embryonic Stem Cell Line Derived from a Cloned Blastocyst." Published online Feb 12, 2004; 10.1126/science.1094515 (Science Express Reports).

Hwang, Woo Suk (Hwang, U-sôk). *Saengmyông konghak kisul kwa in'gan saenghwal. Saengmyông konghak tonghyang* [Biotechnology and Human Life: The Tendency of Biotechnology], 5/1. Seoul, 1997.

Jung, Kyu Won (Chông, Kyu-wôn). "A Legal Discussion on Human Embryonic Cell Cloning by Somatic Cell Nuclear Transfer Technique." *Journal of the Korean Bioethics Association*, Vol. 1, No. 1, May, 2000, 21-39 (article in korean).

Kim, Ch'ang-ki. "Human Cloning Test Succeeds." In *Digital Chosun. The Chosun Ilbo,* Dec 14, 1998.

Kim, Ch'ôl-chung. "Bioethics of Human Embryonic Cloning." [An Interview with Woo-suk Hwang and Dong-ho Lee], *Chosun Ilbo*, March 1, 2004.

Koehler, Angela. "Klon-Pionier: ‚Heilung mit koerpereigenen Transplantaten'." *Wirtschaftswoche*, Feb 18, 2004.

Kollek, Regine. "Klonen ist Klonen - oder nicht? Warum der erste Menschenklon nicht die Gestalt ist, an der sich die Urteilsfindung orientieren muss." In *Hello Dolly? Über das Klonen*, Johann S. Ach, Gerd Brudermueller and Christa Runtenberg (eds.). Frankfurt am Main, 1998, 19-45.

Koo, Young-mo (Ku, Yông-mo) and Sang-ik Hwang. "Some Ethical Problems Concerning the Cloning Research and its Application." *Korean Journal of Medical Ethics Education*, Vol. 3, No. 2, November, 2000, 199-209 (article in korean).

Ku, In-hoe. *Saengmyông yulli-ûi ch'ôrhak* [Philosophy of Bioethics]. Seoul, 2002.

Lee, Pil-ryul (Yi, P'il-ryôl). "Is the 'Pre-embryo' not Equivalent to a Human Being?" In *Proceedings. IV Asian Conference of Bioethics. 'Asian Bioethics in the 21st Century',*" 22-25 November 2002, Seoul National University. Seoul, 92-95.

Park, Hee-joo (Pak, Hûi-chu). "Cloning Controversies in S. Korea." *Journal of the Bioethics Association*, Vol. 3, No. 1, June, 2002, 47-73 (article in korean).

Park, Un-jong (Pak, Ûn-chông). "Research Ethics and Policy Involving Vulnerable Subjects and Human Body Components." *Journal of the Korean Bioethics Association*, Vol. 1, No. 1, May, 2000, 99-122 (article in korean).

Potter, Rensselaer van. "Bioethics." B*ioscience* 21, 1971, 1088.

Rager, Günter. "Embryo - Mensch – Person." In *Fragen und Probleme einer medizinischen Ethik*, Jan P. Beckmann (ed.). Berlin and New York, 1996, 254-278.

Schneider, Ingrid. "‚Reproduktives' und ‚therapeutisches' Klonen." In *Bioethik. Eine Einführung*, M. Düwell and K. Steigleder (eds.). Frankfurt am Main, 2003, 267-275.

Siep, Ludwig. "Ethische Probleme der Gentechnologie." In *Fragen und Probleme einer medizinischen Ethik*, Jan P. Beckmann (ed.). Berlin and New York, 1996, 309-331.

Simerly, Calvin, Tanja Dominko, Christopher Navara, Christopher Payne, Saverio Capuano, Gabriella Gosman, Kowit-Yu Chong, Diana Takahashi, Crista Chace, Duane Compton, Laura Hewitson, Gerald Schatten. "Molecular Correlates of Primate Nuclear Transfer Failures." *Science* 300, Number 5617, Apr 11, 2003, 297. 10.1126/science.1082091.

Sô, Chông-sôn *Saengmyôngch'e Pokche kisul-ûi munjejôm. Kwahak kwa kisul* [Problems of Reproductive Cloning], 6, 1997, 52-54.

Soh, Ji-young (Sô, Chi-yông) and Tae-gyu Kim (Tae-kyu Kim). "Ministry Considers Creating Embryo Research Complex." *The Korea Times*, Feb 27, 2004.

Song, Du-yul. *Schattierungen der Moderne. Ost-West-Dialoge in Philosophie, Soziologie und Politik*. Köln, 2002.

Tikhonov, Vladimir. "Residues of this view of the modern world order are strong." *BJOAF* 27. München: Iudicium, 2003.

Some Observations on Buddhist Thoughts on Human Cloning

Jens Schlieter

Abstract: So far, many Buddhists appear to have taken a fairly neutral attitude to cloning. It seems that, provided no harm is done to any of the beings involved, the procedure of cloning as such does not offend religious feelings or basic value sets of Buddhists. This is evidenced by the rather relaxed way in which the issues of therapeutic and reproductive cloning are discussed by South and South-East Asian Buddhists. After identifying four main characteristics of the Buddhist ethical discourse (considering the intention of a deed and its 'karmic' consequences for the perpetrator; the aspects of 'contextual consideration' and 'moral self-cultivation'), the cases of reproductive cloning and cloning for research are discussed. Finally, the article considers the impact of Buddhist ethics on Thailand's regulative procedures. It seems that hitherto Buddhists have not formulated specific principles of 'Buddhist' medical ethics or bioethics; they, therefore, appear to have less influence than Western biomedical ethics. Yet, the positions taken by Buddhists are as manifold as those by specialists of other religious traditions.

Key Words: Buddhism, Thailand, Southeast Asia, Reproductive cloning, Cloning for research, Intentionalism, Legal regulation.

1. Introductory Remarks: Hwang's Buddhist Justifiction of His Cloning Research

The following observations focus on Buddhist estimations of the ethical legitimacy of human cloning. So far, many Buddhists appear to have taken a fairly neutral attitude to cloning, i.e. the biomedical procedure of artificial, non-sexual reproduction. It seems that, provided no harm is done to any of the beings involved, the procedure of cloning as such does not offend religious feelings or basic value sets of Buddhists. On the whole, Buddhists adopt much the same range of positions as those held, for example, by Christians of Catholic, Orthodox or Protestant denominations.

Before I proceed to examine the foundations and circumstances of Buddhist reasoning,[1] I would like to refer to the most recent cloning research of the Korean research team led by Woo Suk Hwang, published on the February 12, 2004 in the prestigious journal *Science*. The article describes the team's successful application of the Cell Nuclear Transfer

Method to human eggs. 242 eggs from 16 volunteers[2] - who donated their eggs for the purpose of experimentation - were denucleated, and a cell nucleus was inserted. Thirty living blastocysts were established, but died more or less immediately. One stem cell line was actually extracted out of one of the cloned embryos.[3]

Of significance for our topic is the fact that the leading researcher Hwang tried to underline the ethical legitimacy of his experiment with a reference to his Buddhist faith: "I am a Buddhist, and I have no philosophical problem with cloning. And as you know, the basis of Buddhism is that life is recycled through reincarnation. In some ways, I think, therapeutic cloning restarts the circle of life."[4]

Indeed there are some rather unique aspects in Buddhist attitudes towards human life which will be analysed below. Yet Hwang's definition of 'therapeutic cloning' is quite unusual. It has commonly been defined as the direct opposite to 'reproductive cloning' - the latter characterised as the attempt to create a new being, or, more generally, a new independent form of life. Astonishingly, in his quotation above Hwang seems to take the position that "therapeutic cloning" should not be condemned because it is just another 'appearance' of a life cycle, i.e., a new phase of a cyclic life series.[5] Can this statement, which equates "therapeutic cloning" with the beginning of a new life form, be reconciled with Buddhist thought? And is it consistent with the current use of the two terms "therapeutic" and "reproductive" cloning? Apart from the critique of the concept "therapeutic cloning" - it should be more adequately termed 'cloning for biomedical research'[6] - the concept of a strict separation of the two has lately attracted much criticism. Not surprisingly, therefore, Hwang himself also admits that his experiments cannot be separated from the initial procedure of cloning human beings. "Yes, this technique cannot be separated from reproductive cloning," Hwang said in an interview,[7] and he added, that every country should prevent any reproductive cloning experimentation with this technology.

He seems to have no "philosophical problem" with cloning because for him it is just an asexual way of triggering a new life-cycle. Furthermore, the question of ending the life of the embryo becomes irrelevant (if I understand his remark correctly) with his perception that stem cells in themselves form a new 'life' phase. He thereby avoids the central question whether it is justifiable to stop the development of the embryo, or, in other words, to deliberate end its current life phase. However, scrutinized more closely, Hwang does not offer an ethical evaluation of his cloning experiments. He simply attempts to describe 'ontologically' or 'dogmatically' the procedure as such. His "philosophical" problems vanish by declaring that the production of an embryo - in order to create stem cells - is nothing less than the inception of a new phase of a 'being-to-be-reborn.' In other words: The ethical question whether the destruction of the embryo should be allowed, does not precede the experiment but rather follows the accomplished cloning experiment. There is quite a strong Buddhist presupposition of the "ethical neutrality" of existence: If scientists are able to manipulate 'nature,' they can do so only by developing a hidden quality or capability in 'nature' itself. Generally, this presupposition creates a permissive environment for the manipulation of genes or genomes etc., but, of course, it does not provide ethical solutions. The ethical sphere is simply ignored.

What makes matters more complicated is Hwang's statement on another occasion that he fails to see that a cloned human embryo should be acknowledged as human life at all: "'That requires the egg from a woman and the sperm from a man,' he said. 'We used no sperm'."[8] And he concludes: "Nothing in Buddhist teachings raises precise ethical questions about the next step - inserting that cloned embryo from a test tube into a women's womb to clone an infant."[9]

I doubt if it is possible to harmonise the statements cited above. Of course, we have to bear in mind the possibility that one or more of these quotations are distorted. But, as Phillan Joung notes, the confusion of the two categories of cloning tends to hide an important fact: Indeed, the cloning procedure is - for the first initial steps - the same in both cases. But due to the ethical difference of the underlying intentions the cloning procedures cannot be lumped together. Joung suspects "biopolitical tactics" on the part of the reporting press, initiating confusion in order to distract attention from the main problem: the unresolved conflict about the legitimacy of "therapeutic cloning."[10]

Anyhow, if Hwang had a serious interest in advancing or even applying Buddhist ethical reasoning to human cloning research, he would have made sure that the audience got a clear picture of his idea of human life. Or, at least, of what he sees as the moment from which human life must be protected. Hwang never mentions that Buddhist texts contain quite clear definitions of the beginnings of human life, and, moreover, most of them argue, generally speaking, that to kill embryonic life is an unwholesome act. A parallel to Hwang's disregard can be found in some of the articles that zealously adopted his attitude in all respects as an example of the 'Buddhist' attitude as such. Actually, Korean Buddhist critics of cloning-for-research play an important role in the movement 'People's

Solidarity for Participatory Democracy,' a fact seldom mentioned in Western reports.[11] From the viewpoint of an external observer, Hwang's reference to his Buddhist opinions nevertheless has to count as a testimonial to the Buddhist perspective, or, more precisely, the perspective of a lay Buddhist. One may ask, though, if it is a coherent position, if it is in line with the dogmatic positions expressed in central Buddhist texts, or, if and in what respect it differs from interpretations by religious specialists.

A specific teaching of Mahāyāna Buddhism (predominant in East Asian countries) is based on the ideal of helping others out of compassion, with special regard to their suffering. Hwang, by asserting in an interview in the *New York Times* that there is an "obligation" for scientists "to do this kind of research because it is for a good purpose,"[12] evokes associations with the Buddhist self-commitment to help others who are suffering. Indeed, this ideal of "great compassion" (skt. *mahâ-karunâ*) may, according to some central texts, override all other Buddhist 'ideals,' which will be discussed below.

I would like to end my introductory remarks on Hwang's line of argument with a comment on the role of the observer. There is a well-known methodological difference between the normative approach of ethics and the descriptive approach of cultural studies. We have to draw a distinction because a combination of the two will lead to various methodological difficulties which I am unable to discuss here. Anyhow, the descriptive approach chosen here also has his disadvantages. For example, it is impossible to exclude certain personal presuppositions and assessments. Various methodological pitfalls endanger comparative ethics. Since I am citing scholars and religious specialists whom I was able to interview, I - like anybody else in the context of communication - had a certain influence on the answers by the choice of the wording, the way of asking, by simply being a Westerner, an 'outsider' etc. These factors are of some considerable significance, since many Buddhist scholars are, for example, very much aware of the recent efforts by UNESCO and the UN legal committee to draft an international ban on cloning.

2. How Should 'Buddhist' Bioethics Be Defined?

Let me proceed with a question: May we speak of 'Buddhist' bioethics as we speak of 'Christian' or 'Muslim' bioethics? This is not to say that well-grounded opinions on bioethical issues are lacking in Buddhist cultures. But I agree with Damien Keown who suggested in his *Buddhism & Bioethics*, that a theoretical formulation that is specifically Buddhist is still in its nascent stage: "Despite the contemporary importance of issues such as abortion and euthanasia, there has been comparatively little discussion of them from a Buddhist perspective. [...] [P]roblems such as embryo research [...] have scarcely been raised."[13] Keown's own work is most likely the first systematic attempt to apply Buddhist ethics to biomedicine and biotechnological procedures. It is not too surprising that this work was written by a Westerner. Even today there are only few circles in which Buddhists discuss bioethical principles from a theoretical point of view. Most confine themselves to the discussion of cloning and stem cell research at a regional level.

What are the reasons for this? First and foremost, Buddhists have a pronounced interest in the possible doer of deeds but, generally speaking, less interest in legal regulations. Indeed, there are only very few countries in which Buddhist teachings are an integral part of the constitution or civil law. Traditionally, Buddhist monks and nuns have left the field of legislation to lay people. Of course, they hold their own ethical opinions based on their interpretation of classical Buddhists texts and regulations (the Vinaya, or codices of monastic discipline). But they seldom feel the need to announce their positions to a broad (lay) public audience.

It seems to be widely, though not universally, held by Buddhists that ethics must be combined with the spiritual well-being of the actor. Every action will have its effects, and it is hoped - and expected - that the individual will refrain from unwholesome actions. Thus, Buddhist ethicists usually do not claim that the perspective of the victim as such has to be considered. This is an important point when it comes to the question of embryo research or cloning.

Apart from encouraging individuals to examine their motives, it is rare indeed for Buddhists to engage in heated exchanges on ethical matters. I may also add, as a personal observation, that in their publications very few Buddhist scholars of bioethics discuss positions held by other scholars. Normally there is a tolerant and liberal attitude toward the opinions of others. Even today, in most Buddhist monasteries it does not matter what kind of *view* or philosophical argumentation a Buddhist embraces - it does not lead to his exclusion from the monastic community ('disrobing'). Only certain *actions* normally have that effect. One may add that the term "Buddhism" itself is a Western notion and was coined in around 1830 to denote different teachings,[14] schools and traditions, which for their part seldom felt the need to unite in doxological perspectives. In effect, there is a wide range of Buddhist teachings in respect to the major schools of Theravāda, the dominant tradition in South-East Asian countries, and Mahāyāna Buddhism of Central and East Asia. This diversity can also be observed in bioethical reasoning.

Can the statement therefore be justified that there is no such thing as 'Buddhist bioethics'? If defined in its context of formation - the US of the 60s and early 70s - 'bioethics' may be described as a reaction and reflection by theologians (in the early phase most of them Protestants like Fletcher and Ramsey) on new medical practices such as artificial insemination, organ donation/brain death, the 'pill,' and so on. If defined as an attempt by religious specialists and moral thinkers to cope with ethical problems of scientific medicine, there is certainly something we may call 'Buddhist bioethics.' This attempt, if made in a traditional manner, will turn to the classical sources of Buddhism in order to find ethical guidelines relating to birth, death, or illness that may be transferred to modern medical applications. There are indeed early medico-ethical regulations in the Buddhist canon, for example, on abortion or euthanasia. Taken together, they form a kind of general ethos. But, on the other hand, most of these passages on "ethics" in Buddhist scriptures have a narrow sense and application because they were tailored to the lifestyle of monks and nuns, who were rarely involved in the ethical dilemmas posed by medicine. Thus, it seems for them a difficult task to apply pre-modern worldviews and terminology to modern scientific medicine and biomedical research. It is quite understandable, therefore, that a straightforward 'Buddhist' bioethics has not been published, or, if it has, it has not received much attention in the Buddhist world.

Far more common and promising than a reconstruction of premodern Buddhist medical ethics is the attempt to develop, on broader Buddhist grounds, a systematic approach to bioethics that may answer questions of embryo research or human cloning. Let us take a work by Pinit Ratanakul as an example:[15] In his book *Introduction to Bioethics* (1986) he combined principles of the deontological ethicist W.D. Ross and the moral action guides of Beauchamps and Childress with the ethical principles of Buddhism. He thereby introduced a systematic bioethical approach that transcends the limited scope of those ethical guidelines either intertwined with a specific belief system or based on the views of indigenous medicine. One advantage of such an attempt is that it is no longer bound to a specific life style such as that of a monk or nun.

But bioethics of specific religious traditions is, for sure, more than just a modern reformulation of traditional moral action guides. Most religious contexts consist of 'formative stories' which speak about the meaning of birth and death or the spiritual goal of human life. Let me take the example of human cloning and Buddhism: Obviously the formative stories of Karma and rebirth, the final liberation from the world of suffering, and the great narrative of a human being becoming a Buddha, have a profound influence on ethical reasoning on cloning. Also, the lack of other formative stories, like that of man as created by god, or the creator-god himself, has an observable influence on Buddhist bioethical thought. Ultimately, we may conclude that a direct equivalent to the Western concept of 'ethics,' and 'bioethics' in particular, is about to emerge and that it will reflect the influence of Western ethics to a considerable degree.

3. Four 'Core Elements' of Buddhist Ethics

In the following I shall consider Buddhist ethical thoughts under four European philosophical headings and try to show how these Buddhist principles guide the Buddhist discourse on different hypothetical human cloning scenarios («a» and «b»).

'Intentionalism': Buddhist ethics are strongly based on (the examination of) the intention of the doer of deeds. Intentions can be wholesome or unwholesome. If the intention governing a deed is to harm or kill a living sentient being, that deed is seen as unwholesome. Other unwholesome intentions are ignorance, greed, fanaticism and delusion. Wholesome intentions, on the other hand, are characterised by the intention of "non-harming" (*ahimsâ*), compassion, and the attitude of loving kindness, a love that embraces all beings. Schools differ in their understanding of the application of these principles, e.g., whether compassion implies "action" or just a mental attitude towards others.

'Considering consequences' (for the doer/the one who is treated): Buddhists also consider the outcome of actions. Important here are not the consequences for possible victims, but the consequences for the doer of deeds. Unwholesome actions, it is believed, will assemble bad Karma and therefore imply an effect on the next rebirth. So it has a direct effect on the perpetrator's next life. A consequentionalist (not synonymous here with utilitarian) approach naturally allows a certain 'weighing' of anticipated effects: A deed may have more than one effect or, combined with the core element of 'intention' (see above), it can be expected to produce a main good effect that may involve certain bad side-effects.

'Contextualism'/'situative ethics': Buddhist ethics can also be characterised as 'contextual' (situative ethics), due to the fact that instructions differ in relation to the spiritual status of the people involved.

Monks and nuns observe a large number of rules, and in Mah□y□na schools there are further distinctions in ethics according to the spiritual progress achieved by the doer. Buddhist ethics generally follow a "gradualist approach,"[16] and its application is quite flexible. Several programmatic advices describe a situative selection of possible means. Sometimes the advice is given not to stick to dogmatic decision or to adhere slavishly to "views."

'Self-cultivation': Last but not least, Buddhist ethics originally serve as a means to liberation. To act 'skilfully' and 'wholesomely' is a necessary precondition for achieving this goal. By their nature, Buddhist ethical codices of monastic discipline were formulated as training rules. Of course, they express the goal to be generally pursued by Buddhists (see also 'Contextualism'), but nevertheless the conception of ethics as self-cultivation implies that 'immoral,' faulty behaviour may occur. Its occurrence, indeed, may lead the perpetrator to intensify his or her personal efforts, because, according to the Buddhist standpoint, the doer of wrong deeds harms him or herself first and foremost, whereas good deeds contribute to the accumulation of merit.

Obviously these characteristics of an 'intentionalist,' 'consequence-orientated,' 'contextual,' 'offender-centred' and 'self-cultivationist' ethics will have a considerable effect on the Buddhist assessment of human cloning. The most important difference to Western ethics, according to my understanding, may be seen in the lack of a universalist reasoning based on the idea of 'dignity.'[17] I will now present some Buddhist evaluations of cloning. For the sake of clarity I will refer to the aspects of the scheme above where they seem relevant.

4. Buddhist Positions on Human Cloning

A. Case «a»: Cloning-for-Biomedical-Research ('Therapeutic Cloning')

The case of so-called 'therapeutic cloning' differs significantly from cloning-to-produce-children as regards the criterion of *intention*; the first steps of all cloning procedures are, indeed, the same. Because in 'cloning-for-research' there is no direct advantage in bringing to life a new being, but rather it involves taking (an embryo's) life with rather an uncertain effect on others, some Buddhists tend towards a strict application of the first precept, that is, to abstain from killing altogether.[18] But actually some monks and scholars argued recently that an embryo may not count as a full human being. Is this interpretation in line with classical Buddhist texts? And what about a position like that of Hwang, who said he is not sure if a cloned human embryo is 'human' at all?

In the most common procedure of 'cloning-for-research' the cell nucleus of an oocyte is extracted and replaced by the somatic cell nucleus of a donor. This initial procedure is, according to most Buddhists, not a problematic one since the transfer does not take place after a 'conception' (i.e., descent of consciousness). To change the pattern of sexual reproduction implies - at least for most Therav□da Buddhists - nothing of ethical significance. Any destruction of the embryo (after the descent of the developing life force of consciousness), however, is a problematic deed. But when exactly does this descent take place?

According to one important description, the consciousness principle seeking rebirth descends into the womb and finds 'a halt' in the 'bodily matter' (*rûpa*), thereby sparking a 'new' life.[19] The strict position, prevalent in classic texts as well as in modern commentaries,[20] is expressed, for example, in Perera's Buddhist interpretation of the "Universal Declaration of Human Rights": "It is the Buddhist view that the right to life commences at the very first embryonic stage of a being, since *maitrî* or love, according to the *Mettasutta* (Sn [=*Suttanipâta*], vv. 143-152) should be extended even to the embryo or 'one seeking birth' - *sambhavesî* (Sn v. 147)."[21]

But according to Bhikkhu Mettânando,[22] a modern interpretation should take the emergence of the brain as the 'bodily' basis of consciousness.[23] Other scholars, too, suggest a later date for the 'descent' of the consciousness principle. In this case, Buddhist scholars like Mett□nando have to define two kinds of consciousness: first the 'consciousness' principle which serves as the life force (p. *jivitindriya*) and second, the 'consciousness principle' (p. *viññâna*) which serves as the basis for full personhood.

However, Bhikkhu Huimin, fully ordained Buddhist monk and Professor at Taipei National University, argues that any action resulting in the death of an embryo is a serious misdeed (2002, § 4.2.). An 'embryonic sacrifice' for the sake of better treatment of severe diseases is something that many Buddhists do not support. At first glance there might be - for Mahâyâna Buddhists - a possible parallel to the idea of sacrifice for the sake of others: the Mahâyâna ideal of the compassionate Bodhisattva. Spiritually developed Mahâyâna Buddhists might perceive themselves as acting in the role of a Bodhisattva, helping the sick by offering themselves in self-sacrifice (see Durt 1998).[24]

But in crucial aspects the two situations are not comparable. The Bodhisattva is able to choose to sacrifice himself. He knows the recipient of his donation, which makes this intentional action a noble one. But if a human clone is created only for the purpose of being sacrificed, he or she is not intentionally choosing to be a donor for the sake of others. And without good intention on his or her part, this does not, for Buddhists, constitute a good deed, even less so, if a clone is used merely to provide spare organs or tissues for an existent individual.[25] A Buddhist will certainly express his or her rejection of such use of a human being.

B. Case «b»: Reproductive Cloning to Produce Children

When Buddhists speak of cloning, they think primarily of cloning as a means to help infertile couples. If human cloning "could satisfy the parenting desire of childless couples and if it does not cause pain and suffering to all parties concerned nor destruction of life, Buddhism will have no difficulty in accepting it."[26] A quite 'liberal' attitude towards the general idea of cloning human beings (which, I may add, seems to be based in some cases on a meagre knowledge of the biological procedures on the part of modern Buddhist commentators) has already been noted in various studies.[27] In some Asian countries infertility is, it seems, a problem complicated by the religious significance of symbolic 'duties.' Medical help here is much appreciated. Somparn Promta emphasises the fact that if the intention is directed towards creating a new being, the whole act is governed by the motive of bringing something to life,[28] which should be seen as positive. If Buddhists consider the aspects mentioned above, the *intentions* of the parties involved (parents who want to have children; physicians who help them fulfil this wish) are - on first sight - not to be condemned as consciously intending harm or suffering. But would this statement stand up to more thorough investigation? If one is to abstain from doing harm to whomsoever involved, then far more than just the direct "intentions" of the acting parties are of concern. As P.D. Premasiri points out,[29] the "golden rule" must be applied to the question of cloning.[30] He argued: "But if there is any objection [to cloning, J.S.], it has to be on consideration of other facts like: If we produce cloned human beings, are there likely to be certain problems, emotional problems? Under normal conditions we have parents that care, parental relationships which are connected with the emotional development of the present; now, without these will a person produced in this artificial manner become a misfit, have emotional problems - could it result in disastrous consequences later on?"

On the other hand, if the 'golden rule' is introduced with reflection on the amount of 'suffering' of a being-to-be, it seems to be truly applicable only if a clone is born and his or her emotional situation can be observed. It is not a reasoning based on principles a priori, but confined to an assessment of empirically observable 'signs.'

Apart from some further anthropological and soteriological conceptions that affect Buddhist evaluations of human reproductive cloning,[31] the main objective is the principle of non-injury, which is to avoid killing. If no killing is implied, then it is the extent of harmful emotional consequences which should guide ethical reasoning. And if such negative consequences would also prove to be absent, 'reproductive cloning' would most probably not be subject to fundamental Buddhist rejection. According to the latter two characteristics of Buddhist ethics in the scheme above, i.e. contextualism and self-cultivation, a Buddhist might argue that to be cloned or not is not a relevant fact in the field of ethics at all - if, as Premasiri points out, it is assured that the cloned human is not prevented from accumulating merit (which, for example, takes time - thus, to live a long life may be a crucial point here). But this will be at risk, if current animal reproductive cloning techniques are transferred one-to-one to humans. There is a high loss of embryos and foetuses, and further epigenetic malformation in a certain number of individuals is expected. I doubt whether Buddhists who have taken a permissive attitude to reproductive cloning, including Nakasone, Guruge, Nabeshima, Promta[32] would stick to their views if they were aware of the implications of current practices.

But there are also voices from the Buddhist world that have criticised the means of cloning. To give an example from Thailand: In 2000 a Thai writer, Wimon Sainimnuan, published a novel entitled *Amata*,[33] a Pali word meaning "immortal," "deathless." This word (*amata*, skt. *amrta*) denotes in classical Indian mythology a drink that implies immortality. It was used by the Buddha in his first sermon (*Dhammacakkappavattana-sutta*) to describe the goal of his teachings, the way to *Nirvâna*: "Listen! Immortality is found" (p. *amatam adhigatam*, Vin I.9,15ff.).[34]

The novel portrays a business tycoon, Prommin, who is obsessed with the idea of having several clones of himself. These clones are to be used as organ donors for himself, the ageing tycoon. A Western scientist, Dr. Spencer, helps him to realise his plan. But one of his 'clone-brother-sons,' Arjun, a faithful Buddhist, persuades Dr. Spencer not to take single organs, but to transplant Prommin's brain into his, Arjun's, young body.

Dr. Spencer does so; Prommin's brain is transferred to the body of his younger 'brother-son,' while Arjun's brain is transplanted into the body of Prommin. Prommin's body, after the procedure little more than waste in the eyes of the others, is then deep-frozen. Arjun's body is believed to contain the spirit of Prommin. We now come to the climax and turning point of the novel. According to Buddhist sources, the heart (p. *hadaya*) is the seat of thought and feelings (*mano-viññâna-dhâtu*),[35] equal to "mind" (p. *citta*). Accordingly, Arjun's mind takes control of Prommin's implanted brain. Since everybody believes that he is Prommin and the real Prommin is put away in the freezer, Arjun is finally free.

This subtle "Buddhist" plot portrays the close alliance between big business and biotechnology as a close alliance. Their erroneous search for immortality, based literally on a 'wrong' perception of the brain, ends up in a freezer, while the Buddhist clone-son triumphantly takes over the business and returns to a traditional Buddhist lifestyle. Incidentally, he is not hindered in any respect by actually being a clone.

We might see in this plot an attempt to revolt against globalisation - indeed, it is a foreign scientist who carries out the cloning service for the tycoon. On the philosophical level, however, we may reflect upon the fact that from the Western scientific point of view the battle of the heart against the brain is lost. But is it a hopeless revolt somewhere in Asia, too?

In the end, Wimon Sainimnuan's novel demonstrates that the question of cloning cannot be answered in isolation from other relevant aspects of changing societies that have to cope with the coexistence of Western science and methods and more traditional Buddhist world views.

5. The Legal Regulation of Cloning in Thailand

As an example of the relationship between the Buddhist bioethical discourse and legal regulation in a country influenced by Buddhism, I will now try to give an impression of the situation in Thailand.[36] Some aspects of the situation (regulation procedures, information politics, business and research structures) are similar to those in China, Korea, and Japan, where a certain Buddhist influence can also be observed. However, I doubt whether these similarities are attributable to Buddhism. Rather, it is most probably these 'aspects' themselves which make legal regulation and its current status comparable.

Thailand has some rapidly growing domestic biomedical centres, as well as laboratories run by foreign companies or as joint ventures. Thai scientists succeeded in Cloning cows as early as 2000, and they have announced plans to clone rare and endangered species of the Siamese cat or the buffalo. Scientific research is generally highly recognised in Thai society. Some Thai scientists and intellectuals celebrated the birth of Dolly as a breakthrough. Even the King of Thailand, His Majesty Rama IX (Bhumiphon), as such also Head of the Thai Buddhist Community, has been quoted as saying that the cloning of animals, where useful, should be employed systematically.[37] In June 2003 the Deputy Prime Minister Suwit Khunkitti announced the approval of a life science research centre that will focus on therapeutic cloning technology for organ transplantation. It will be funded by the Thai government with initially 400 million Bht. According to plans, this new life science centre will bring together US American and domestic companies. It may start working in 2004.

However, the current status of human embryo research is difficult to ascertain. There are at least 3.000 surplus embryos from IVF procedures stored in more than 25 Thai centres of assisted reproduction, and there are at least some centres engaged in stem cell research. However, in comparison to other Asian countries, e.g. Korea, the current status of biomedical treatment and research in Thailand is comparatively modest.

At present, no central state authority is officially responsible for the ethical aspects of cloning research. However, different institutions have presented proposals for legal and ethical regulation: First, there are ratified guidelines by the Medical Council of Thailand, but these are binding only for physicians in hospitals and scientists in governmental institutions. Second, there are ministerial decrees of the Ministry of Science, Technology and Environment, but they have not been ratified by parliament, and thus have a limited scope of application. Third, there are a number of initiatives of which the most prominent is that of the Bioethics Advisory Committee of the National Health Foundation of Thailand (NHF) in cooperation with the National Centre for Genetic Engineering and Biotechnology (BIOTEC). But, again, their recommendations are not yet enforced by law. And there are further Research Ethic Committees in the Universities. The Bioethics Advisory Committee tries to implement international standards, for example, the WHO/UNESCO guidelines and declarations,[38] and has organised hearings, at which Buddhist positions have also been presented.

In August 2001 Pramual Viruttamasen, director-general of the Medical Council of Thailand, announced newly adjusted ethical guidelines: In order to prevent controversial embryo research, medical personnel should limit their research on surplus embryos from IVF procedures up to the 14th day, assuming that the ability of the nervous system to process sensations or thought has not yet developed. The decision of a sub-committee of the Medical Council also backed cloning research for just one purpose, namely, that of breeding organs. Pramual said: "Even though the sub-committee unanimously agreed not to allow human cloning for reproductive purposes in our country, the majority of committee [members have] expressed their support for that kind of cloning [organ cloning]. That is the basis for backing stem cell research."[39] But the scope and meaning of the decision seemed to be less than clear. In December 2001 Prof. Anek Areepak, chair of the Human Research Ethics Club of Thailand, declared that the Ministry of Health had prepared to draft a bill to ban human cloning. It enforced more or less the above regulations of a permissive approach, partly adapting British regulations, and it was enacted in 2002. During my visit to Bangkok in August 2003, I was told that all companies wishing to do stem cell research now have to submit an application to the Ministry of Health and the BIOTEC. Since there have been no applications for permission to do research on human cloning, there has so far been no rejection. But one interview partner of the NHF told me that it is most difficult for the few specialists in the Ministry to check whether research is being done in the proposed way or not.

On the international level, however, the Thai delegate at the UN LegalCommittee for the Possible Convention Draft against Human Cloning, Manasvi Srisodapol (Minister Counsellor), voted in favour of a ban on reproductive cloning on the one hand, and the formulation of 'appropriate guidelines' for therapeutic cloning on the other hand. This position, he said, was consistent with the position taken by the BIOTEC/NHF committee in his home country.[40]

Interestingly, compared to German bioethics committees, there is a rather small number of 'religious specialists,' namely Buddhists, participating in the institutions mentioned above. There were no representatives of the Thai Sangha, the Buddhist Community, to officially introduce a Buddhist perspective in the committee sessions. But some sessions were attended by Buddhist scholars from the Universities, medical doctors, and lay Buddhists and monks conversant in medical ethics. On September 23, 2002, a great assembly of more than 100 representatives of the main religious traditions in Thailand - Buddhism, Christianity, and Islam - again organised by BIOTEC and the NFH, brought to light that all major traditions oppose human cloning. Reportedly, Buddhists were less strict in their objection. A quite liberal attitude was also the outcome of a survey questioning 2,516 people in six different provinces, again conducted by the NFH and BIOTEC and published in December 2002. According to that survey 18.3 % of the Thai people said that they agreed with human cloning. This figure shows that a considerable number of people have far-reaching confidence in new technologies that would possibly vanish if certain conditions of actual research procedures received more attention. Be that as it may, this figure of 18% might also point to the fact that Buddhists are more likely than other religious groups to abstain from making a blanket judgement in these matters.

6. Concluding Remark

Buddhists of all traditions seem to feel an increasing need for fundamental orientation in the field of biomedicine; it is my impression that right now a fresh interest is evolving in the discussion of these questions across the boundaries of schools and cultures, sparked, for example, by Wang's comments on Buddhist attitudes. The formulation of philosophical bio-ethical principles based on Buddhist principles, as done by Ratanakul, Taniguchi, and Keown, seems to be an important step in that direction.

Notes

1. For technical reasons a simplified transliteration for Pâli and Sanskrit words will be used in this paper.
2. Lately, criticism was raised that the consent of the donators was not received in the usual manner, but as there is still an ongoing process of examination this is still an unsettled dispute.
3. See www.sciencexpress.org [10.1126/science.1094515]; February 12, 2004.
4. Hwang in Dreifus, 2004.
5. Traditionally, the continuity of the life-death-life...-cycle is conditioned by karma, and, first and foremost, a sorrowful experience of endless circulation (*samsâra*) - so starting a new life may not necessarily be a positive act.
6. see L. Kass et al. 2002.
7. Interview-transcript of the Australian Broadcasting Corp. (ABC), Michael Vincent (reporting), Friday 13, 2004, see http://www.abc.net.au/am/content/2004/s1044228.tm.
8. Faiola 2004, A01. The article continues: "Yet Hwang vehemently vows not to take that step and says he would lobby against those who would. Not for religious reasons - but on medical and philosophical

grounds." Moon, a colleague of Hwang, also subscribes to the view that a cloned human being might deserve to be human: "We have always had the idea that life begins after fertilization - by sperm and the egg. But cloning is a totally new idea. We have no definition for cloned human beings …" (Bernton, 2004).

9. Hwang is, for sure, correct in asserting that in early Buddhist scriptures descriptions of technical procedures or new biomedical practices cannot be found. Yet some clear embryological and normative descriptions can be found that indeed demand some further ethical deliberation: Buddhists widely believe that a human being comes into existence by a consciousness-spirit (Sanskrit: *gandharva*; *vijñâna*) that enters into the intermingled fertile substances at the time of conception (the 'coming together,' i.e., intercourse). Thus, Buddhists regard *three* conditions as necessary preconditions for conception: Semen, an ovum (called 'blood' in pre-modern Buddhist embryology), and a consciousness awaiting rebirth (compare Keown, 1995a, 90). The idea that an 'artificially' produced human being is not a human, insofar and as long it is outside of the mother's womb, has been reported to the author as a position taken by some "conservative Buddhists" in Thailand (Somparn Promta, in an interview recorded September 1, 2003, Bangkok). Anyhow, it is worth mentioning that the majority of Therav□da Buddhists believe "that however premature and small this foetus is compared to an adult, in this foetus all physical and psychic attributes are already present." (Taniguchi, 1987b, 76) In Mah□y□na Buddhism, there is a more complex view, but even in the "human-like" phase of the "embryo" the consciousness principle is already present.
10. see Joung, 2004.
11. Kim Byoung-soo, a speaker of PSPD, commented on Hwang's experimentation: "But to extract the stem cells, researchers should kill the cloned human embryos. Given that the embryo is an early stage of human life, the behaviour is obviously unethical." (Tae-gyun, 2004)
12. Dreifus, 2004.
13. Keown, 1995a, XVIII.
14. A term that tries to combine all aspects of the autochthon threefold scheme: *Buddha*, *Dharma* (the teachings, the Buddhist "law," skt. *dharma*), and *Sangha* (the Buddhist monastic community).
15. Director, *Religious Studies Department*, Mahidol University, Bangkok, Thailand. Compare his contribution in this volume.
16. Harvey, 2000, 51.

17. If it comes to 'dignity,' there might be a Buddhist equivalent in the idea of the preciousness of human life, which is appreciated for the rare possibility to attain Buddhahood. In the words of the Burmese democracy movement leader: "Buddhism [...] places the greatest value on man, who alone of all beings can achieve the supreme state of Buddhahood. [...] Human life therefore is indefinitely precious." (Aung San Suu Kyi, 1991, 174) But this 'spiritual functionality' seems not to be equivalent to Western notions of human dignity. Buddhist ethics in this respect come closer to the general notion of 'sanctity of life,' including thus all sentient beings in the scope of ethics. However, Somparn Promta offered as an interpretation (if we look for possible equivalences) that this potential should be seen in the embryo, too - and, if it is destroyed, it might take a large number of lower rebirths before this being gets back into a human rebirth and regains its specific chances to proceed spiritually (Interview, 1.09. 2003, Bangkok).
18. "Whatever monk should intentionally, with his own hand, deprive a human or one that has human form of life, supply him with a knife, search for an assassin for him, instigate him to death, or praise the nature of death [...], this monk is *pârâjika*, expelled." (Prebish, 1996, 51f.) This precept to abstain from killing is part of the "Pr□timoksha," the central 'rules' of what should be morally binding. It is one of the initial four; if a violation is confessed, immediate expulsion from the Buddhist monastic community should take place. But lay Buddhists must also *idealiter* abstain from killing.
19. see *Mahânidâna Sutta* and Rhys, Davids, 1951, 60.
20. The classical description can be found e.g. *Mahâtanhâsankhaya Sutta*, Majjhimanikâya, Nr. 38 [M I.265f.], cp. Haldar 1977, 27. I will refrain here from interpreting the Buddhist classical sources on the beginning of human life which will be done elsewhere.
21. This passage, nevertheless, is ambiguous, as Buddhaghosa declares in his commentary (*Paramattha-Jotikâ*): Either *embryos* (or birds in the egg) pushing to get into existence, or *un-released humans*, 'worldly' humans, who will be reborn (cp. Nyanatiloka, 1955, 256), could be depicted here.
22. Thai monk of *Wat Rajaorasaram*, former physician and member of the Bioethics Committee of Chulalongkorn University; counsellor of Buddhist affairs to the UN.
23. (Interview transcript:) "I think the arising of the consciousness of a human being does not take place immediately after conception, [...] what do we learn from science? If you fertilise [...] a test-tube baby,

you put an ovum and fertilise it, it becomes a little zygote, you submerge in minus 75° Celsius, for how long? Infinitely. It remains a zygote. So, let's say, one hundred years, or, 120 years, that zygote, according to *Samantapâsâdikâ* [*] is no longer a human being. But if you thaw it out, it grows, you put into the womb - it grows, it becomes a baby! So [the] description according to *Samantapâsâdikâ* cannot be applied here." [**Samantapâsâdikâ,* ascribed to Buddhaghosa, being an extensive commentary (5th c.) on the *Vinaya*].

24. By now this position has no strong support by Buddhists, but some are probing its applicability (see Promta, 2004 and Yoon, 2004).
25. Compare Sainimnuan's novel *Amata*.
26. Ratanakul, 1997, 406.
27. see Campbell, 1998. Buddhists emphasise that they actually do rely on the law of Karma: "Each resulting being from cloning procedures would be a different individual with his or her own Karmic heritage from past lives." (Guruge, 2002, 114) It might be said that many Buddhists have a neutral attitude to man-made artefacts. Whatever can be done technically must be in the scope of nature and is, therefore, - by its status alone - beyond good or evil. Less indifferent to changes of the course of nature are some East Asian Mahâyâna Buddhists, especially in Japan, who subscribe to the view that the realm of the 'Buddhas and Gods' must not be invaded (compare the Japanese commission report *Seimeirinri Iinkai Hokoku* [1993]).
28. See Promta, 2001. Promta was member of the *Bioethics Advisory Board* of the Thai National Health Foundation/BIOTEC [two interviews (recorded by the author), September 1 and 3, 2003 (Bangkok)] at the *Department of Buddhist Studies*, Chulalongkorn University. He also pointed out that for him the intention of finding cures for severe illnesses governs the act of cloning-for-research and is therefore permissible, too.
29. Director, *Department of Pali and Buddhist Studies*, Peradeniya University, in an interview (recorded by the author), February 27, 2004 (Peradeniya).
30. The principle can be found in Buddhist scriptures too, e.g., the well-known *Dhammapada*, v. 129 ff.: "All tremble at violence; all fear death. Comparing (others) with oneself one should not kill nor cause to kill." (Norman 1997, 20) The 'exchange' of one's own position and that of another was later developed into a common meditation procedure.

31. One of these teachings may be addressed in the Buddhist anthropology. The Buddhist teachings of "No-self" give the advice not to identify oneself with the empirical person, the bodily and mental factors of one's own human existence. It should not be assumed or believed that body, feelings, volition, mind or consciousness imply the existence of a "substantial self." Thus, for a cloned person, the fact of 'being cloned' is just a part of conventional existence. It is certainly regarded as 'unwholesome' if someone identifies himself with his 'good genes'; and the same holds true for the cloned human being, if he were to identify himself 'negatively' with being 'just a copy' (compare the *Vasettha-Sutta*, *Majjhimanikâya*, Nr. 98, [M II. 5,8], or the *Sutta-Nipâta*, v. 611, for the idea that there is no such thing as a "discrimination by birth" (p. *jâti-vibhanga*) - the circumstances of birth (genealogy etc.) are purely conventional; people only become what they are by their actions - compare Swaris, 2000, 57-59). However, I have to admit that very few Buddhists argue with the Buddhist nominalist conception in the field of ethics (compare Falls et al., 1999; Barnhart, 2000; Guruge, 2002). Alas, most Buddhist monks or scholars with whom I have recently spoken (Tibetan Mah□y□na, Sinhala and Thai Therav□da Buddhists) told me that they fail to see in what respect a clone should differ in status from any ordinary human being. If they are concerned, it is not about his or her status but about his or her possible emotional situation, or, that a human clone might not be treated as a human.
32. Nakasone, 2001; Guruge, 2002; Nabeshima, 2000; Promta, 2004.
33. Hongladarum, 2002b.
34. cp. Vetter, 1988, 8f.
35. Buddhaghosa, 1975, 256; Smyuttanik□ya, 1884-98, I 199, I 207.
36. It may serve as background information for Pinit Ratanakul's presentation of Thai Buddhist positions included in this book.
37. Suthus Sriwattapong, The Nation, 22. 10. 2001.
38. In a "National Meeting on Bioethics and Human Genetics" held on June 25-26, 2001 - the first of its kind, organised by the NHF and BIOTEC - arguments were exchanged concerning the legal situation, possible genetic discrimination, and the ELSI programme.
39. The Nation, 15.08. 2001.
40. "The Thai National Centre for Genetic Engineering and Biotechnology had appointed a bioethics team to develop appropriate guidelines for therapeutic cloning, which it was believed could provide important answers in the treatment of diseases." This is the wording of a press re-

lease by the 57th General Assembly, 6th Committee, 17th Meeting, Oct. 18, 2002, www.unhchr.ch/huricane/huricane.nsf/0/BCD26DF64D647 DE4C1256C5A004F55FF?opendocument.

References

Barnhart, Michael G. "Nature, Nurture and No-Self: Bioengineering and Buddhist Values." *Journal of Buddhist Ethics* 7, 2000, 126-144.

Bernton, H. "South Korean Researchers at the Center of Cloning Debate." *The Seattle Times* (15.02.2004) (http://seattletimes.nwsource.com/html/ localnews/2001858346_koreanscientists15m.html).

Beauchamp, Tom L. and James F. Childress. *Principles of Biomedical Ethics*, 5th edit. Oxford: Oxford University Press, 2001.

Buddhaghosa. *Visuddhimagga*, C.A.F. Rhys-Davids (ed.). London, 1975.

Campbell, C.S. "A Comparison of Religious Views on Cloning." In *Cloning*, G. McCuen (ed.). Wisconsin: GEM Publications, 1998, 119-139.

Durt, Hubert. "Two Interpretations of Human-flesh Offering: Misdeed or Supreme Sacrifice." *Journal of the International College for Advanced Buddhist Studies* 1, 1998, 236-252.

Faiola, Anthony. "Dr. Clone: Creating Life or Trying to Save It? S. Korean Defends Ethics of His Controversial Research." *Washington Post*, February 29, 2004; Page A01 (http://washingtonpost.com/wp-dyn/articles/ A15746-2004Feb28.html).

Falls, Evelyn, Joy Skeel, et al. "The Koan of Cloning: A Buddhist Perspective on the Ethics of Human Cloning Technology." In *Second Opinion* (Park Ridge Center, Chicago) 1, 1999, 44-56.

Fujii, Masao. "Buddhism and Bioethics." In *Bioethics Yearbook*. Volume 1: *Theological Developments in Bioethics* (1988-1990), Baruch A. Brody et al. (eds.). Dordrecht, Boston: Kluwer, 1990, 61-68.

Guruge, Ananda W. P. "Bioethics: How Can Humanistic Buddhism Contribute?" *Hsi Lai Journal of Humanistic Buddhism* 3, 2002, 86-117.

Haldar, Jnan Ranjan. *Medical Science in Pali Literature*. Calcutta: Indian Museum, 1977.

Hongladarom, Soraj. "Globalization, Bioethics and the Cultures of Developing Countries." *Eubios Journal of Asian and International Bioethics* 12 (2002a), 103-105.

Hongladarom, Soraj. "Human Cloning in a Thai Novel. Wimon Sainimnuan's Amata and Thai Cultural Attitude Toward Biotechnology." (2002b) (www.pioneer.chula.ac.th/~hsoraj/cloning.pdf).

Huimin Bhikkhu. "Buddhist Bioethics: The Case of Human Cloning and Embryo Stem Cell Research." *Chung-Hwa Buddhist Journal* 15, 2002, 457-470.

Joung, Phillan. "Forscher in der Rolle der Schurken." In *Neue Zürcher Zeitung*, March 27, 2004 (http://www.nzz.ch/2004/03/27/zf/page-article9GOEH.html).

Kass, Leon R. (ed.). *Human Cloning and Human Dignity. An Ethical Enquiry*. Washington, 2002 (http://bioethics.gov/reports/cloningreport/index.html).

Keown, Damien. *Buddhism & Bioethics*. London and New York: Macmillan/St. Martins Press, 1995a.
Keown, Damien. "Are there 'Human Rights' in Buddhism?" *Journal of Buddhist Ethics* 2 (1995b), 3-27.

Keown, Damien (ed.). *A Dictionary of Buddhism*. Oxford: Oxford University Press, 2003.

Mahânidâna Sutta, *Dîghanikâya* Nr. 15 (cp. T.W. Nabeshima, Naoki. "Bioethics of Interdependence: A Shin Buddhist Reflection on Human Cloning." (2000) (http://meta-library.net/iftm/nabe-frame.html).

Nabeshima, Naoki. “A Shin Buddhist Perspective on the Application of Human Embryonic Stem Cells for Medical Treatment.” *Religion and Ethics* 2, 2002, 3-22.

Nakasone, Ronald Y. “The Opportunity of Cloning.” In *Ethical Issues in Human Cloning: Cross-disciplinary Perspectives*, Michael C. Brannigan (ed.). New York (Seven Bridges Pr.), 2001, 95-97.

Nudeshima, Jiro. “Human Cloning Legislation in Japan.” *Eubios Journal of Asian and International Bioethics* 11, 2001, 2.

Norman, Kenneth Roy. (Trl.) *The Word of the Doctrine (Dhammapada)*. Oxford: Pali Text Society, 1997.

Nyanaponika. (Trl.) *Sutta-Nipata*, Konstanz: Christiani, 1955.

Perera, L.P.N. *Buddhism and Human Rights. A Buddhist Commentary on the Universal Declaration of Human Rights*. Colombo: Karunaratne & Sons, 1991.

Prebish, Charles S. *Buddhist Monastic Discipline: The Sanskrit Prâtimoksha Sûtras of the Mahâsamghikas and Mûlasarvâstivâdins*. Delhi: Motilal Banarsidass, 1996.

Promta, Somparn. “A Buddhist View of Cloning.” *The Nation*, 14.12. 2001, 5a.

Promta, Somparn. “Moral Issues in Human Cloning. A Buddhist Perspective.” (http://www.thainhf.org/Bioethics/Document/aoy.pdf; English - translation [2002], unpublished).

Promta, Somparn. “Human Cloning and Embryonic Stem Cell Research. A View from Theravada Buddhist Morality.” Bangkok, ASEAN-EU LEMLIFE Project (Working Papers), 2004, 5-11.

Ratanakul, Pinit. *Bioethics: An Introduction to the Ethics of Medicine and Life Science Introduction to Bioethics*. Bangkok: Mahidol University, 1986.

Ratanakul, Pinit. "Buddhism, Prenatal Diagnosis and Human Cloning." In *Bioethics in Asia*, Norio Fujiki, Darryl R. J. Macer, (eds.). Eubios (Christchurch), 1997, 405-407.

Rhys Davids, C.A.F. *Dialogues of the Buddha*, Part II, London, 1951, 60 (D II.62).

Roach, Geshe Michael et al. "Buddhists on Cloning." In *The Human Cloning Debate*, Glen McGee (ed.). Berkeley, Berkeley Hill Books, 2000, 285-288.

Samyutta-nikāya (4 vol.), L. Feer, C.A.F. Rhys Davids (eds.). London 1884-1898.

Schlieter, Jens. "Die aktuelle Biomedizin aus der Sicht des Buddhismus." Max-Delbrück-Center for Molecular Medicine, Berlin, 2003 (www.bioethik-diskurs.de/portal/documents/wissensdatenbank/gutachten/ Biomedizin_Buddhismus.htm).

Seimeirinri Iinkai Hôkoku. "The Report of the Bioethics Committee." In *Indogaku Bukkyôgaku Kenkyû* (IBK) 41, 1, 320-369. (Artikel auf japanisch).

Spranger, Tade Mathias. "The Japanese Approach to the Regulation of Human Cloning." *Biotechnology Law Report* 5 (2001a), 700-706.

Spranger, Tade Mathias. "What is Wrong about Human Reproductive Cloning? A Legal Perspective." *Eubios Journal of Asian and International Bioethics* 11 (2001b), 101-102.

Suu Kyi, Aung San. *Freedom From Fear*. London: Penguin, 1991.

Swaris, Nalin. *Buddhism, Human Rights and Social Renewal*. Hong Kong: Asian Human Rights Commission, 2000.

Tae-gyun, K. "Embryo Cloning Success Sparks Ethics Debate." *Korea Herald*, 21.02. 2004 (http://kn.koreaherald.co.kr).

Taniguchi, Shaniguchi. *A Study of Biomedical Ethics from a Buddhist Perspective*. MA Thesis, Berkeley: Graduate Theological Union and the Institute of Buddhist Studies, 1987a.

Taniguchi, Shaniguchi. "Biomedical Ethics from a Buddhist Perspective." *Pacific World New Series* 3 Fall (1987b), 75-83.

Vetter, Tilmann. *The Ideas and Meditative Practices of Early Buddhism*. Leiden: Brill, 1988.

Yoon, Jong-gab. "Ancient Buddhism and Bioethics in Korea." In *Bioethics and Culture in Korea*, Korea Research Foundation & Deutsche Forschungsgemeinschaft (ed.). Busan, 2004.

Human Cloning: Thai Buddhist Perspectives

Pinit Ratanakul

Abstract: For Thai Buddhists human cloning is not against nature, though it is not the natural way of human birth. Nor does human cloning come into conflict with the Buddhist concept of individual kammic life-force and the insistence on impermanence. With advances in science and reproductive technology moving so fast and the empowerment this technology will give to humankind, it is only a matter of time before human cloning happens. Accordingly, we will have to expand our moral horizon and our idea of human birth and human existence to cope with such a possibility. The ethics of cloning should be focused on the safety of humankind and the minimization of the harm to the clones both physical and psychological.

Key Words: Nature, Individual karmic life-force, Impermanence, Empowerment, Moral horizon, Safety, Minimization.

1. Introduction

Therav□da Buddhism, also known as Hînayâna or Southern Buddhism, has been a prevailing cultural force in Thailand since the establishment of the first Thai kingdom in the 13th century. Even though Thailand has committed herself to religious pluralism and even though the present constitution does not make it compulsory for every Thai to be a Buddhist, 98% of the population is Buddhist. Secularisation, which is taking place in many Southeast Asian countries, is not a dominant force in present day Thai society particularly in the rural areas where most of the population is. Similarly, despite some impacts of globalisation on the lives of Thai people, Buddhism still continues to be the basis for the value system of the majority of the people. Together with the institution of kingship, it forms the core of civic Thai culture. Thai culture cannot be appreciated without having an understanding of Buddhist teachings which mould the Thai world-view and perception.

2. Thai Cultural Orientation

The basic teachings of Buddhism which provide the foundation for Thai cultural orientation may be summarised as follows:

A. Dependent-Origination, Impermanence, Suffering

According to Buddhism all phenomena, physical and mental, are subjected to two predominant natural laws i.e. the law of dependent-origination (Pâli: *paticcasamuppâda*) and the law of change (p. *aniccatâ*).[1] These two laws form the basis of all important Buddhist teachings.[2] The law of dependent-origination, also referred to as the law of causality or conditionality, emphasises that all phenomena are conditioned. They arise and exist due to the presence of certain conditions, and cease once these conditions are removed: nothing is independent.[3] Subjected to the law of change, phenomena are also impermanent, always changing, or, in other words, they have no substances or essences (*dhamma-anattâ*). Even what we call the self is merely a combination of physical and mental aggregates (*khandha*) made up of body or matter (*rûpa*), sensation (*vedanâ*), perception (*saññâ*), mental formation (*sankhâra*), and consciousness (*viññâna*). All these components are causally interrelated and impermanent and never remain the same in any consecutive moments. The fact that all phenomena in the external world and our bodies and minds are conditioned and impermanent has to be accepted as such. Our understanding of this aspect of reality will lead to happiness. To hold on to things as if they had a permanent essence and to regard ourselves as enduring entities is the root-cause of suffering.

B. Kamma, Samsâra, Nibbâna

Buddhism is an atheistic religion that gives humankind total responsibility of its destiny. Everyone is the architect of his or her own life. This emphasis indicates the importance of human action and its effects. In Buddhism action and its resulting effects are called *kamma*. The *kammic* effects may be good or bad depending on the nature and quality of actions performed. One reaps what one sowed in the past. These *kammic* effects are believed to be active across space and time. They act as an individual life-force (*gandhabba*) causing rebirths through successive lifetimes in samsara i.e the cycle of births and rebirths in which all beings are involved perpetually.[4]

In Buddhist cosmology there are 3 realms (*loka*) in sams□ra, where beings in different forms dwell. These realms consist of the realm of desire (*kâma loka*), the realm of forms (*rûpa loka*), and the formless realm (*arûpa loka*).[5] The dwellers in the realm of desire are gods, sentient beings, humans, animals, spirits, and hell-beings. The realm of forms is the abode of the refined and subtle beings. Those with a higher degree of these qualities dwell in the formless realm. Despite difference in forms and life-span dwellers in all realms are impermanent and subject to death and rebirth. There is always a migration of beings from one realm to the other regulated by *kamma*. When a being dies, its *kammic* life-force sparks another form of existence either in a higher or lower realm according to the good or bad *kamma*.[6] Evil actions lead in the worst case to rebirth in hell while good deeds lead to heavenly rebirth. The life-force will completely cease only when there are no more *kammic* actions, i.e. actions motivated by greed, anger and delusion. This cessation is called *nibbâna* or absolute emancipation.[7] Accordingly, it is the end of *samsâra* existence, for there are no longer births or rebirths.

C. The Preciousness of Human Rebirth

Of all rebirths, human rebirth is considered to be the most precious because the human sphere is the only place where there is enough suffering to motivate humans to seek the way to be liberated from misery, and enough freedom for them to act on that aspiration. In the higher realms, the gods are too absorbed in the blissful state to find the way out of *samsâra*, whereas the animals, spirits and hell-beings are in irremediable misery, and have little freedom to do either good or bad. These suffering beings will gain precious human rebirth only when the result of the bad *kamma* that led them to lower rebirth is exhausted. When this happens, the results of their good deeds performed when they were human will lead them to rebirth in other realms and, sooner or later, to the human sphere again.

The difficulty about human rebirth is illustrated in a parable given by the Buddha. It was about a partly blind turtle in the vast ocean where there was a piece of floating wood with a hole in it. The Buddha estimated that the chance of any particular being to obtaining human rebirth is about equal to the likelihood of the turtle pushing its head through the hole in that piece of wood. This is why the first precept in Buddhism prohibits the killing of life, particularly human life, in any stage of its development. Whoever transgresses this precept will suffer the *kammic* consequences, i.e. brevity of present life, ill health, and rebirth in the state of suffering as animals or hell-beings.

D. The Importance of Merit-Making

Believing in the efficacy of *kamma*, the Buddhists regard merit-making as an important means to ensure good rebirth in the higher realms and/or to sever the knot that ties us to *samsâra* existence. In Theravada Buddhism meritorious acts (*kusala*) include selfless giving (*dâna*), keeping the precepts (*sîla*), meditation (*bhâvanâ*), teaching and listening to the Dhamma or the Buddhist teachings.[8] Selfless giving is emphasised not just for good *kammic* results, but the act itself brings joy to the giver and also the means to reduce egoism and attachment to one's own possessions. The observance of the precepts is important for leading a moral life. Precepts are not commandments as in the Judeo-Christian tradition. They are voluntarily self-imposed rules for the development of one's spirituality. In a Buddhist society the precepts are also expected norms to live by. Due to the religious and social significance of the precepts, the Buddhists, in the accumulation of merit, try to avoid violating the precepts as much as possible. For them the religious significance is especially crucial with regard to rebirth. Those who keep the precepts are assured of a good rebirth.

3. Thai Buddhists and Human Cloning

Despite the ongoing process of modernisation in Thai society, Buddhist monks continue to have influence on Thai people, who usually follow their guidance with respect. Given this fact, we deem it necessary to approach the highly respected and learned monks to get their views on human cloning. I was able to contact six leading monks who command the respect of both the clergy and the laity. Their considered views, based on a profound knowledge of Buddhism, represent a Buddhist position on this particular issue which, of course, did not arise at the time of the Buddha. We also interviewed ten well-informed lay people consisting of scientists, physicians and ethicists. It is our hope that these findings will give us some idea about what Thai Buddhists think regarding human cloning.

A. The Possibility of Human Cloning

None of the Buddhists we interviewed was surprised by the possibility of human cloning. Believing in the law of causality, they see human cloning as the natural outcome of scientific research and experiments culminating in the birth of Dolly the sheep. For them human cloning is also possible because of the law of change. Impermanence and change imply that the human condition will not remain the same. With advances in science and reproductive technology moving so fast, it is only a matter of time before this possibility becomes a reality? The question for Buddhists is how to accommodate this change. Perhaps, as suggested by some ethicists, we will have to expand our moral horizon and our idea of human existence. Such expansion is consistent with the Buddhist spirit of openness and flexibility. In Buddhism there is an absence of strong dogmatism. This allows scope for tolerance and flexibility to meet new challenges brought about by scientific progress.

B. Cloning and Human Birth

Thai Buddhists - particularly the monks - do not see cloning as an act of creation. Their views are based on the belief in the law of causality. For them cloning provides only one necessary condition for human birth, namely the genetic make-up. The presence of an individual life-force (*gandhabba*) is another necessary condition. Without the infusion of this life-force into the genetic make-up, cloning is not successful and human birth cannot take place. The emphasis that cloning depends on a multitude of conditions implies that in the process of cloning no one can play God. The scientist is acting only as an agent to bring the clone into being. In Buddhism life cannot be created. It manifests itself when the conditions are right.[9] Without the right conditions no event will occur. A flower, for example, is dependent on multiple causes and conditions, such as the seed, earth, moisture, air, and sunshine. If one of these causal factors is absent, there is no flower.

The Buddhists who have profound knowledge of Buddhism such as the monks we interviewed, believe that human cloning is not against nature, though they accept that the clone is not born in the natural, customary way from parents. For them nature is a mystery, much more comprehensive than our own understanding. If human cloning is against nature it will be unsuccessful. It is only because nature allows anything to happen in the world. With the possibility of cloning we have to redefine the notions of "nature" and "human" and should not cling to one fixed idea. Such clinging will bring suffering to us. Change that comes with the new reproductive technology challenges us to think beyond accustomed notions and norms, as witnessed in the case of IVF babies. We no longer consider the IVF child to be less human than ourselves.

C. The Clones and Impermanence

All the Buddhists agree that we cannot become immortal through cloning. Like the original, the clone has no permanent self. His/her physical and mental components (*nâma-rûpa*) are always in constant flux. No matter how many clones we may have of ourselves, none of them is the same as the other. In light of the concept of *kamma*, each clone will have a separate *kammic* life-force to bring it into being. Even if Einstein were to be cloned, the result would not be his carbon copy. Although genetically the original and the clones are identical, in terms of *kamma* they are different in mentality and character. These are believed to be formed mainly by the individual *kammic* life-force (*gandhabba*) entering the body of the clone.[10] The clones, which themselves are impermanent, cannot perpetuate the original, nor can the replacement of degenerative organs by cloned ones. Human life, like all conditioned existence in sams□ra, is characterised as impermanent (*anicca*), insubstantial (*anattâ*), and suffering or unsatisfactoriness (*dukkha*). We can escape from this conditioned existence only through the practice of morality, meditation, and the cultivation of wisdom leading to the unconditioned *nibbāna*. Cloning, therefore, is in no way a means to perpetuate ourselves.

D. The Status of the Clone

For Thai Buddhists clones, like IVF children, are human like ourselves and should be treated as any other human. Their character and mental tendencies are the result of their past *kamma* and the way they are nurtured. In one social environment there may be a clone who practices Buddhism and can eventually attain *nibbâna*. There may also be a clone who indulges in sensualism and neglects spirituality. Since clones are as human as ourselves, before proceeding with human cloning we should think carefully and altruistically about the benefits and harm the procedure will bring to them. Social stigma and the lack of respect for their rights, for example, will alienate and dehumanise them. We should not embark on this venture just for ourselves or for its own sake.

E. Therapeutic Cloning and Stem Cell Research

Although Thai Buddhists accept the possibility of human cloning, they have reservations about reproductive cloning primarily because it involves the destruction of embryos, and its potential harmful effects seem to outweigh its benefits. With regard to therapeutic cloning they have different opinions. The monks believe it is unethical because in the process the week-old embryos have to be destroyed to obtain the stem cells. For the monks the embryo already has life even in its early formative stage.[11] But for the lay people, particularly the scientists and physicians, the need to relieve those who are suffering from degenerative diseases, such as diabetes and Parkinson's disease, outweighs consideration for the destruction of the early embryo. Of course, as Buddhists they also feel uneasy about setting priorities here: the life of the embryo or the urgent need of the suffering person. Some scientists and physicians believe that the negative *kammic* effect of destroying the early embryo is not so grave because its life is still in a very early formative stage in which it has not yet developed into a sentient being. In Buddhism the gravity of *kammic* effects depends on many factors, such as the intensity of the doer's intention and effort, as well as the size and quality of the life being destroyed. Buddhist ethics emphasises that the motivation and intention (*cetanâ*) behind actions are morally significant factors, and not simply the end result. This is the reason why an action that has *kammic* results is defined as intentional action, and not just action. However, the Buddhist ethicists caution that even though the motivation behind medical research may be good, namely, i.e., to relieve human suffering caused by incurable diseases, much research must be done before proceeding with any therapy. They are also concerned that if non-reproductive cloning is permitted, scientists may not stop at stem cell research, but may go on to reproductive cloning. The argument often heard from scientists engaged in new research and testing new capabilities is, "if we can do it, we ought to do it" or "if we don't do it, someone else will." Human weakness and temptation will also make scientists want to go on with this venture to gain recognition ahead of others. In order to prevent this, Thai ethicists see the need for timely regulation and control of stem-cell research.

F. The Ethics of Human Cloning

From the Thai Buddhist point of view the conclusion whether human cloning is immoral or not should be drawn by considering its impact on men, women, family, society, and humanity at large. Thai Buddhists show sympathy for infertile couples who cannot have children by other means and also for those who have lost an only child through a traffic accident or suchlike. But none of them recommends reproductive cloning to ease the suffering of these people because of the possible harmful effects this venture will bring to humanity. Once we have started the process, it will be impossible to reverse it. It is like sliding down a slippery slope with no means of arresting the fall. Therefore, they want the scientists to think carefully before initiating the procedure. The potential negative effects of human cloning for Thai Buddhists are the trans-cultural issues that have been under vigorous debate at international forums.[12] These effects include upsets in the social order (for example, by confounding the meaning of parenthood and confusing the identity and kinship relations of any clone child), the abuse of cloning, the commercialisation of cloning, and the fostering of injustice in society where only the rich can benefit (from costly cloning). These negative effects seem to outweigh the benefits cloning will bring to humanity. Such effects will add more suffering to what we already have. Besides, Thai Buddhists also feel that the cloning technique at present is still imperfect and if it is done at this stage there would be more harm.

The ethics of cloning should be considered in terms of what is involved in the procedure and its potential effects both on men, women, family, society and humanity, as well as on the clones themselves. These clones are as human as we are and we must not manipulate them as we please. It was reported that there had been 277 failed experiments before Dolly came into being. And yet, Dolly was not as healthy as other sheep. We cannot be sure how many failures there may be and how many embryos will be destroyed in the case of human cloning. Nor are we certain of the health of the clone. There have been cases in which cloned animals were born with some kind of deformity or suffering from some immune deficiency.

Nevertheless human cloning will happen in the near future, and we cannot ban it forever. Taking into account all the relevant ethical issues, the health of the clone and the relief of suffering, the ethics of cloning requires us to proceed with great circumspection.

4. Concluding Remarks

Thai Buddhists are open to change and new possibilities. They consider human cloning to be a natural result of scientific advances in genetics and reproductive technology, and regard it as a new challenge that we have to cope with. They are chiefly concerned about the potential consequences which will profoundly affect the lives and well-being of all people, society and humanity. They also realise that we cannot close our eyes to new technological development, and that time is usually in favour of technological innovation that empowers us to fulfil our desires. Since there are moral limits as to what man ought to do to present and future generations of humankind, the possibility of someone carrying out reproductive cloning, particularly at present when the technique is still imperfect, must be discouraged.

Human cloning is now a new challenge for all cultures. Ethical issues regarding human cloning are trans-cultural and cannot be settled only in a given cultural environment. It is essential that they be examined urgently in a wider context. Accordingly, all concerned people from different cultures will need to pool their insights and experiences to develop regulations to prevent the abuse of human cloning and to minimise the harm it may bring. An international agreement which can be enforced globally will need to be reached in a not too distant a future. Time is running out. Our actions or our failure to act will have serious consequences for the future of humanity.

Notes

1. In keeping with Theravâda tradition, Pâli rather than Sanskrit terms are used throughout the paper.
2. For implications of the law of Dependent-Origination for ethics, see Lovin/Reynolds, 1989.
3. This law is also referred to as "The Twelve-linked Chain of Dependent Origination," beginning with ignorance (*avijjâ*) and followed consecutively by action (*sankhâra*), consciousness (*viññâna*), name and form (*nâmarûpa*), the six organs (*salâyatana*), contact (*phassa*), feeling (*vedâna*), craving (*tanhâ*), grasping (*upadâna*), becoming (*bhava*), birth (*jâti*), old age and death (*jarâ, marana*). All these constituents are causally inter-related and lead to suffering. Nibbâna is attained only when the links in this chain are severed through the development of wisdom (*vijjâ*).
4. The Buddhist concept of rebirth is different from that of re-incantation which implies the transmigration of a soul and the material form it takes. The Buddhist doctrine of no-self *(anattâ)* implies that there is nothing that transmits except the individual *kammic* life-force, which creates the new psycho-physical aggregates *(nâma-rûpa)* in accordance with its precious *kamma*. In Buddhism life is compared to a flame. Rebirth is the transmitting of this flame from one *nâma-rûpa* to another. The flame of life is continuous although there is an apparent break at the so-called death which, in Buddhist understanding, is the

state of passing away, while birth is the state of rising again, regulated by the law of *kamma.*

5. Reynolds, 1982.
6. See footnote 4.
7. There are two kinds of *nibbâna* viz. (i) *saupadisesa-nibbâna* attainable during a life-time when there are no longer all forms of craving *(tanhâ)* and (ii) *anupadisesa-nibbâna* attained with complete cessation of the *kammic* life-force.
8. For lay Buddhists there are 5 most important precepts to keep i.e. refraining from killing, stealing, sexual misconduct, lying, and imbibing intoxicants. There are also 8 precepts which the lay Buddhists like to observe on special days such as Sabbath (*uposatha*). The additional three precepts concern abstaining from eating beyond noon; dancing, singing, wearing ornaments, garlands, perfumes; and use of luxurious beds. Concerning meditation there are 2 kinds i.e. calm meditation (*samatha*) to make the mind tranquil, and insight meditation (*vippassanâ*) to cultivate wisdom.
9. According to Buddhism in order for conception to take place there have to be three conditions: (i) the egg and sperm must meet; (ii) the egg and sperm must be fertile; (iii) there must be a *kammic* life-force *(gandhabba).*
10. This individual *kammic* life-force manifests itself as *gandhabba or* "being-to-be-born." In Buddhism the term *gandhabba* is used only in this particular context and must not be mistaken for a permanent soul.
11. There are 5 periods of embryological development viz. first week after conception (kalala), second week (arbuda), third week (pesi), fourth week (ghana), fifth week (pasâkhâ). There is life already from the first period. The question whether or not the embryo is a person in the earliest stage does not arise for the Buddhist, because fertilisation happens only when there is infusion of the life-force to form the psychophysical elements of life (nâma-rûpa).
12. See Brannigan, 2001.

References

Brannigan, Michael (ed.). Ethical issues in Human Cloning - Cross Disciplinary Perspectives. New York: Seven Bridges Press, 2001.

Harvey, Peter. *An Introduction to Buddhism*. Cambridge: Cambridge University Press, 1993.

Heinberg, Richard. *Cloning The Buddha: The Moral Impact of Biotechnology*. Wheaton: The Theosophical Publishing House, 1993.

Keown, Damien. *Buddhism & Bioethics*. London: MacMillan, 1995.

Keown, Damien. *Buddhism and Abortion*. London: MacMillan, 1998.

Lovin, Robin and Reynolds, Frank (eds.). *Cosmogony and Ethical Order*. Chicago: The University of Chicago Press, 1985.

Ratanakul, Pinit. "To Save or Let Go: Thai Buddhist Perspectives on Euthanasia." In *Contemporary Buddhist Ethics*, Damien Keown (ed.). Richmond, Surrey: Curzon Press, 2000,169-182.

Reynolds, Frank. *Three Worlds According to King Ruang*. Berkeley: Asia Humanities Press, 1982.

The Ethics of Human Cloning: With Reference to the Malaysian Bioethical Discourse

Siti Nurani Mohd Nor

Abstract: The article outlines the ethical discourse on cloning with special reference to the Public Conference on Reproductive Human Cloning held in Kuala Lumpur in 2002. This conference o brought together speakers from the diverse communities in Malaysia, above all the Islamic, Buddhist, Hindu and Christian communities. The four main communities agreed on the sanctity of the marriage institution and regarded it as the only legal means of producing a child. Reproductive cloning was unanimously opposed. The article discusses in detail the ethics of human cloning as viewed from the Islamic perspective. The author believes that the concept of the beginning of life as discussed among scientists and Islamic authorities in Malaysia has not yet been conclusively determined and should be dealt with more urgently in light of the new issue of human cloning. The methodology of ijtihad, a reasoning process that allows for the derivation of new regulations on current problems not addressed in the Quran and the Sunnah, is outlined.

Key Words: Islamic ethics, Shari'a, Fatwa, Cloning, In vitro fertilisation, Embryo, Sanctity of human life, Human dignity, Family lineage, Warnock, Public interest.

The 23 million multiracial and multicultural population makes Malaysia the richest country in Asia in terms of ethnic and cultural diversity, and she has earned the reputation of 'melting pot' of some of the major organised religions and oldest philosophical systems of the world. The various ethico-religious traditions can provide interesting and practicable resources in worldviews, rituals, institutions, styles of education, and simple ideas of human relations.

Bioethical deliberations in Malaysia are often significantly influenced by Islam as the dominant religion.[1] However, other groups are regularly invited to put forward ideas for a common stand on several bioethical issues. In fact, these groups have contributed to the development of the Malaysian culture, making it a distinct Asian culture in its own right, one that is an alternative to the Western European and North American exemplification of the Enlightenment mentality. For example, the conspicuous absence of the sense of community is absent in the Enlightenment tradition. Consciously aware that traditional Greek philosophy, Judaism and

Christianity were instrumental in giving birth to the Enlightenment mentality, there is an urgency to transcend the exclusive dichotomy of matter/spirit, body/mind, sacred/profane, man/nature and self/society with other values such as the sanctity of the earth, the continuity of the being, the beneficiary interaction between the human community and nature as an ideal philosophy.

> The Greek philosophical emphasis on rationality, the Biblical image of man having 'dominion over fish of the sea, and over the fowl of the air, and over every living thing that moveth upon the earth,' and the so-called Protestant work ethics provided necessary if not sufficient sources for the Enlightenment mentality. However, unintended negative consequences of the rise of the modern West have undermined the sense of community implicit in the Hellenistic idea of the citizen and the politics, the Judaic idea of the covenant, and the Christian idea of universal love and fellowship.[2]

The ideal philosophy of ethics would incorporate the disciplined reflection and recognition of the religious pluralism and multiculturalism that is eminent within the Malaysian peoples. It would invoke not only a re-examination of spiritual values and resources such as self-identity, humility, family cohesion, communal solidarity, social integration, and cross-cultural cooperation that has been lost as a consequence of modernity, but also the reconsideration of Enlightenment values such as liberty, equality, human rights, and privacy.

Bioethics, a discipline of study that has arisen due to the gravity of human sciences upon traditional ideas regarding life and death, provides the opportunity to rediscover these lost spiritual roots. As pointed out by Schuster, the proposal to draft the cloning treaty under consideration by the United Nations offers contemporary bioethics a unique opportunity to create a bioethics framework and an international forum to deal democratically with all kinds of bioethics issues such as reproductive cloning and embryo cloning research.[3] Subsequently, a bioethics that would truly be the exclusive product of the Malaysian people would ideally be one that has given due and proper consideration to the diverse cultures present in that society in its absolute totality. In a stronger language, the beliefs of these differing communities should not be compromised. However, most of the local interfaith discourse on bioethics has shown relatively few conflicting viewpoints. For example, the discourse on the ethics of organ donation and transplantation, for example, reached a consensus that organ donation can suitably be regarded as an act of charity. However, the dialogue on the ethics of human cloning only ostensibly struck a common dissonance on the sanctity of the marriage institution.[4] Public dissent and consent are therefore of relevance when considering national policies for the conduct of research in biomedicine in any society.

A public Conference on Reproductive Human Cloning jointly organised by the Ministry of Foreign Affairs and the Institute of Strategic and International Studies of Malaysia was held in Kuala Lumpur in February 2002.[5] I will now present objectively the various views put forward by representatives of diverse quarters of Malaysian society. Their main aim is to explore the ethics of human cloning and to develop a national consensus to be submitted to the United Nations.

My discussion will often be juxtaposed with the ethics of in vitro fertilisation and the ethics of organ transplantation in order to present a fair impression of the nature of ethical deliberations in Malaysia.

My main focus will be on the ethical discourse in the Muslim community. However, I will also briefly look at the fundamental values inherent in each religion. For Islam submission to God, doing what is approved and avoiding evil, is as central as the giving of alms or *zakat* and self-discipline by way of fasting and prayer. Christianity focuses on loving others, even enemies, compassion and, mercy, forgiveness and reconciliation. Hinduism is about understanding the self as being made up of body, mind and soul, in addition to serving others as service to God and accepting everything as being part and parcel of the whole. Buddhists believe in the 'middle path,' right understanding and detachment. It is my novice understanding that generally the Buddhists of Malaysia do not really adhere to, nor are they today in touch with, classical Buddhist teachings. The majority of the modern Chinese community is more inclined to speak of values provided by Taoism and Confucianism (if not Christianity), even though they declare their faith as essentially Buddhist. Taoism is about yin and yang or the balance between negative and positive energies, harmony with nature, gentleness and seeing beyond the obvious. Confucianism deals with treating others as one would like to be treated, advocating values such sincerity, earnestness, modesty, humility, and learning through observation. Other minorities consist of the Sikhs who emphasise community spirit, purity of thought and action, and courage in the face of tyranny. Finally, of course, there are the indigenous Orang Asli communities who have advanced socio-economically and educationally but who are known to have retained great and even ethereal respect for the environment.

The diverse religious and cultural beliefs in Malaysian society and strategic considerations, therefore, make the task of reaching a consensus on the ethics of human cloning a challenging one. The question posed at the opening of the conference was 'Does cloning undermine human dignity and human rights?' This conference is most relevant to our discussion on culture and the ethics of human cloning in Malaysia because participants included scientists and medical doctors, academicians as well as leaders and representatives from various religious and non-governmental organisations.

Before we discuss the findings of the conference. an understanding of what human cloning means would be most useful. When scientists in the 1950s sought to understand why only genetic material from gonadal cells could result in the formation of a new organism or individual, the interest in cloning began. Early failures to clone organisms using genetic material from differentiated or specialised cells led scientists to believe that cell differentiation resulted in the loss or permanent inactivation of genes. Subsequent attempts at cloning began to utilise genetic material from 'young unspecialised' cells. Through the 1970s and 1990s advances were made in animal cloning by using techniques such as embryo splitting, blastomere separation of human embryos, and the transfer of embryonic nuclei into enucleated unfertilised eggs. However, the birth of Dolly revolutionised the idea that cell differentiation was irreversible. Dolly was produced via nuclear transfer from the cell of an adult mammary gland. Scientists were able to starve the cell so that it could be 'reprogrammed' to behave like a fertilised ovum. Since then the idea that fully-developed humans can be created became a plausible scientific possibility.

Cloning is defined as the production of a cell or organism with the same nuclear genome as another cell or organism. Reproductive cloning refers to the production of a foetus without the process of normal sexual reproduction. Therapeutic cloning refers to the production of stem cells, tissues or organs.

Scientists at the conference, therefore, presented detailed information on the science of human cloning and attempted to 'clarify the misconception' that cloning creates 'copies' of existing human beings. Cloning only produces a genetically similar twin much later than an already existing individual. As with ordinary identical twins, the resultant individual will develop differently, both psychologically and intellectually, according to his or her environment.

As highlighted before, the cloning debate in the Western world is almost always connected to the issue of human dignity. Hence the question arises: Does cloning undermine human dignity and human rights? The issue of human dignity is known to be significantly influenced by the philosophy introduced by German philosopher Immanuel Kant. Kant's second formulation of the categorical imperative emphasises the idea that every person has an inherent dignity by virtue of his or her humanity, that is, his or her rational nature:

> Act in such a way that you always treat humanity, whether in your own person or in the person of any other, never simply as a means, but always at the same time as an end.[6]

All persons are rational creatures and, therefore, are entitled to respect not only from others but from themselves as well. Human life has intrinsic value, and the moral obligation that is owed to humans is one of respect. The human individual produced as a result of cloning would be designated as a mere 'means.' The art of human cloning technology is seen as the violation of the individual's right to a unique genetic identity.

The February conference was informed that the United Nations General Assembly had accepted the proposals of France and Germany to adopt a resolution to establish an Ad Hoc Committee to consider a mandate for an international convention *against* reproductive human cloning. A Ministry of Science, Technology and Environment spokesman stated that 'Malaysia supports the establishment of the Ad Hoc Committee in the light of the far-reaching implications of reproductive cloning for mankind, *including its integrity and dignity* as well as the urgency for the international community to deliberate and take a considered position on the issue.'[7] The decision was deemed important to serve 'as an effective and feasible deterrent to overzealous researchers.'

A list of international declarations that highlighted the concept of human dignity was consulted. Article 1 of the Universal Declaration of Human Rights reads, 'All human beings are born free and equal in dignity and rights.' The term 'human dignity' is also included in the United Nations Charter. The World Health Organization 1998 Resolution of the Ethical, Scientific and Social Implication of Cloning is given as, 'Cloning for the replication of human individuals is ethically unacceptable and contrary to human dignity and integrity.' Cloning technology is believed to open up the possibility of the commodification of human beings, and this is agreed to be dehumanising and degrading.

Hence, from the very start the public was provoked and forced into a debate that focused on the technological infringement of human dignity and integrity. A diverse representation of the various communities was invited to the conference where the issue of human dignity was also inevitably discussed.

Christian leaders quoted relevant verses from biblical texts in support of the principle of human dignity. Islamic scholars, however, selected to argue from the perspective of the violation of God's role as Supreme Creator rather than the violation of human dignity. Both, however, shared other compelling arguments, including the risks to health from the mutation of gene pools and the risk of abuse of the technology.

While there were religious concerns about the abuse of cloning, scientists generally felt that 'any science can be abused and used for evil purposes.' And this fact, they argued, cannot be used to justify the termination and 'criminalisation of research in this area.' They outlined the potential of stem cell research and suggested that it be allowed to advance, giving due consideration also to the ethical problem that is raised when it involves the production of cells from human embryos. Scientists discussed the view that stem cells do not possess 'past or present consciousness' and therefore 'cannot be regarded or treated as humans,' and this view received cautious and quiet attention. Scientists continued to express their reservations should research be banned altogether: to discourage research in the area of cloning would be to 'deprive the welfare and well-being of humankind.' The words of Baroness Mary Warnock were also brought into the discussions. According to her conviction, 'Human cloning should never be allowed, except perhaps in case of complete male infertility when all other remedies have failed.'

It is thought that the formulation of a well-balanced guideline, one that corresponds to the strategic importance of science and technology, being a significant tool for development, is what the country needs. Citing the UK Human Cloning Act 2001 as an example, some parties lamented the lack of local legislation in the field. The UK Act allows experimentation in human cloning, but forbids the implantation of the cloned embryos into the uterus. The offence is punishable by up to 10 years in prison; Japan, too, carries the same penalty for the crime. The difficulty in drafting national laws is underlined here: legislation on experimentation with embryos produced through in vitro fertilisation is still not in place after three decades.

1. Christianity

The Christian perspective on human cloning is based on the belief that every human being has innate dignity and rights. There is further the idea that a basic equality exists among all human beings. The idea that the human form reflects the image of God gives him or her an ontological sense of dignity.

> Being in the image of God the human individual possesses the dignity of a person, who is not just something but someone.[8]

The Book of Genesis Chapter 1, verse 27 states that 'God made man to his own image and likeness, male and female, He created them.' A Malaysian bishop reflects that the Catholic Church rejects human cloning and that 'it must be judged negatively with regard to the dignity of the person cloned, who enters the world by virtue of being the 'copy' (even if only a biological copy) of another being.' The practice is believed to pave the way to the 'clone's radical suffering,' because his psychic identity would be jeopardised by the real or even by the merely virtual presence of his 'other.' In addition, the bishop warns that

> ...this exploitation of human beings, sought by certain scientific and industrial circles, and pushed forward by underlying economic interests, retains all ethical repugnance as an even more serious offence against human dignity and the right to life, since it involves human beings [embryos] who are created in order to be destroyed.

He emphasises that 'halting the human cloning project [on the other hand] is *a moral duty* which must be translated into cultural, social and legislative terms.'

However, differing sentiments exist among Roman Catholics regarding the sanctity of human embryonic life. Some believe that all acts that cause the termination of the life of the embryo and foetus constitute murder, and others believe that it may be carried out for a justifiable motive.

2. Hinduism

Scholars convey the idea that cloning is not new in Hinduism. Hindu mythology and folklore suggest that Devi Parvathi created Lord Ganesha from a small fragment of her skin and that a spark from Shiva's third eye created Lord Murugan. In another anecdote, a drop of Mahisa's blood produced a replica of himself so much so that it was difficult to distinguish between it and the original. It is believed that the creation of the replica of Mahisa had an evil purpose: to deceive. This became the basis of the Hindu objection to human cloning. The words of Premukh Swami Maharaj, who advocates the sanctity of the marriage institution, were recalled:

> Human cloning would mean a parentless society, full of surrogate mothers, careless donors and loveless children; society will face unknown consequences.

Hindus regard human birth as 'a rare gift from god through the accepted norms and traditions of marriage.' Cloning therefore creates humans 'against the order of nature.' The Hindu leader ended his speech with the following thoughts:

> Hinduism places great importance on the soul. A soul is a self-existing entity from eternity. It may take any form, with the grace of god, in accordance with its previous Karma… The *Essentials of Hinduism* reads, 'It is possible to clone a human being, as we know it now but the scientists were not able to say whether they can clone the soul or not.'

In reference to the manipulation of human embryos in cloning technology, Hinduism as well as Buddhism and Jainism abide by the rule of *ahimsa* which advocates that any act of violence or aggression upon living things would constitute a crime. The sacredness of life, including non-human life, is particularly upheld by believers of Jainism.

3. Islam

Muslim-Malays in Malaysia are bound by a common ethic deriving from the Quran and the Prophetic traditions. Some social scientists have observed that after the world-wide resurgence of Islam in the 1970s and also because Islam has always been part of the cultural landscape of the Malays, the Muslims of Malaysia have looked to Islam as the ultimate source of ethics, morality, civilisation, and everything else of value, and in addition, a source of Malay unity and dignity.[9]

Life or social ethics for the ordinary Muslim is based essentially on the *Shari'a* which acts both as a code of law and a code of morals. The *Shari'a* provides the foundation of social morality. It means more than simply 'good behaviour' or *akhlak* or *adab*, as it has been termed in the *Encyclopaedia of Islam*.[10] It is an ethic that governs social relations in that the Muslim is not only asked to act well in private life but also in public and in relation to others around him. It also becomes the responsibility of the Muslim (if he is able), to decide whether an action is good or otherwise.

Ijtihad or critical reasoning of an ethical problem is applied by the method of analogical deductions and community consensus. *Ijtihad* is also defined as the method that involves thorough moral reasoning on issues that have not been expressly regulated by the *Shari'a* (Quran and the Prophetic traditions, the Sunnah). *Ijtihad*, which literally means 'striving,' designates the self-exertion by a *mujtahid* (the expert who performs the moral deliberation) to the best of his ability in order to deduce the ruling of a particular issue from the 'evidence.' 'Evidence' here means the revealed sources of the *Shari'a*, namely the Quran and the Sunnah.

In Malaysia, however, responsibility for exercising *ijtihad* does not lie with the Muslim public, nor does it devolve upon professionals, such as the community of doctors and scientists who may be 'experts' on complex matters such as the science of human cloning. Rather, *muftis* are given the authority for interpreting disputed points of doctrine and issuing *fatwas*. *Muftis* head the religious council of each of the thirteen states in Malaysia, but there is also a National Council of Islamic Affairs headed by the Chief Mufti of the country. *Muftis* are entrusted with the task of *ijtihad*, to seek appropriate resolutions to a contemporary ethical problem, but with regard to bioethics they nonetheless draw on the advice and consider the proposals of scientists and doctors.

The reasoning process as inherent in the *ijtihad* gives the *Shari'a* dynamism to respond to novel challenges and concerns that are constantly arising within the modern Muslim culture.[11] There are precisely ten normative sources of guidelines for the derivation of fatwas in Malaysia. They are laid down in order of importance as follows:

1. The Qur'an
2. The Sunnah
3. Ijma (general consensus)
4. Qiyas (analogical inference)
5. Istihsan (public interest)
6. Maslahah Mursalah (unrestricted public interest)
7. Sad al-dhariah (preventing hardships)
8. Syar'u man Qablana (following existing rules)
9. Istishab (presumption of existing rules)
10. Emulating the practices of the people of Medina

Ijtihad usually begins with careful evaluation of the problem by referring to the Quran and/or the *Sunnah*. Next the *Ijma*, or a collection of the traditions of the Prophet's companions, is examined. It is claimed that during the early days of Islam, there were instances when the Prophet acted or said something that was not expressed in the Quran or revealed directly by God.[12] His companions would form their own judgements which were sometimes at variance with the Prophet. Often the Prophet accepted their opinions. If the *Ijma* failed to have any bearing on the problem, the method of *Qiyas* follows next. This designates the analogical inference to a relatively similar incident in the past.

Istihsan, which basically means public interest, is contemplated next. It involves avoidance of hardship or difficulties and choice of an action that brings comfort or benefits.

Maslahah mursalah is the unrestricted consideration of public good and interest, and this principle essentially means that decisions should be taken with respect to the preservation and maintenance of a good life.

Literally *maslahah* means benefit or interest; when it is qualified as *maslahah mursalah*, it refers to unrestricted public interest in the sense of not having been regulated by the Lawgiver and no textual authority can be found on its validity or otherwise. It is synonymous with *istislah* and is occasionally referred to as *maslahah mutlaqah* on account of its being undefined by established rules of the *Shari'a*.[13]

The problem is then examined from the principle of Sad al-dhariah (preventing an evil before its occurrence). The precept of istishab follows next, which essentially means that a fatwa or a rule of law or a custom remains in force, in so far as it has not been denied (or discontinued) due to lack of evidence that requires it to be changed.[14] The principle of emulating the practices of the people of Medina is taken as a final reference point. The people of Medina are those who have had the experience of seeing and witnessing the Prophet (pbuh) handle matters of practical relevance. The Medina society typifies a system of law that is derived from a mechanism of communal or collective deliberation or *musyawarah*. The Medina Constitution includes a clause: 'whenever you differ about a matter it must be referred to God and Muhammad.'[15] This essentially enforces adherence to and therefore the return to the two primary scriptural documents, the Quran and the Prophet's Sunnah.

Fatwas are generally viewed only as guidance on controversial matters for Muslims. They provide temporal or 'earthly' judgements on many issues, and deviation from them does not always entail punishment. Criminal cases such as theft, murder or abortion are usually tried in the common courts of law. Accordingly, *Shari'a* courts have limited jurisdiction over the administration of justice in Malaysia. This is confined largely to areas in the domain of banking and insurance and family matters such as adultery and the division of property.[16]

The ethical discourse in Malaysia on the question of the moral status of the human embryo has been limited to news media coverage and public lectures by the medical and scientific community on the subject of in vitro fertilisation (IVF). Much of the discourse is confined to discussions of the Islamic rule that IVF is ethically legitimate for Muslims if the gametes used are obtained from legally married couples. There has been no exclusive deliberation on the moral status of the human embryo, which would have an ethical bearing on problems relating to the fate (manipulation, disposal and preservation) of the embryo within an IVF procedure, despite the fact that the first IVF baby was born over fourteen years ago.[17] This may also perhaps rest on the fact that public lectures on IVF by professionals are meant to give factual information and exposition of the novelty of the technique as a state-of-the-art solution for infertility. Lectures are therefore designed to be customer-oriented. Relatively little attention has therefore been given to the sanctity of human life that may be endangered in the treatment of infertility, an ethical problem that has embedded itself within Western discourse over the past two to three decades.

In order to examine the Islamic ethical position on the sanctity of life of the human embryo, the writer began to search for national discussions on the subject of abortion that may shed light on the moral status of early human life. One article published in a local law journal attempted to deliberate on the moral status of the human embryo.[18] However, because discussions are based largely on non-Malaysian resources, they do not represent a local contemplation of the beginning of life. I am inclined to submit that there is no original independent reasoning or *ijtihad* exclusively on the subject of the moral status of the embryo.

I recall that Iqbal once observed that Muslims in the contemporary world do not engage in rigorous exercise of *ijtihad*.[19] Likewise, Nagata in 1984 claimed that since Muslims in Malaysia belong to the Shafii sect, the most orthodox tradition among the Sunni schools, they 'generally believe that the doors of *ijtihad* are closed forever by Shafii's system.'[20] Nagata attributes this to two observations. Firstly, there is a tendency among Muslim-Malays to return to the method of judging a controversial subject by relying exclusively on interpretations of the Quran. Secondly, the mechanism of the *ijma* underscores a particular requirement for the production of a fatwa: the unanimous vote of all members of the fatwa committee is essential. Certain contemporary problems, such as those regarding medicine, are sometimes too complex to achieve a clear consensus from each and every member of a particular fatwa committee. Fatwas therefore take years and sometimes decades to be drafted, depending on the complexity and possibly also public interest as well as the political urgency of the matter or matters at hand.

Interestingly though, other Islamic law scholars point out that in some instances the derivation of fatwas in Malaysia can deviate from the absolute requirement of the *ijma*.[21] The state of Selangor, probably the most developed state in Malaysia, takes the view that if the Shafii school of jurisprudence appears prohibitive on a matter of public interest, other schools can consequently be consulted. This is endorsed by the Selangor Islamic Administrative Laws of 1952, Clause 41(2) which states that

> [i]n the process of developing and producing a fatwa the Council and the Islamic Religious Committee should follow the absolute kul or provisions of the Shafii school. However, if the kul is found to steer away from the interest of the public, then the Council and Committee can follow the kuls from other Sunni schools of Islamic Law.[22]

Shah in 1986 also submits that public interest is a valid reason for contemplating a problem from the tenets of other schools of law.[23]

When the ethics of organ transplant technology were deliberated in the country, the principle of *Maslahah Mursalah* or 'unrestricted public interest' was chiefly applied in combination with the principle of *Sad al-dhariah*, which means 'preventing the onset of evil or hardships.' Therefore, on the basis of preventing subsequent hardships to a person with a malfunctioning organ, an organ transplant may be viewed as satisfying this principle. A verse in the Quran (1: 185) is also invoked in the context of the case:

> Allah intends you comfort and ease and He does not want to burden you with difficulties.

The purpose of the Shari'a is to secure the welfare of the people by promoting their benefits or by protecting them from harm.[24] Islam also lays down a principle whereby the deviation from an established norm must be preceded by a burden of proof. Muslims generally believed that Islam prohibited the dissection of the dead body (the established norm): the Prophet (pbuh) is believed to have said, 'Breaking the bones of a dead man is similar to breaking the bones of a living man.'[25] It was on the basis of this principle that Islamic authorities in Malaysia proposed that organ transplant as a medical technique should be accepted because both local and international ventures in the field of organ transplantation have proven beneficial in saving lives. It then became permissible to remove organs from the dead. The Malaysian public is also informed that this medical procedure should be allowed on the precept that 'the needs of the living have priority over the dead.'[26]

The Sanctity of Human Life In Islam

Let us reflect more closely on the discussion of what differentiates 'the embryo' and 'the foetus.' This is important because the embryo and foetus are presumably both protected under the Malaysian Criminal Law Section 315 which states:

> Any person or persons found voluntarily causing a pregnant woman to miscarry will be punished by a jail term of up to three years, and if the woman aborts the baby in its complete form, she or any other person responsible will be liable to imprisonment up to seven years and liable also to be fined.

Under this law medical practitioners are excluded, to allow the act of abortion for 'therapeutic reasons.'

> Exception: This Section is not extended to the medical practitioner registered under the Medical Act of 1971 who terminates the pregnancy, because he believes, in good faith, that the continuation of the pregnancy will involve risks to the life of the woman or cause a greater degree of harm, whether mentally or physically.

Malaysian law scholars tend to accept that the term 'embryo' denotes the conceptus from the moment of fertilisation to the eighth week, while the term 'foetus' commonly refers to the conceptus after week eight until the end of the pregnancy. Abortion is defined as the 'act of terminating a pregnancy by removing the foetus and other contents of the uterus.' An abortion is regarded as a crime, and a 'complete act of abortion' constitutes the event whereby the foetus emerges from the womb, alive or dead.[27]

On the basis of several other passages in the Quran and of the sayings of the Prophet, most Muslim scholars believe that ensoulment occurs about 120 days after conception.[28] Some scholars hold that it occurs much earlier, at around 40 days after conception.[29] A verse in the Quran explains various stages in the development of the human foetus:

> Man we did create from a quintessence (of clay). Then we placed him as (a drop of) sperm in a place of rest, firmly fixed [or the nutfa stage in the womb]. Then we made the sperm into a clot of congealed blood [alaqa]; then of that clot we made a (foetus) lump [mudgha]; then we made out of that lump bones and clothed the bones with flesh; then we developed out of it another creature. So blessed be God, the Best to create! (The Holy Quran, 23: 12-14).

Malaysian Islamic law scholars also believe that ensoulment and, simultaneously, 'life' begins 120 days after fertilisation in reference to the *hadith* below,

> Everyone of you is formed within your mother's womb for a period of 40 days as the nutfah, then as the alaqah

> over the same period of time, then the mudghah which also develops over 40 days, before the angels are sent to blow your souls into yourselves.[30]

Maliki jurists took the extreme viewpoint, maintaining the notion that life begins the moment the semen has settled in the womb, and prohibited abortion right from the moment of conception.[31] Notably though, some Malikis do not object to abortion when it is performed within forty days of conception. They declare that abortion done within this period does not make the action *haram* or sinful, but that it is looked upon as *makruh* or an action that is not recommended in Islam. Similarly, law scholars in Malaysia also reflect that the 40-day term can be equated to the embryonic stage as described in medical science.[32] At this stage it deserves respect, but its termination (for whatever reason) is decided as *makruh tanzihiyyah*. At this point the transgressor is not liable to be fined the *ghurra*.

The Shafii school of thought believes that the earliest form of life includes stages described in the Quran as 'the clot of blood' and 'the lump' that clings. Other interpretations profess that the Arabic word 'janin' or foetus literally means 'that which is concealed' and 'that which contains life.' Ebrahim notes that Shafii's analysis of the human life form irrefutably points out that it begins at the eighth week after fertilisation, and he additionally qualifies that this definition closely resembles the definition offered by modern science.[33] Such derivation has since also been made by our scholars when they compare the definition in Black's Medical Dictionary.[34] They note that the word 'foetus' refers to the living child in the womb and that life has already begun, even though 'quickening' can only be felt by the mother during the fifth month of pregnancy.

Sheikh Mahmoud Shaltut, the late Grand Imam of the University of Al-Azhar, mentions 'quickening' in his deliberation on abortion, and it is worth mentioning here. After considering the discussions by the various schools of Islamic jurisprudence, he agrees that 'after quickening takes place, abortion is prohibited.' He declares that abortion would then denote a 'crime perpetrated against a living being.'[35]

In accordance with the *Shari'a*, a person who causes a pregnant woman to abort is liable to be fined the diya or 'blood-money.' The amount of the diya imposed increases as the foetus matures. If the foetus is aborted alive but dies outside the mother's body, the full diya is imposed. This judgement is based on the view that the foetus would have been deprived of the vital nutrients and protection provided while within the womb.

A lower fine or the ghurra is imposed if the foetus is aborted dead. The *ghurra* is about one-fifth the value of the *diya*. According to the *Shari'a*, the *ghurra* is equivalent to 'a male or female slave, or five camels or 100 sheep or the monetary equivalent at the time, which was 50 dinars or 600 dirhams.'[36] This judgement on differing levels of fines is related to the belief that the foetus has legal rights in Islam, namely, the rights to paternal inheritance.

This is also the first instance of Islamic ethics using the language of rights. Rahman claims that it is also precisely in connection with infanticide and vengeance (*jihad*) that Muslim jurists first began to lay down their theory of the five basic human rights which must be protected by the state or the government. These rights are:

1. protection of life,
2. protection of property,
3. protection of faith or religion,
4. protection of family lineage and dignity, and,
5. protection of one's mental integrity or the search of knowledge.[37]

These five rights now correspond to the objectives or the general purpose of the Islamic law.[38]

It has been noted that the *Shari'a* accords moral rights to the foetus in the womb by alluding to its rights to inheritance. As Al-Ghazali has pointed out in his deliberations on contraception and abortion, the crime becomes increasingly serious as the foetus matures.[39] There is, therefore, recognition of an entity with rights which increase in value as it matures. Human cloning technology in research would seem immoral in Islam, because it would involve manipulating and destroying a human entity with rights.

It is worth noting that national deliberations on the moral status of the human embryo have indeed involved the contemplation of a greater anticipated harm such that it became imperative for an older or established norm to be displaced by a more stringent rule. For example, Muslim authorities once deliberated whether the 120-day limit for abortion should be brought forward both to cope with both the problem of illegal abortion and to control ambitious researchers in the field of human embryo experimentation.[40] For this purpose the opinion of the Maliki school, which states that life begins at the moment of fertilisation was nearly adopted so as to protect the sanctity of early human life.[41] However, in early 2002, the authorities issued a fatwa that tallied with most countries of the Islamic world and put the beginning of human life as the 120th day and no earlier. This indicates an acute and almost uncanny caution by authorities not to deviate from the established norm of the majority or the Muslim world consensus.[42] Even so, it must be acknowledged that the fatwa includes the reminder that the human entity is to be respected from the moment of conception and that abortion may be allowed before or after that period if and only if the mother's life is in danger and for no other reason.

Perhaps due to the dissatisfaction in some quarters regarding the deliberations on the beginning of life, some Malaysian physicians began to acclaim the value of biological facts in determining the precise moment at which life begins. They declare that 'scientific evidence concerning fertilisation and the development of the very early embryo [will] make a purely philosophical or theological stand untenable.'[43] It is divulged that scientists currently use the 14-day rule for the permissibility of human embryo research, and such limits have been incorporated in guidelines concerning human embryo research.[44]

A National Committee spearheaded by the Ministry of Health was established to address the ethics of human cloning and eventually develop policies and regulations pertaining to the issue. The Committee includes the Ministry of Science, Technology and Environment, the Ministry of Health, the Malaysian Department for Islamic Development, Islamic Studies Departments, and the Foreign Ministry. It was reported that the Committee would especially study regulations formulated in the UK and Australia.

On May 8, 2002, the First National Life Science Conference 2002: 'Genomics and Health Implications for Malaysia' was organised by the College of Physicians, the Malaysian Academy of Medicine, and the Ministry of Health. The Conference concluded with a resolution that demonstrated firm support for existing international declarations which prohibit reproductive human cloning. These include the Universal Declaration on the Human Genome and Human Rights adopted by the General Conference of UNESCO (1997) and the Convention for the Protection of Human Rights and Dignity of the Human Being with regard to the Application of Biology and Medicine: Convention on Human Rights and Biomedicine of the Council of Europe (1997). The soul-searching statements in the preamble to the latter were considered by the Conference:

> Conscious of the accelerating developments in biology and medicine; convinced of the need to respect the human being both as an individual and a member of the human species and recognising the importance of en-

> suring the dignity of the human being; conscious that the misuse of biology and medicine may lead to acts endangering human dignity; resolving to take such measures as are necessary to safeguard human dignity and the fundamental rights and freedoms of the individual with regard to the application of biology and medicine.[45]

The findings of the World Health Organization (WHO), which had addressed the issue in several forums, were reiterated, '...that the use of cloning for reproductive purposes was deemed ethically unacceptable' '...and harmful to the dignity of the human being.' Similarly, Article 11 of the Declaration of the Human Genome and Human Rights was referred to: 'practices contrary to human dignity, such as reproductive cloning of human beings, were not permitted.' The Intergovernmental Committee (May 2001) reaffirmed that human reproductive cloning was contrary to human dignity and therefore asked member states to establish controls to prohibit human reproductive cloning.

In the Islamic world, every Islamic seminar, *Fatwa* council and individual scholars have reflected that cloning procedures aimed at producing human clones are not legitimate. Declaring the procedure to be *haram* or forbidden is thought of as a way to prevent harm. Aly Mishal of the Federation of Islamic Medical Associations, however, recommended a reversible ban on human cloning 'in case new information about benefits become clearer and approved by the *Shari'a.*'[46]

Local scientists continue to support research in human cloning and show an inclination to agree with Panayiotis Zavos, who once commented that cloning is not an ethical issue but a medical issue. They claim 'We live in a world that keeps evolving. We can solve problems with high technologies. If we simply say no to it, then it is a tragedy, it is the end of our progress. We think cloning opens up the possibility to many things.' Most scientists, however, agree that reproductive cloning may only be used as a last resort to overcome infertility and recommend that a degree of regulation be put in place.[47] On the question of the uniqueness of human identity, the responses included 'I think the problem of losing individuality is not a big concern' and 'our society should not be overly concerned about the issue of individuality.'[48]

As mentioned before, Islamic scholars did not particularly speak on the grounds of human dignity but rather argued that human cloning violates the power of God. The Quran states 'And He creates humans from liquids with a genetic trait and family ties; truly God is the most powerful who creates anything that He wishes.' (Al-Furqan: 54) Cloning is therefore described as the act of playing God and therefore against the order of nature. 'Truly We have created men in the best of forms.' (Al-Tin: 4) And 'God knows what lies within the mother's womb and whether the foetus is perfect or otherwise. For every creation has a predetermined fate. And He knows of things that are unseen or seen. For He is the Greatest and the Highest of all things known.' (*Quran,* Al-Ra'd: 8)

The *Fatwa Mu'asarah,* a book containing a collection of *fatwas* or Islamic rulings on various moral and social issues by Yusuf al-Qardhawi, was quoted:

> 'God's knowledge regarding the maternal womb is utterly detailed. This includes knowledge regarding the nature of the conception, whether viable or not, intelligent or otherwise, weak or strong, happy or suffering. Man only knows how to determine whether the sex of the foetus is male or female.'

This adds to the belief that the ability to produce human clones intervenes in the works of God. Such actions not only take away the humility of human scientists but also undermine their faith.

Others have advanced the idea that the very knowledge of the possibility to clone humans ought to remind Muslims of the supremacy of God's powers and the impending life in the Hereafter. If an entire organism can be replicated from a single adult cell, then God's powers to resurrect the dead in the Hereafter is a reality that would not be too difficult to comprehend.[49] This knowledge ought to further instil in Muslims a sense of humility, obedience and respect for God's laws. Muslims are, in addition, asked to reflect on a cautionary statement in the Quran

> ...(establish) God's handiwork according to the pattern (*fitrah*) on which He has made mankind: no change (let there be) in the work of God: that is the standard religion: but most among mankind understand not.[50]

Cloning is regarded as the reduction of the status of humankind to that of animals. Humans are elected as 'God's vicegerents and therefore the most noble of all creations with the duty to manage the world as well as possible.'[51] I am not too sure here whether the designation of humans as 'the most noble of God's creations' allows humans a sense of dignity. But according to the *Shari'a*, one of the general purposes of the law is to protect the individual's family lineage and hence his dignity. The individual's dignity derives from his or her being a certified member of a 'legal' family. An illegitimate child [born out of wedlock], for example, is generally believed to have been deprived of his/her dignity in the sense that there is uncertainty about his/her family ties.

Through the 1980s, when the country was confronted by the moral controversies surrounding *in vitro* fertilisation and surrogacy, *fatwas* were quickly formulated to outlaw surrogacy. Surrogacy 'clearly' fell outside the stipulations of a legal marital relationship. Associated *fatwa*s[52] were also produced to address the status of babies produced from IVF techniques. It was held that IVF babies produced from married couples have legal rights to paternal inheritance. The swift formulation of such *fatwa*s and the prohibition of surrogacy seem to illustrate that the safeguarding of Islamic Law regarding the protection of family lineage and dignity were matters of prime concern to the Malaysian religious authorities. They greatly overshadowed issues regarding the establishment of rules to regulate the handling and disposal of embryos during research and infertility treatment, not to mention respect for the moral status of the human embryo.

Human cloning eliminates the 'paternal factor' in the child produced in a different way, and most scholars concluded that there is, therefore, more harm in allowing human cloning and that it must ultimately be banned. Such harm would include the disruption of lineage and family relationships. The social fabric of humanity would be thrown out of balance, as cloning is expected to interfere with male-female population dynamics. The fear caused by reports associating cloning with mortality, deformities, genetic disorders, and premature ageing further underpins the belief that endless harm can result from human cloning. It is also anticipated that certain genetic disorders and mutations might proliferate. The representative of the Islamic Medical Society therefore proscribed human cloning on the basis of the above 'anticipated social, moral, psychological and legal implications of human copies.'[53]

The view that cloning research shows disrespect for the human embryo is not echoed within national borders, as evidenced by the bold suggestion that 'fertilised ova in the laboratory do not achieve the same status and therefore do not deserve consideration as human foetuses, unless they are restored to the mother's uterus.' This statement was made in reflection of a 1987 Muslim ethical consideration of the moral status of the embryo, which had begun to incorporate the notion that there are intrinsic differences between the embryo in the womb and the embryo grown in the Petri dish or test-tube. Al-Khateeb argues that the embryo that is artificially kept in a scientific medium has no rights to inheritance in the way the embryo in the natural womb does. It naturally follows that experimentation on human embryos, including therapeutic cloning, is considered legitimate.[54]

So far, no *fatwa* has been issued in Malaysia, but the Department of Islamic Development has considered the *fatwa* developed by the Singapore *Fatwa* Committee which allows cloning technology in the development of replacement organs and tissues. The use of the DNA of the person undergoing treatment minimises tissue rejection, and this is viewed as a 'good' that can be gained from cloning technology.

It is also agreed that it is permissible to utilise cloning technology in gene therapy and genetic engineering to produce and introduce genetic material into a human being, but the intention must be strictly to cure illness. Cloning is not permissible if the intention is, say, to enhance the physical features of the person. It is prohibited to change God's creation in any form. Cloning may only be used to introduce healthy genetic material into ova, fertilised ova or foetuses in order to avoid diseases. The DNA material of the somatic cell of the husband may be fused with an enucleated ovum of the wife.

Scientists and religious scholars alike observe with interest the potential that cloning offers in the field of biomedicine and have adopted a wait-and-see attitude. In the light of the principle of *dharuriyyat* or 'necessity allows the forbidden,' cloning technology may be allowed to proceed.

Scholars have also pondered that the principle of *maslahah* or public interest may be extended to the ethics of human cloning: it is acceptable if it demonstrates clear benefits or maslahah for the public, with the minimisation of harm.[55] It has been said that it may be permissible to offer cloning technology as a treatment for infertile patients because it assures continued family lineage. However, the application of the principle of *maslahah* is dependent upon the concept of *dharuriyyah* or utter necessity. This means that the commissioning of an act is only applicable where the livelihood of the person or group of persons is absolutely dependent on that act. Thus, condoning the act would mean 'harm' to the person/persons involved. At the same time primary consideration must be given to the 'harm' deriving from the act of cloning, in that it would create confusion in family ties which may, in turn, throw the whole of society into chaos. It is precisely in this regard that a spokesman for the Department of Islamic Development disapproves of human reproductive cloning;

> A clone that develops from the female cell with no genetic contribution from the male partner will create confusion in family lineage. This is because the child would be produced from the somatic adult cell which means that he or she is produced from the twin sister cell and the family hierarchy will result in a state of disorder. Islam emphasises and *encourages marriages between couples of different germ-lines* and cloning is anticipated as causing disarray in the family tree.[56]

According to the Shafii school of jurisprudence, in the interest of certifying a child's paternal lineage, a woman is required to observe the so-called 'period of retirement' or 'period of iddah' on the event of her husband's death or their divorce, before she can consider marrying another.[57] The iddah period varies according to the schools. If the woman is already pregnant at the time of her husband's death, the Shafii School takes the position that she must observe an iddah period covering the whole duration of her pregnancy. This would ensure the child's paternal lineage. The importance of protecting family lineage is exemplified in the Quran. A relevant verse states

> 'If any of you die, and leave widows behind, they shall wait concerning themselves, four months and ten days. When they have fulfilled their term there is no blame on you if they dispose of themselves in a just and reasonable manner and God is well acquainted with what you do.' (al-Baqara: 234)

For example, because of the importance attached to paternal lineage, the use of stored embryos and gametes in infertility treatment outside the confines of marriage (for example, after divorce or the death of a spouse) would be deemed to be *haram*.

To highlight a classic example of the application of the iddah concept in deliberating the ethics of assisted human reproduction, I shall take the case of Libya. The traditional understanding of foetal gestational age in that society is used as the chief premise for prohibiting the practice of artificial insemination, and this prohibition is embedded within the Libyan Penal Law of 1972:

> Anyone who impregnates a woman by *artificial insemination* by force, threat, or deceit is to be punished by imprisonment not to exceed ten years, [and] the punishment [is] to be for a period of not more than five years if insemination was with her consent.[58]

The Libyan tradition observes an extraordinarily long gestational period of up to twelve years, well beyond that observed by other Sunni schools of Islamic jurisprudence.[59]

It is argued that the maximum period of gestation attached to the foetus is 'not entirely due to the ignorance of the medieval jurists on matters of embryology' but rather 'a desire to avoid attributing the status of illegitimacy to children born after the normal period of gestation.' It appears that an illegitimate child not only has no claims to inheritance, but he or she would also be looked upon as a social misfit. Traditional Muslim society also submits that the birth of a child out of wedlock and outside recognised periods of gestation establishes prima facie evidence of fornication.[60] As such, the law was drawn up not only to maintain public morality but also to avoid the 'severe hadd penalties' on fornication.

Mayer notes that the Libyan observance of the exceptionally long period of gestation raises concern that a child born years after divorce or the death of a husband may claim to be his legitimate offspring.[61] This, in turn, raises the possibility of the abuse of stored gametes and embryos in assisted human reproduction. The Libyan government eventually ruled that artificial insemination is prohibited absolutely.

The basis for the prohibition of human reproductive cloning remains strictly that of safeguarding family lineage and not human dignity, which is so much debated in the Western world. In conclusion, the ethics of human cloning from the Malaysian perspective can be summarised as follows :

1. The basic concept in reproduction is abidance by the *Shari'a* approved system of legally binding marriage.
2. Human cloning is against the natural *fitrah* of the human relationship of marriage and reproduction.
3. The natural balance between males and females may be disturbed.
4. Serious harm exceeds any clearly foreseen benefits.
5. Fear of the unknown potential and dangers such as deformities, genetic disorders and premature ageing.

On the last point it is worth recalling arguments by Dame Mary Warnock, chairman of the Warnock Committee, who translated the moral apprehension regarding the unknown potential of human embryo research into acceptable moral principles. Although she acknowledges that such moral principles may be used to 'counteract utilitarianism' that often underlies policy decision making, she elaborates:

> It is difficult to persuade members of committees that feelings or sentiments can have a central role to play in moral decision making. Such people tend to believe that a moral judgement must be rational, or else it must be based on religious dogma. Otherwise, it will not count as a proper moral judgement. They find it shocking to accept that, as Hume put it, 'morality is more properly felt than judged of.' Yet, I believe it is to offend against the concept of morality itself to refuse to take moral feelings or sentiments into account in decision making.[62]

It is therefore necessary at the present time to curb all activities that may lead to the production of human clones. Legislation must be established to assuage these fears of the unknown. Warnock once remarked

> A society which had no inhibiting limits, especially in the areas with which we have been concerned, questions of birth and death, of setting up of families, and the valuing of human life, would be a society without moral scruples. And this nobody wants.

Finally, I believe it would be most appropriate to reproduce the words of Professor Schweidler below to capture the importance of legal instruments in managing the problem of human reproductive cloning.

> The legitimacy of political institutions is primarily at stake when they are confronted with the demand for *help*, when they have to *defend* the rights of their citizens against actual or foreseeable violations. The rule of law has its criterion of legitimation in the abstract power to guarantee equal protection for all human beings, and that power is the more demanded the more certain human beings are unable or disabled in their capacities to

defend themselves against violations. The procedural sphere in which the legal system has to give the paradigmatic proof of its ethical responsibility is the sphere of *legislation*.

Notes

1. Muslims make up approximately 12 million of the population, and Article 3 of the Federal Constitution establishes Islam as the official religion of the Federation. Most legislation in the country is designed according to the Islamic moral and ethical standards which are more acceptable for all Malaysians (Zawawi, 2001; see also Chan, C. K., 2001).
2. Tu Weiming, 1997.
3. Schuster, 2003.
4. For example, Hindu delegates noted that 'human cloning would mean a parentless society, careless donors and loveless children' and that 'human birth is a rare gift from God throught the accepted norms of marriage' but translated such thoughts as the undesirable consequences of human cloning. Muslim delegates believe that disturbing the family lineage through cloning would later become a problem in establishing claims of paternal inheritance.
5. It was the proposal to draft the cloning treaty under consideration by the United Nations that led the organisers to hold the Conference on Reproductive Human Cloning in February 2002.
6. Kant, 1981.
7. *National Conference* 2002.
8. Address of the Bishop of the Catholic Church in *National Conference* 2002.
9. Mutalib, 1990, 32; see also Voll, 1982, 332.
10. Gibb et al., 1979, 326.
11. The following is based on *The Regional Seminar on Muftis and Fatwa* (1997). Kuala Lumpur: Institute of Islamic Understanding. 23-24 September. See also Kamali, 1994.
12. Faruki, 1961.
13. Kamali, 1989.
14. Kamali, 1989, 377.
15. See Guillaume, 1980, 231-233 and Yaacob, 1997.
16. Although civil courts (which practice the Common Law of England) still have wide jurisdiction in matters related to criminal law, the decisions of the *Shari'a* courts have been found significantly in the Com-

mon Law reports. Ibrahim observes, in addition, that 'Islamic law is administered in the *Shari'a* courts but unfortunately the jurisdiction of the *Shari'a* courts is limited and they were neglected and subordinated to the civil courts during the time of British control in Malaysia. Even after independence it took a long time before the position of the *Shari'a* courts and their officials was improved.' See Ibrahim, 1994.

17. To date *fatwas* are produced mainly in reference to the legality of the technology on Muslims, the legitimacy of babies produced from IVF and prohibitions on sperm banks.
18. Tengku Zainudin, 2001.
19. Iqbal, 1951, 149.
20. Nagata, 1984, 21.
21. Ishak, 1980.
22. This statement epitomises the importance given to matters that are seemingly restricted by old Shafii rulings. Some people believe that new *fatwas* are needed to keep up with changing times and the influx of new problems. (Nagata, 1984, 121; see also Ibrahim, 1998, 100).
23. Shah, 1986, 85.
24. Scholars emphasised that the principle of *maslahah mursalah* had been the basis for the *Shariah* law permitting post-mortem. While the integrity of the body of the deceased has to be preserved as projected by the *hadith*, post-mortem is, however, allowed in cases of murder, death by poisoning, or accident-related death. It is held permissible on the basis of the more urgent requirement to seek justice for an accused, for the victim or for his/her relatives (Yaacob, 2000).
25. Many ethical discussions on organ transplant begin with a reflection of this *hadith* from the Sunan Adu Dawud collection. See, for example, Gatrad & Sheikh, 2001.
26. Abdul Rahman, 1998, 67 and 86.
27. Tengku Zainuddin, 2001.
28. Al Bar, 1995.
29. Kamel, 1983.
30. Interview carried out on July 13, 1999 at the Academy of Islamic Studies of the University of Malaya by Tengku Zainuddin (2001), *op. cit.,* 138. The *hadith* quoted is taken from the collection of *hadiths*, see Al-Nasyaburi, *Muslim ibn al-Majah.* Volume 5, 496.
31. Al-Mazkur et al., 1985.
32. Tengku Zainuddin, 2001.
33. Ebrahim, 1993. Ebrahim is an African Islamic scholar who is regularly invited to give lectures on bioethical issues in Malaysia.

34. Tengku Zainuddin, 2001, 96.
35. Shaltut, 1959.
36. Al-Bahansi, 1964, 99-107 as cited by Rispler-Chaim, 1993, 9.
37. Rahman, 1982.
38. The Shari'a puts forth the notion that the objectives of Islam comprise the protection of five fundamental basic elements, namely, the protection of religion, the protection of life, the protection of the intellect, the protection of lineage, and the protection of property (Kamali, 1989, 338).
39. Al-Ghazali, 1984, 110.
40. Tengku Zainuddin, 2001, 146.
41. The four Sunni schools of Islamic Law are divided on the subject of the beginning of life. Although there is consensus among the Shafii, Hanafi and Hambali schools that life begins at 120 days, the Maliki school affirms that life begins at the moment of fertilisation.
42. It took a period of ten years (1992-2002) for the local *fatwa* on 'the beginning of human life' to be drafted.
43. Abdullah, 2002.
44. The National Standing Committee on Assisted Reproductive Technology was established in the year 2000 to study, in addition, the ethical implications of cloning technology. One example is the Malaysian Guidelines On The Conduct Of Assisted Reproductive Technology (2000).
45. Convention 1997.
46. Mishal, 2002.
47. Siok, 2002.
48. Ibid.
49. Zawawi 2001, 27.
50. Ibid., 27.
51. Ibid. 2001.
52. *Fatwa* on test-tube babies (1982 and 1983). Kuala Lumpur: JAKIM, 16-17 November 1982 and 10 October 1983.
53. The President of the Islamic Medical Society at the 2002 Conference quotes the views of the Chief Mufti of Egypt, Nasr Farid Wasil, delivered on June 30, 2001.
54. Al-Khateeb, 1987.
55. Zawawi, 2001.
56. Spokesman for the Department of Islamic Development (2002). Conference on Human Reproductive Cloning.

57. Considering that the woman menstruates and has regular periods of 'purity,' the *iddah* is three periods of purity, and if the woman does not menstruate, the *iddah* for her is three months. (Shafii, 1961, 338; see also Ibrahim, 1975, 188).
58. Mayer, 1980.
59. According to Ghanem, that gestational age is determined from the personal experience of women during the reign of the Prophet's Caliphs and Companions. See Ghanem, 1982, 36.
60. Coulson, 1964, 103.
61. Mayer, 1980.
62. Warnock, 1983.

References

Abdullah, M. F. "Reproductive human cloning: medical, scientific and strategic perspectives." *Conference on Reproductive Human Cloning.* Kuala Lumpur: Ministry of Foreign Affairs & ISIS Malaysia. 6-7 February, 2002.

Abdul Rahman, AF. "The role of religious officials in the realm of organ transplantation." In *Islam and Organ Transplantation*, Ibrahim, I. (ed.). Kuala Lumpur: Malaysian Institute of Islamic Understanding, 1998.

Al-Bahansi, Ahmad Fathi. *Al-Qisas fi Al-Fiqh Al-Islami.* Cairo, 1964.

Al Bar, M. A. "When is the soul inspired." In *Contemporary topics in Islamic medicine.* Jeddah: Saudi Arabia Publishing & Distributing House, 1995, 131-136.

Al-Ghazali, Imam Abu Hamid Muhammad. *Marriage and Sexuality in Islam.* Farah, M (trans.). Salt Lake City: University of Utah Press, 1984.

Al-Khateeb, E. "General comments on sexual assault, fate of bank-deposited embryos and surplus fertilized ova." *Third Symposium on Medical Jurisprudence-Islamic Vision of Some Medical Practices. 18-21 April, Jeddah*, 1987.

Al-Mazkur, K., Al-Saif, A., Al-Gindi, A. R. & Abu Guddah, A. A.-S. (eds.). *Human Life: Its Inception and End as Viewed by Islam.* Kuwait: IOMS Publication Series, 1985.

Chan, C. K. "Ethical aspects of health care reform." *Eubios Journal of Asian and International Bioethics*. 8, 2001, 115-118.

Convention for the Protection of Human Rights and Dignity of the Human Being with regard to the Application of Biology and Medicine: Convention on Human Rights and Biomedicine. Strasbourg: Conseil de l'Europe, 1997. Reproduced in *Journal of Medicine and Philosophy (2000)*. 25(2), 259-266.

Coulson, N. J. *History of Islamic Law*, Edinburgh: Edinburgh University Press, 1964.

Ebrahim, A. F. M. *Biomedical Issues: Islamic Perspective*. South Africa: Islamic Medical Association of South Africa, 1993.

Faruki, K.A. *Islamic Jurisprudence*. Karachi, 1961.

Gatrad, A. R. & Sheikh, A. "Medical Ethics in Islam." *Archives of Disease in Childhood*. 84, 2001, 72-75.

Ghanem, I. *Islamic Medical Jurisprudence*. London: Arthur Probsthain, 1982.

Gibb, H.A.R., J.H. Kramer, E. Levi-Provencal and J. Schacht (eds.). *Encyclopaedia of Islam*, Vol. 1. Leiden: E.J. Brill, 1979.

Guillaume, A. *The Life of Muhammad: A Translation of Ibn Ishaq's Sirat Rasul Allah*. Karachi: Oxford University Press, 1980.

Ibrahim, A. *Islamic Law in Malaya*, S. Gordon (ed.). Kuala Lumpur: Malaysian Sociological Research Institute, 1975.

Ibrahim, A. "The introduction of Islamic values in the Malaysian Legal System." *Journal of the Malaysian Institute of Islamic Understanding (IKIM)*.2 (1), 1994, 27-45.

Ibrahim, A. "Processes in the development of fatwas by muftis." In *Mufti and Fatwas in Asean Countries*, Yaacob, A. M. & Abdul Majid, W. R. (eds.). Kuala Lumpur: IKIM, 1998.

Iqbal, M. *The Reconstruction of Religious Thought*, Lahore: M. Ashraf, 1951.

Ishak, Othman. "The place of Fatwa in Islamic Law and its development in the State of Selangor." In *Tamadun Islam di Malaysia (Islamic Civilization in Malaysia)*, Khoo Kay Kim (ed.). Kuala Lumpur: Historical Society of Malaysia, 1980, 113-123 (article in malayan).

Kamali, M.H. *Principles of Islamic Jurisprudence.* Petaling Jaya, Selangor: Pelanduk Publications, 1989.

Kamali, M.H. "Shari'a and the challenge of modernity." *Journal of the Malaysian Institute of Islamic Understanding (IKIM),* 2(1), 1994, 1-26.

Kamel, Abd Al-Aziz. "Abortion." In *Seminar on Reproduction in* Islam, Al-Gindi, A. R. (ed.). Kuwait: Islamic Organization for Medical Sciences, 1983.

Kant, I. *Grounding for the Metaphysics of Morals.* Transl. by J.E. Ellington. Indianapolis: Hackett Publishing Co, 1981.

Mayer, A. E. "Libyan legislation in defense of Arabo-Islamic sexual mores." *American Journal of Comparative Law,* 28, 1980, 287-313.

Mishal, A. "Cloning and Advances in Molecular Biotechnology: Islamic Shariah Guidelines." *FIMA Yearbook 2002.* Federation of Islamic Medical Associations & Medico Islamic Research Council (MIRC): Islamabad, 2002, 33-47.

Mutalib, H. *Islam and Ethnicity in Malay Politics.* Singapore: Oxford University Press, 1990.

Nagata, J.A. *The Reflowering of Malaysian Islam: Modern Religious Radicals and Their Roots.* Vancouver: University of British Columbia Press, 1984.

(The) National Conference on Reproductive Human Cloning: Medical, Scientific and Strategic Perspectives. Ministry of Foreign Affairs and In-

stitute of Strategic and International Studies. 6-7 February 2002. Kuala Lumpur.
Rahman, F. "Human Rights in Islam." In *Democracy and Human Rights in the Islamic Republic of Iran.* Chicago: Committee on Democracy and Human Rights, 1982. Mimeo. Cited in idem. *Health and Medicine in the Islamic Tradition.* Kuala Lumpur: S. Abdul Majeed & Co, 1993.

Rispler-Chaim, Vardit. *Islamic Medical Ethics in the Twentieth Century.* Leiden, New York & Cologne: E.J. Brill, 1993.

Schuster, E. "Human Cloning: Category, Dignity, and the Role of Bioethics." *Bioethics Journal.* 17(5-6), 2003, 517-525.

Shafii, M. *Shafii's Risala: Islamic Jurisprudence.* Transl. by M. Khadduri. Baltimore: John Hopkins Press, 1961.

Shah, Raja Azlan. "The Role of Constitutional Rulers in Malaysia." In *The Constitution of Malaysia: Further Perspectives and Developments - Essays in Honour of Tun Mohamed Suffian*, Trindade, F. A. and Lee, H. P. (eds.). Singapore: Oxford University Press, 1986.

Shaltut, Sheikh M. *Al Fatawa.* Cairo: Matba'at al Idarah al-Ammah li al Thaqafah of Al Azhar, 1959, December. Also found in *Topics in Islamic Medicine.* Cairo: Council of the International Organization of the Medical Sciences.

Siok Gee, See. *A Study of Issues In Human Cloning: Ethics, Legislation and Human Rights*. Master of .Science thesis, 2002 (unpublished).

Tengku Zainudin, TNA. "Abortion and the right of the foetus to live." *IKIM Law Journal.* 5(2), 2001, 93-146.

The Regional Seminar on Muftis and Fatwa (1997). Kuala Lumpur: Institute of Islamic Understanding. 23-24 September.

Tu Weiming. "Towards a Global Ethic: Spiritual Implications of the Islam-Confucianism Dialogue." In *Islam and Confucianism: A Civilisational Dialogue*. Kuala Lumpur: University of Malaya Press, 1997.

Voll, J.O. "The Resurgence of Islam." In Idem. (ed.), *Islam: Continuity and Change in the Modern World*. Essex: Longman, 1982.

Warnock, Mary. "In vitro fertilisation: the ethical issues." *The Philosophical Quarterly,* 33, 1983, 238-249.

Yaacob, A. M. "The Development of the Mufti Institution of Malaysia." *Regional Seminar on Muftis and Fatwas*. Kuala Lumpur, 23-24 September, 1997.

Yaacob, A. M. "Islam and organ transplant." *Proceedings of the First Regional University of Malaya Medical Centre Conference on Management in Medicine - Islamic Perspective*. Kuala Lumpur: University of Malaya, 2000.

Zawawi, M. *Human Cloning: A Comparative Study of The Legal And Ethical Aspects of Reproductive Human Cloning*. Institute of Islamic Understanding Malaysia: Kuala Lumpur, 2001.

The Charm of Biotechnology: Human Cloning and Hindu Bioethics in Perspective

Heinz Werner Wessler

Abstract: India is proud of its fast growing pharma industry, which invests heavily in biotechnological research. The biotechnology sector is commercially successful and highly competitive. Private concerns are reaching out for the globalised market and fast developing their research capacities. Official statements relate biotechnology research not only to the discourse on worldwide competitiveness, but to development issues and the "welfare of the poor". On the domestic market, human genetic research is related to the market demands of culturally conditioned procreation technology. For the sake of male offspring, couples are prepared to invest a lot of money in fertility and gender determination technologies. A survey clearly demonstrates that cloning is already the focus of many research projects in private and public research institutions. Human cloning is officially forbidden. Human stem-cell research, however, is well established, particularly in private laboratories, restrictions notwithstanding. Commercially offered human stem-cell lines from India are recommended by the US National Institute of Health. With its human biotechnology-friendly public awareness, its strong and commercially successful biotechnology industry, and its loose state control on privately run laboratories, India may soon turn out to be "the world's cloning superpower".

Key Words: India, Hindu ethics, Human cloning, Stem-cell research, In vitro fertilization, Modernisation and tradition.

India's biotechnology appears to be blossoming, both in scientific research and in the commercial field. The research institutions of Indian universities and of the big Indian-based private pharmacological concerns are members of the globalising scientific community. They contribute to international research projects, adapt new research results to domestic conditions, and participate in the commercial transformation of biotechnological progress into market demands - at home and in the "globalising world."

Official statements relate biotechnology research not only to the discourse on worldwide competitiveness, but to development issues and the "welfare of the poor." The *Indian Ministry of Science and Technology* hosts a *Department of Biotechnology* since 1986,[1] which defines its "vision" as "Attaining new heights in biotechnology research, shaping biotechnology into a premier precision tool of the future for creation of wealth and ensuring social justice - specially for the welfare of the poor."

Indian society adapts easily to new technological solutions for its culturally conditioned demands, particularly when they are related to human procreation issues. As is well known, the imperative to beget male progeny is not only part of the mentality of a largely patriarchal society, but it is a religious duty in normative Hindu ethics. Hindu religious literature in general and Dharmashastra literature in particular quite clearly declare that the purpose of the sacrament of marriage is to produce male progeny.

India's population is largely Hindu - depending on definition between about two-thirds and 82 percent. The Muslim population is 12.12 percent, and Christians account for 2.34 percent.[2] Then there are Sikhs, Buddhists, Parsees, Jews - forming a unique and traditionally pluralistic society.

Besides the normative ethics arguments, there is the usual pragmatic rationale behind the production of male offspring, particularly the question of economic security for elderly parents in a country whose social security systems are rather limited in scope. But even in the growing Indian middle class, consisting of roughly 250 million citizens, the pressure on couples to produce male offspring is immense. Childless parents and particularly childless women are stigmatised, and childlessness is among the prominent reasons for divorce.

The market demands of childless couples for procreative technology are therefore strong. In general, people are positive about private investment in applied human biotechnology, and it is commercially attractive for clinics in India to invest in new technology. At the same time, donors of tissue, eggs, and any other raw material used in human genetic research are generally easy to find. For the poor segments of Indian society, the donation of blood, kidneys and other organs, and tissue is already an attractive additional income option. This is obviously also true of more sensitive donations. Egg donation is becoming a "career option" for Indian women, writes *The Telegraph* newspaper of Calcutta in a recent article.[3] The Indian Council of Medical Research has approved guidelines for doctors which allow them to buy eggs from women, usually for between 10,000 and 25,000 rupees (200-500 €).

1. The Charms of Cloning

During a visit by Iranian President Mohammad Khatami to the Southern Indian metropolis of Hyderabad early in 2003, the director of the Centre for Cellular and Molecular Biology based there, Lalji Singh, asked Iran to loan a pair of cheetahs or to provide some cells to clone an animal that had been extinct in India for about half a century.

The cheetah is a leopard-like member of the cat family and is the fastest animal on land, running at speeds of about 140 km/h. It disappeared from India following large-scale hunting during British rule that ended in 1947 but is still found in parts of Iran. According to a Reuters' report, Singh said the institute was setting up a large laboratory to revive endangered species such as the cheetah in the framework of a 110 million rupee ($ 2.3 million) project.

It is of course much easier and cheaper to import and breed cheetahs from Iran than to clone them. But the message of this statement by the director of a prominent biotechnological institute conveys another meaning. I suspect it was launched simply to catch attention. PR is important in the prospering market of genetic engineering in India.

None of the newspapers which reported the cheetah issue gave any information on the epigenetic character of cloning, which means that cloning is unlikely to be the basis of the creation of a reproductive population because of the successful cloning of few individuals. It was taken for granted that the cloning of cheetahs means the revival of the cheetah in India, and it was obviously enacted for and perceived in a research-friendly and biotechnology-friendly public sphere demanding news from the cloning frontline. The secular, English-speaking and largely Western-orientated elites in India are mostly optimistic about scientific progress. Applied modern technology, such as e.g., civil and the military nuclear technology, are widely perceived as proof of India's capacity for progress. Progress and technical modernisation are largely seen as something positive per se in a developing country. This partly explains, by the way, the astonishingly positive echo to the experimental atomic bomb tests in 1999, the military rocket programmes, and the national consensus on the ongoing nuclear armament of the Indian army.

2. Pharmaceutical Concerns

In general, India thinks positive about new technology. The economic, social and political environment in India for research and application of new technology in general and of biotechnology in particular is encouraging. Economically, the pharmaceutical industry is "the sunrise sector" of India's globalising industries. "With branded drugs worth 55 billion US dollars going off patent in the US in the next five years, Indian pharma exports are set to grow five-fold by 2010."[4] Doctors all over the world prescribe medicine produced by the Indian companies Ranbaxy, Dr. Reddy's, Orchid and Wockhardt.

The blossoming pharmaceutical industry is perhaps the most impressive proof of India's growing potential and increasing competitiveness as a key player in a global market. The export success story, which has contributed to the continuing economic growth of the Indian economy since the 1980s, is not merely based on generic products, but also on patented original products. Remember: Thirty years ago, medical doctors were exported in large numbers to secure the National Health Service in Great Britain. Nowadays, Non-Resident Indian (NRI) doctors prescribe Indian pharmaceutical products in Great Britain, Canada and the US - three of the economically most important countries, in which a high percentage of practising medical doctors stem from the Subcontinent.

Since the late 1980s, India has progressed far beyond its earlier self-reliance strategy (*svadeshi*) in its industrial production. After decades of industrial self reliance, the economic liberalisation policy of the last 15 years has demonstrated that Indian industry can be competitive in the world market. Even now, however, anything that is produced in India is perceived as Indian (i.e. *svadeshi* inheriting its former anti-colonial connotation) on the old folio of the self-reliance doctrine, which in turn conforms to Gandhian ideas and overtones in favour of manual labour and a preference for home-industry production to overcome the in-built "estrangement" inherent in the modern production process.

I do not wish to overstate this argument of a hidden correspondence of the spirit of self-sustainability, including of course an anti-consumerist notion of the ideal of simple living, and the booming high technology in India. But the spirit of *svadeshi* is nevertheless there, encouraging Indian creativity and the country's enormous efforts to create and secure the continuing economic growth. At one time, *svadeshi* meant limiting imports. Today, it has come to mean expansion of exports and competitive production for the global market, including a highly efficient elite education as a prerequisite for competitiveness. There is something of a correspondence between the *svadeshi* ideal and the globalisation of Indian industry.

The crucial question in this context is how far Indian concerns are perceived correctly, how far their "Indianness" in fact goes. They may be partly owned by foreign international conglomerates, and it is also open to question how far licence production based on international patents is "Indian." It is also open to question how far biotechnological research is strategically delegated by internationally operating concerns to private Indian institutions funded from anywhere in the world - not only because excellent research facilities including highly specialised personal, may be available at a lower cost and therefore economically competitive in the Subcontinent, but also because the political and legal environment appears to be easier to handle with regard to the highly sensitive issues in human biotechnology.

3. The World's Cloning Superpower?

The Department of Biotechnology in the Indian Ministry of Science and Technology presents (12/2003) 38 projects on its web page relating to cloning in the fields of animal, medical and plant biotechnology, among them projects with somehow ridiculous titles like "establishment of embryo cloning by somatic cell nuclear transfer: application in generation of transgenic animals," "establishment of cloning technology in farm animals," "cloning and characterization of the estrogen receptor gene expressed in the mouse brain cortex," "biotechnological investigations for cloning of mango." The Ministry follows the moratorium on the use of federal funds for human cloning recommended by the *National Bioethics Advisory Commission.*

The most important tool for bioethical control are the "Ethical guidelines for biomedical research on human subjects" issued in 2000 by the *Central Ethics Committee on Human Research (CECHR)* of the *Indian Council of Medical Research.*[5] The guidelines are firm on human cloning: "...research on cloning with intent to produce an identical human being, as of today, is prohibited."[6] These are, however, guidelines, not (yet) law.

The guidelines clearly allow human in vitro fertilisation (IVF) for research as well as for clinical purposes. The Department of Biotechnology proudly presents its research with animal embryonic stem cells, but it does not support human embryonic stem cell research, at least not officially, even though the guidelines do not restrict embryonic research within the first 14 days, i.e. until the primitive streak appears.[7] Beyond that, the moratorium is binding for state institutions only. This is probably the most important reason why human stem cell research appears to be located somewhere else in India - in private research laboratories.

India's private pharmaceutical industry supports not only research with animal stem cells and their production, but India is also one of the key players in the human embryonic cell market. Following news reports on cloning in India, we might even ask whether India - and not China[8] - might turn out to be the world's cloning superpower in the near future.

Two Indian research institutions are included in a list of 10 identified by the U.S. National Institutes of Health in August 2001 as having at least one of the 64 possible human embryonic stem cell lines that American researchers can use for research under US-American federal grants.[9] They offer a total ten cell lines to the scientific community, even though they are registered as unavailable for the time being on the internet (see table 1).

In interviews, the founder and director of Reliance Life Sciences, Firuza Parikh, estimated that research would continue in India, even if the funding from the US came to a halt for ethical or other reasons. In an email to the Washington Post,[10] Parikh said "We are for embryonic stem cell research. Religious, cultural and political circumstances here are not in conflict with our work."

The laboratory is owned by the Reliance Group.[11] Founded by Dhirubhai H. Ambani (1932-2002), the Reliance Group is India's largest and most respected conglomerate with total revenues of Indian Rupees 800 billion (US$ 16.8 billion), net profit of over Rupees 98 billion (US$ 2.1 billion), net profit of over Rupees 47 billion (US$ 990 million), and exports of Rupees 119 billion (US$ 2.5 billion). This is seriously big business! The group's activities span exploration and production (E&P) of oil and gas, refining and marketing, petrochemicals (polyester, polymers, and intermediates), textiles, financial services and insurance, power, telecom and infocom initiatives. Beyond that, Reliance is the 'Most Admired Business House' for the third consecutive year in the 'Business Barons - TNS Mode Opinion Poll' for 2003. The results of this exclusive opinion poll appeared in the latest issue of *Business Barons*.

National Centre for Biological Sciences/Tata Institute of Fundamental Research	
Cell Lines	**Contact Information**
NIH Code: **NC01*, NC02*, NC03***	Dr. Mitradas M. Panicker University of Agricultural Sciences

Provider's Code: **FCNCBS1*, FCNCBS2*, FCNCBS3*** * These lines are not yet available for shipping. **Information on Cell Characteristics**	GKVK Campus GKVK P.O., Bellary Road Bangalore, 560065 India Phone: 91-80-363-6420 Fax: 91-80-363-6662 E-mail: panic@ncbs.res.in

Reliance Life Sciences	
Cell Lines	**Contact Information**
NIH Code: **RL05*, RL07*, RL10*, RL13*, RL15*, RL20*, RL21*** Provider's Code: **RLS ES 05*, RLS ES 07*, RLS ES 10*, RLS ES 13*, RLS ES 15*, RLS ES 20*, RLS ES 21*** *** These lines are not yet available for shipping. Cell characteristics not available from provider.**	Dr. K.V. Submaraniam Maker Chambers IV Naariman Point Mumbai 4000021 India E-mail: K_v_subramaniam@ril.com Web site: http://www.ril.com

Table 1: Online-advertisment of two indian research institutes, which, according to the U.S. National Institutes of Health, possesses human embryonic stem cell lines that American researchers can use for research.

4. In Search of Arguments From Tradition

Prima facie, the discourse language on cloning and embryonic research in general in India is not religious language. It is the language of science and technology, of progress and globalization. Modern India is a secular state, and the primary language of secularism is English. The preamble to its constitution does not refer to God, but to the secular and socialist character of the federal state: "We, the people of India, having solemnly resolved to constitute India into a sovereign socialist secular democratic republic…" The normative "socialist secular" characterisation of the republic was introduced by the *Constitution (Forty-second Amendment) Act, 1976* under Indira Gandhi's extraordinary authoritarian emergency government in 1976.

Generally speaking, the language of research is almost exclusively English, and the public debate is much dominated by this implanted language. At the same time, only a very small section of Indian society is able to use English creatively. English in India in general, and the language of bioethical discourse in particular, is to a large extent rather an antiseptic language. Even though the subcontinent inherited English from the Raj (as it had inherited Persian from the Moghuls before), the language of education of a small percentage of the Indian population comprises an idiom of estrangement, of an elite, somehow cutting itself off from the rest of society.

English in India is more than simply a medium of communication: It symbolises class, modernisation, westernisation, globalisation, honesty, gentlemanliness, science, anti-traditionalism etc. Disguised in the English enlightened "new speak," deeper dimensions and particularly cultural conditionings become visible when words and phrases are put into their proper context in Indian society.

The announcement that two private Indian laboratories offer human embryonic cell lines commercially for research purposes met with a mixed response from religious authorities in India. The most pronounced statement came from one of the five Shankaracharyas, Jayendra Saraswat, one of the highest authorities on Hindu issues. He is strongly opposed to the use of human embryonic cells for any purpose. He said, according to news reports, "this is clearly against the traditional Hindu perception of abortion, artificial insemination, test tube babies and human cloning."

As Julius Lipner's analysis of the moral status of the unborn in classical Hinduism shows, the human being or perhaps the personhood - if this term is applicable in the Hindu context - starts with conception according to the main sources of *Dharmashastra* (Hindu law). From an orthodox Hindu view based on the traditional codes, abortion at any stage of pregnancy starting from conception is, therefore, a case of violence against a being, a case of murder (*brunahatya*). The classical tradition does not believe in the ensoulment of the embryo or foetus at some stage of pregnancy. Starting from conception, according to *Dharmashastra*, the embryo is to be defined as an ensouled living human being and should therefore be treated under the paradigm of *ahimsa* (non-violence).

Hindu ethics are very much based on the *ahimsa* principle. Mahatma Gandhi therefore argued from within the tradition when he said: "It seems to me clear as daylight that abortion would be a crime."

In general, according to the common brahmanic tradition, the relationship between body and soul is loose. The soul is incarnated over and over again according to a person's moral performance in his or her former existence, in accordance with the law of *karma*. Good deeds lead to a good reincarnation, immoral deeds lead to a decline at least during the next existence. The status of the "person" is always conditioned by and related to the chain of former existences. The residues of the past existence (*vasana*) are always there, unless the soul attains the state of *moksa* (liberation) through ascetic practices.

We could scan Hindu myths in search of non-sexual forms of procreation for anything relating to cloning: mythical stories of gods and sages being created out of male semen without the female egg, from a diversity of body parts, from plants etc. These issues, which can be identified and related to cloning issues, deserve closer study. Altogether, it seems that arguments drawn from the mythic tradition may be used to point out a permissive attitude towards cloning as such, but tradition is strict on the moral status of the unborn and can hardly be anything but prohibitive on embryonic research.

The production of a human embryo outside the body and the production of a human clone may, however, not necessarily be a taboo as such. In short, the conclusion could, for example, be that it is natural to have parents, but non-natural reproductive techniques are not necessarily strictly forbidden. In the Hindu perspective, the focus is on the survival of the embryo. My impression is that the crucial issue for brahmanic orthodoxy is the calculated "death" of embryos as a side effect of the successful transference of one living embryo into the female womb. From the perspective of religious law, the calculated "death" of the embryo must be considered similar to abortion. The traditional Hindu legal perspective cannot be but restrictive on human reproductive cloning issues.

5. Modernising Tradition

I would like to look more closely at the interpretation of the idea of man's spiritual perfection in the philosophy of Sri Aurobindo (1872-1950), since this idea can be related to the cloning issue. Neo-Hindu reformism, starting from the early 19th century, is usually interpreted as a result of the interaction between tradition and Western modernity. Aurobindo somehow fits into this pattern. He was a political revolutionary in early adulthood, later became a teacher of his so-called integral yoga and an outstanding representative of modern Hindu philosophy. His innovative conception of the spiritual evolution of mankind includes the idea of so-called *supermental manifestations upon earth*, and the idea of a physical transformation of man following his spiritual transformation. "A divine life in a divine body is the formula of the ideal that we envisage…The process of the evolution upon earth has been slow and tardy - what principle must intervene if there is to be a transformation, a progressive or sudden change?"

This statement on "sudden change" is a remarkable statement, and, from a 21st century perspective, a somehow truly prophetic uttering. The idea of "new man," so crucial to the project of Western modernity, is translated into an Indian context and is given a highly individual, and somehow terrifying, interpretation. Of course, cloning was never an issue for Aurobindo. Nor is abortion to be found in the index of his collected works.

In recent years, particularly after the spectacular news of the birth of the cloned sheep Dolly on July 5, 1996, some Hindu authorities have uttered clearly positive statements about human cloning in general. Crawford quotes Mata Amritanandamayi,[14] the so-called president of Hinduism, at the second parliament of religions in Chicago 1993: "The idea of cloning, though implemented only recently through modern science, was in the minds of the ancient saints and sages of India." Others, like Tiruchi Mahaswamigal, a prominent living saint, founder of the monastical *Kailasa Ashram* (Bangalore) and "Hindu of the year 2003" - awarded by the conservative "Hinduism today" magazine -, rejects cloning for pragmatic reasons: "It is not new to our cultural history… We did not call it cloning, but there were other methods of procreation. At present we do not require any such alternative methods of procreation." Paramhans Swami Maheswarananda, the founder of a worldwide "yoga in daily life"-movement,[15] who passed away in October 2003, however, opposed cloning for ethical and moral reasons, because it is "against nature." Others, like Swami Omkarananda Saraswati (Omkarananda Ashram) and Chidanand Saraswati (Parmath Niketan), reject the copying of human beings explicitely. Human individuality has to be preserved.

Religious authorities in present-day Hinduism have differing opinions on cloning. Some are permissive, others prohibitive.[16] Beyond that, cloning is not or is not yet a key issue in modern Hindu thinking. This may change quickly, once the "human clone" catches public attention, perhaps in near future.

6. Research in Its Environment

Of course, the languages of social reality and of religious commandments in general and orthodox Hindu legal statements on procreation issues in particular are different. As many as 35 million girls have been killed before, during or after birth in India over the last 100 years, according to the Indian census commissioner, J.K. Banthia. Modern Indian law allows abortion under certain indications. The misuse of sex determination for the purpose of pre-birth elimination of females (PBEF) is forbidden. Forbidden or not - it is common practice and easily available for cash, at least in urban areas. The issue is, however, one of the great taboos in Indian society. Even though everyone knows about it, the topic is mostly avoided both in official and private communication.

The impact of sex determination and PBEF on the population of India is easily demonstrated. India's population grew by 21.34 percent between 1991 and 2001. Although it has gone well above the one-billion mark, the gap between girls and boys is widening in an alarming way, particularly in India's wealthier states and districts. A recent report, Missing: mapping the adverse child sex ratio in India, issued by the UN Population Fund,[17] shows that the sex ratio, which is calculated as the number of girls per 1000 boys in the 0-6 age bracket, declined from 945 in the 1991 census to 927 in the 2001 census. The normal ratio is about 950 to 1000.

The steepest declines have taken place in the prosperous Hindu-dominated northern states of Punjab, Haryana, Himachal and Gujarat, where there are now less than 800 girls per 1000 boys for the first time. The lowest ratio, 754, was in Fatehgarh, in Punjab. The less developed Hindu-dominated states like UP, Bihar, and Orissa also display uneven gender balances, but the shortage of females is not as dramatic as in the economically better off states. Interestingly, the ten districts with healthy sex ratios are largely in non-Hindu areas like Jammu and Kashmir and in the north eastern tribal regions.

Observers agree that the decline is largely due not only to the neglect of girls and infanticide, but to a large extent to pre-birth elimination of females (PBEF). According to the *Campaign Against Female Foeticide*,[18] 90% of the estimated 3.5 million abortions in India each year are of girls. Once parents decide to have the test, either at the suggestion of a doctor or for personal reasons, and the foetus is found not to be of the desired male sex, abortion follows almost automatically. According to a recent study, "there is universal awareness about [prenatal] sex determination."[19]

At the heart of the decline are numerous doctors offering ultrasound scans to check the sex of the pregnant woman's baby. Since 2001, the government has reportedly confiscated 16,000 illegal ultrasound machines.[20] Yet many private clinics still offer sex determination more or less openly in connection with selective abortion.

Sex determination has been banned since 1994, but the booming of ultrasound checks, which can be observed since the late 1980s, amniocentesis and the blossoming *in vitro fertilization* (IVF) market remains largely unchecked in India, largely due to the desire and the pressure on Indian couples to produce male offspring. This is in some way to be considered as a kind of rational choice, since dowry demands have become more and more extraordinary, not so much - as might be expected from an outside perspective - in the more traditional milieu, but predominantly in the context of the modern urban middle class. As everybody knows, dowry demands can easily turn into a nightmare for the brides parents. The option to avoid future problems of this kind is naturally attractive. In the recent study mentioned before, "71% of the respondents went for female foeticide because they already had daughters and now they wanted a son."[21]

With the advent of IVF, it is no longer necessary to resort to abortion to ensure the sex of a child. Undesirable embryos are simply disposed of and the desired male embryo is implanted. The pressure on parents to produce male offspring could easily be transformed into a commercial demand for reproductive human cloning technology.

Students at India's leading medical school, the *All India Institute of Medical Sciences*, recently refused to cooperate in an awareness drive. According to media reports,

> "They are not taught enough about medical ethics," quoting Dr M.K. Bhan, a paediatrician at the Institute. "There is a large vacuum in the medical curriculum. The students are young. They are under a lot of pressure. In liberal arts, you are taught about ethics. In medical science, you are not."[22]

The statistical data clearly indicate that the problem of PBEF persists and that the law is unable to eliminate it, despite *The Pre-Natal Diagnostic Techniques (Regulation and Prevention of Misuse) Act, 1994* and its *Amendment Act, 2002*:

> "An Act to provide for the prohibition of sex selection, before or after conception, and for regulation of pre-natal diagnostic techniques for the purposes of detecting genetic abnormalities or metabolic disorders or chromosomal abnormalities or certain congenital malformations or sex-linked disorders and for the prevention of their misuse for sex determination leading to female foeticide and for matters connected therewith or incidental thereto."

The gap between the law and social reality is wide. Therefore, the question has to be raised whether India is ready to control or contain human cloning, if it at some time becomes technically possible.

During a visit to Mumbai in February 2003, the Indian President Abdul Kalam - himself a noted nuclear scientist and one of the fathers of the nuclear bomb in India - publicly opposed the idea of human cloning. During a session organised by the *Sri Shanmukhananda Fine Arts and Sangeet Sabha* and the *South Indian Education Society* he said, according to an article in *The Tribune*,[23] "By 2009 a personal computer that will cost Indian Rupees 25,000 will be able to perform one trillion calculations per second. By 2019 the computing ability will be that of a human brain and by 2029 it will be 1,000 times faster than the brain. But we cannot replace the human brain. … I believe that human cloning should not be done."

This is the kind of diplomatic statement one would expect from a state president aware of UNESCO's *Universal Declaration on the Human Genome and Human Rights*, which clearly forbids reproductive human cloning in article 7. The following randomly chosen prominent opinion from within the Indian research community, contrasts with the presidential statement and illustrates the other end of the scale. K.K. Verma, retired Professor of Zoology, writes in his article in the *Eubios Journal of Asian and International Bioethics*,[24] in a note on another article arguing against reproductive cloning, contributed by Prof. Tade Matthias Spranger of Bonn University,

> Almost all nation states have banned human cloning. As Spranger has rightly pointed out, the ban is mostly guided by ethical rather than legal considerations. … But in some cases practising human cloning would not be contrary to human rights and dignity. To bring home this point Spranger cites the case of a childless couple. In such a case to deny cloning would be violating individual rights; hence cloning in this situation should be permitted.

The inevitable conclusion is that, as with other forms of procreative techniques, human reproductive cloning has a good chance of realisation in India. Opposition on ethical grounds is not univocal, and it is weak. The Indian scientific community is divided on bioethical principles. Opinions in favour not only of therapeutic, but also of reproductive, human cloning are pronounced and are stated publicly.

The quoted personal opinion of the Indian President may be representative to some extent, and it may have some effect on public opinion and legislation. It is, however, only one of many opinions prevailing in public discourse in India. It is open to question whether the political will is strong enough to effectively control human cloning or to develop mechanisms to restrict it to therapeutic applications. There is a distinctly research-friendly atmosphere in relation to biotechnology in general and to cloning issues in particular.

My conclusion is that efforts to control and to restrict human cloning will remain as loose as they are now. Therapeutic cloning, being more acceptable than reproductive cloning, may very well serve as a door opener: Once that door is open, it will not be easily closed. The potential is there. There is an atmosphere of permissiveness in bioethics.

Let us just consider the example of the nuclear bomb issue in South Asia: A first experimental bomb was detonated in 1974, and for 25 years Indian governments argued in favour of total nuclear disarmament worldwide, continually assuring its international partners that India's nuclear competence was being used for civil purposes only. In 1999, a series of five experimental explosions in India and six in Pakistan inaugurated the epoch of the nuclear arms race in South Asia. Is it possible that one day, all of a sudden, some form of a living human clone may appear in India to take the world by surprise?

Secular India as well as modern Hinduism, at least in part, takes a permissive view on human cloning, including reproductive cloning. The research environment is well developed. The technological imperative "can implies ought" is strongly felt in a country that is optimistic about science and technology. The private pharmaceutical industry is commercially strong, and is funding excellent research. The country has well-equipped laboratories, and politics and public opinion are proud of them.

At the same time, neither politicians, nor public opinion, nor the scientific community, nor religious institutions actively seek to put the human cloning issue on the political priority agenda. The market demand for clinical reproductive cloning is there. Sri Aurobindo's vision of the new man and his divine body may well become real in a hitherto unforeseen way.

Notes

1. http://dbtindia.nic.in
2. Census 1991 data, see http://www.censusindia.net/ for more details.
3. *The Telegraph*, 17.8.2003.

4. *India Today International*, December 1, 2003, 21.
5. http://icmr.nic.in/ethical.pdf.
6. Ethical guidelines for biomedical research on human subjects, Chapter VII, issued in 2000 by the *Central Ethics Committee on Human Research.*
7. Ethical guidelines for biomedical research on human subjects, Chapter VII, issued in 2000 by the *Central Ethics Committee on Human Research.*
8. Charles J. Mann in http://www.wired.com/wired/archive/11.01/ cloning_pr.html
9. http://stemcells.nih.gov/registry/unavailable.asp
10. *Washington Post*, 27.8.2001.
11. http://www.ril.com
12. *Navajivan*, 9.12.1928.
13. *The Divine Body*, Complete Writings, vol. 16, 20.
14. Crawford, 2003, 161ff.
15. http://www.yoga-in-daily-life.org
16. Crawford, 2003, 163.
17. http://www.unfpa.org.in/publications/16_Map%20brochure_English.pdf
18. compare http://www.indiafemalefoeticide.org
19. Agarwal, 2003, 73.
20. compare http://www.guardian.co.uk/india/story/0,12559,1037144,00.html
21. Agarwal, 2003, 75.
22. *Calcutta Telegraph*, Oct 24; *The Hindu*, Oct 21; BMJ.com, Nov 1; *New York Times*, Oct 26.
23. *The Tribune*, 16.2.2003.
24. *Eubios Journal of Asian and International Bioethics* 12 (2002), 30.

References

Agarwal, Anurag. *Female foeticide: myth and reality*. New Delhi, 2003.

Coward, Harold G., Lipner, Julius J., Young, Katherine K. *Hindu ethics: purity, abortion, and euthanasia.* Delhi: Sri Satguru Publications, 1991 (first New York 1989).

Crawford, S. Cromwell. *Hindu bioethics for the twenty-first century.* Albany: State University of New York Press, 2003.

Das, Rahul Peter. *The Origin of the life of a human being: conception and the female according to ancient Indian medical and sexological literature*. (Indian medical tradition; vol. VI). Delhi, 2003.

Ghose, T. K./Ghosh, P. (eds.). *Biotechnology in India*, T. K. Ghose (ed.), 2003.

Kane, Pandurang Vaman. *History of Dharmasastra: (ancient and mediaeval, religious and civil law)*. Poona : Bhandarkar Oriental Research Institute, 1975 (5 vols. in 7 bindings).

Lohray, Braj B. "Medical biotechnology in India." In *Biotechnology in India*, Vol. 2, (Advances in biochemical engineering, biotechnology, 85), Ghose, T.K. and P. Gosh (eds.). Berlin: Springer, 2003, 215-281.

The Cultural and the Religious in Islamic Biomedicine: The Case of Human Cloning

Abdulaziz Sachedina

Abstract: The present study investigates Muslim opinions about human cloning in the context of Muslim culture that provides the moral presuppositions and formal normative framework for moral choices and justifications. Every ethical system is ultimately a synthesis of intuitive and rational assertions, the proportion of each varying from culture to culture. There is also in every culture an admixture of the ethnocentric and the universal, of that which is indissolubly bound to a particular geography, history, language, and ethic strain, and that which is common to all humans as humans. In the case of human cloning, moral judgements reveal an amalgam of relative cultural elements derived from the particular experience of Muslims living in a specific place and time, verified by the timeless universal norms derived from the scriptural sources that possess common elements applicable to all humans as humans.

Key Words: Islam, Cultural relativism, Istinsakh (making copy), Public good, Natural-artificial, No harm no harrassment, Technically assisted reproduction

1. Introduction

Islam, as a comprehensive religious-moral system, does not divide the public space in terms of spiritual and secular domains with separate jurisdictions. Rather it strives to integrate the two realms to provide total guidance about the way human beings ought to live with one another and with themselves as citizens, or professionals, or workers of one kind or another, or simply as human beings. Various human institutions, cultural and religious practices, and political systems have one major goal, namely, to serve God, the Merciful, the Compassionate. Muslim ethics tries to make sense of human moral instincts, institutions, and traditions in order to provide a plausible perspective on the making of moral judgments, the fashioning of rules and principles, and devising of a virtuous life. Its judgments are ethical in the sense that they deal with the sense of what reasonable people count as good and bad, praiseworthy and blameworthy, in human relationships and human institutions. Ultimate questions connected with human suffering through illness and other afflictions, reproduction and abortion, death and dying, are within the purview of its religiously based ethics. Human beings are essentially God's creatures, and, hence, their total welfare related to this and the next world is God's concern.

How do Muslims solve their ethical problems in biomedicine? Are there any distinctive theories or principles in Islamic ethics that Muslims apply in deriving moral judgments in bioethics? Is the revealed Law, the Shari'a, as an integral part of Islamic ethics, the only recognized source of prescriptive precedents in Islam? Can it serve as a paradigm for the moral experience of contemporary Muslims living under different circumstances? What is the role of human experience/intuitive reasoning, culture in moral justification?

I begin with these questions in the hope of verifying the validity of a poignant observation and its applicability in the Islamic case made by Edmund D. Pellegrino that:

> Culture and ethics are inextricably bound to each other. Culture provides the moral presuppositions, and ethics the formal normative framework, for our moral choices. Every ethical system, therefore, is ultimately a synthesis of intuitive and rational assertions, the proportion of each varying from culture to culture. There is also in every culture an admixture of the ethnocentric and the universal, of that which is indissolubly bound to a particular geography, history, language, and ethic strain, and that which is common to all humans as humans.[1]

In this paper I will delve into the validity of the above observation in the Islamic case. I will examine the nature of Islamic ethical discourse in order to demonstrate that ethical judgments in Islam are an amalgam of relative cultural elements derived from the particular experience of Muslims living in a specific place and time, verified by the timeless universal norms derived from the scriptural sources that possess common elements applicable to all humans as humans.[2] In the recent years attempts have been made to engage Muslim scholars in the ongoing debate in the area of biomedical ethics in the West. At the center of this debate is the role of ethical principles and rules in the moral assessment of an action from many different perspectives including those of the agent, the act itself, the end, and the consequences. There are different viewpoints about the meaning and nature of moral principles and rules as they are applied to different types of moral dilemmas. In fact, many disputes in biomedical ethics in the West stem from disputes about the generalizability and applicability of more than one of these normative principles that function as action-guides, categorizing actions as morally required, prohibited, or permitted. Moreover, there exists the great variety of principles-oriented approaches that take into consideration general principles as sources of rules, and rules that specify type of prohibited, required, or permitted actions before any particular judgments can be derived in bioethical cases.[3]

2. The Nature of Islamic Juridical-Ethical Discourse

For every ethical situation Islamic juridical tradition seeks to address and accommodate the demands of justice and public good. Without adequate training in the legal sciences, especially legal theory, one cannot pinpoint the principles and the rules that Muslim jurists utilize to justify and assess moral-legal decisions within their own cultural environment. Legal doctrines and rules in addition to analogical reasoning based on paradigm cases enable a Muslim jurist to resolve ethical dilemmas that face the community in dealing with immediate questions about autopsy, organ donation, dignity of the dead, and so on. The practical judgments or legal opinions reflect the insights of a jurist who has been able to connect cases to an appropriate set of linguistic and rational principles and rules that actually provide keys to a valid conclusion of a case under consideration.

Undoubtedly, the enunciation of underlying ethical principles and rules that govern practical ethical decisions is crucial for making any religious perspective an intellectually insightful voice in the ongoing debate about a morally defensible ethics of biomedicine cross-culturally. Given all cultures share certain moral principles like beneficence or nonmaleficence, all require rules like truthfulness and confidentiality as essential elements in regulating responsible physician-patient relationship, yet there are major issues that generate controversy on a global scale. What kind of ethical-cultural resources do different faith communities possess to provide internationally collaborative efforts in creating a cross-cultural ethical discourse to resolve issues in biotechnology?[4]

In Muslim community there is a raging debate about the normative status of inherited classical juridical tradition. Some scholars maintain their immutability and sacredness and, hence, when it comes to critical assessment of the normative resources by means of which major ethical problems created by the introduction of modern technology in the field of health care can be resolved, they refuse to entertain changed circumstances to extrapolate fresh decisions. This uncritical approach to the normative sources has deep roots in the theology of revelation in Islam. Briefly stated, there are two major trends about the meaning and relevance of the revelation for Muslims. According to the one, Islamic revelation, in its present form, was 'created' in time and space and, as such, it reflects historical circumstances of that original divine command. According to the other, the revelation was 'uncreated' and hence, its current form is not conditioned by place and time. Most Sunni Muslims reject any hints that the revelation's interpretation is a cultural or historical variable. In the wake of both quantitative and qualitative change in the modern Muslim world, the question arises as to how far traditional readings of the revelation are relevant to their present situation? It is this critical theological question with drastic ramifications for the overall status of normative tradition that is usually resisted by the traditional centers of Islamic jurisprudence.

3. The Question of Cultural Relativism in Ethical Values

Culture and ethics are, as observed by Pellegrino, inextricably intertwined. Human experience is the main source of moral reflection and ethical decision-making. Moreover, in order for any ethical value to attain acceptability, it seeks cultural legitimacy in prevailing economic and political circumstances. Accordingly, these values even they attain universal application can hardly be expected to be free from cultural relativity. Since human reason depends on the data of experience to make correct ethical judgments, moral presuppositions operative in society interact with the specific experience to provide culturally conditioned moral justification.[5] In fact most objectivist ethical theories, which affirm that value has a real existence in particular things or acts, regardless of the opinion of any judge or observer, include a certain aspect of social or conventional relativism. In the highly politicized debates about the applicability of the International Bill of Human Rights, cultural relativism figures prominently in the arguments made by the non-Western nations against the charter's ethnocentric language that defies its claim to absolute application.[6] Similar arguments against a single universally acceptable bioethical theory in the inherently pluralistic ethical discourse are commonly heard in national and international biomedical ethics conferences.

However, there is a need to be cautious in overemphasizing cultural relativism in the area of biomedical ethics. This is indispensable because today biomedical technology has become universal by its being adopted globally by health care providers. There is a growing consensus in the international community to adopt a more or less consensually derived framework of ethical principles and rules that could engage theologians, scholars and policy makers in the health profession in a dialogical mode to search for solutions to the ethical problems in biomedical technology and research cross-culturally.

Although the historical and cultural context of most of the present scholarship in bioethics is Western, the moral reflections and justifications covered in these studies suggest their applicability across cultures, provided particular experiences and cultural expectations of the non-Western societies are meticulously accounted for in testing the coherence between the principles and the moral judgments. More pertinently, by understanding common human conditions cross-culturally, it is possible to speak about universal morality for ethical assessment and judgment in other societies. The need for substantive cultural consensus founded upon common concerns to address some of the issues like human cloning that confront humanity today, require us to look into biomedical ethics that can speak to the world community across socio-political and cultural barriers to communicate its fears and hopes.

In this paper, then, my purpose is not merely to search for Islamic responses to the possibility of human cloning in the context of the success of animal cloning. Rather, I intend to make a strong case for distinctly Islamic, and yet cross-culturally communicable, deontological-teleological ethics that is operative within the Muslim social-cultural context in assessing moral problems of cloning.[7]

Moral assessment of cloning technology has already begun in Egypt and Iran, where religious scholars, medical professionals and the government are searching for ontological foundations of Islamic religious law to enable them to make authentic choices of what is morally and legally justifiable conduct in biomedical research and practice and its application in the Muslim society. I mention Egypt and Iran only because these are the only Muslim countries where religious scholars are engaged in formulating national policies related to health care.

The important thing that deserves serious consideration in Islamic context is that even when the source of normative life is regarded as having been revealed by God in the Shari'a, the procuring of a judgment and its application is dependent upon reasons used in moral deliberation. This moral deliberation takes into account particular human conditions in their cultural context, which affect the way Muslims justify an action to be moral. In other words, Islamic law developed its rulings within the pluralistic cultural and historical experience of Muslims and non-Muslims living in the different parts of the Islamic world. It recognized the autonomy of other moral systems within its sphere of influence, without imposing its judgments on peoples whose cultural beliefs and practices were at variance with its own. More importantly, it recognized the validity of differing interpretations of the same revealed system within the community, thereby giving rise to different schools of legal thought and practice in Islam. In the absence of an organized 'church,' or a theological body that speaks for the entire tradition or the community, as a source for the normative and paradigmatic religious system, Islam was and remains inherently discursive and pluralistic in its methods of deliberation and justification of moral actions. Hence, on the basis of particular application of principles and rules to emerging ethical issues, like human cloning, it is possible to observe differing judicial opinions toward which speculation over the interpretation of revelatory sources that preserve paradigm cases and the principles that were applied to discover them has led.

4. The Paradigm Case of Human Cloning

Since the cloning of Dolly, the sheep, in 1997 a number of Muslim scholars have, on the one hand, deliberated on the ethics of human cloning and, on the other, on the relationship between religion and science, and religion and culture. The urgency and even crisis situation created by the cloning debate in the Muslim world has led to an unprecedented interfaith cooperation in formulating a proper response to the possibility of human cloning and the adverse ways in which this scientific advancement will affect human society in general, and human relationships in particular. Majority of the Muslim legists' ethical-legal rulings studied for this paper show that these concerns are centered on the cloned person's hereditary relationship to the owner of the cell and the egg, and the relational ramifications of that to the other individuals in the child's immediate families. It is not difficult to see that religious-ethical questions are spurred by cultural sensitivities regarding an individual's identity within familial and extended social relationships. In addition, there are questions about the ways human cloning will affect the culture of intense concern with a person's religious and social distinctiveness. It is precisely at this juncture that cross-cultural communication between Muslim and Christian scholars assumes a front position in highlighting cultural variations among their communities. Whereas, individuality of a cloned human being is central to much discussion in the Western-Christians cultures, it is the concern with a child's lineage, familial and social relationships that dominates the Muslim cultural sensitivities.

One of the most important studies dealing with the subject in Arabic is: *al-Istinsakh bayna al-islam wa al-masihiyya* (Cloning in Islam and Christianity). The study aims to demonstrate plurality as well as mutuality among the cultures of the people in the Middle East. Leading Christian, Sunni and Shiite scholars, representing shared cultural concerns while holding pluralistic opinions in their respective traditions, have contributed to the debate on the way in which human cloning will adversely impact upon the future of the institution of marriage and parent-child relations. The interfaith discourse is based on a common concern in these communities, namely, concern with the negative impact of human cloning on the culture of human interaction.

To be sure, the guiding principle regarding any scientific advancement in Islam is the cautious note in the sacred law of Islam, the Shari'a. The precautionary note in the Shari'a takes into consideration the norm that there is seldom a thing of benefit without some inherent disadvantage affecting people's religion, life, lineage, reason and property.[8] Islam's concern to combine noble ends with noble means rules out the idea of good end justifying a corrupt means. Taking the specific case of human cloning, the most important rule is avoidance of anything that might adversely affect human nature and human relationships. Islam forbids any tampering with human nature in any way other than legitimate methods of correction. Anything that is done for prevention or as treatment is legitimate. Ethical judgment on medical procedure is made on the basis of predominance of benefit (*istislah*) that requires rejection of probable harm (*daf' al-darar al-muhtamal*).

5. Scholarly Opinions in their Cultural-Religious Context

Following the euphoria over the latest success in animal cloning in 1997 prominent Muslim scholars representing both Sunni and Shiite centers of religious learning in the Middle East, mainly Cairo, Beirut and Qumm, expressed their opinions on human cloning. Some of these opinions are regarded as official Sunni and Shiite positions.[9] The Arabic term used for this process in the legal as well as journalistic literature is indicative of the widespread speculation and popular perception regarding the goal of this technology, namely, *istinsakh*, meaning 'clone, copy of the original.' This interpretive meaning is not very different from the fictional cloning portrayed in *In His Image: The Cloning of Man* by David Rorvik in the 1970s when cloning by nuclear transplantation was the topic of the day in North America. It is also because of the popular misperception about human "copies that can be produced at will through cloning" that the leading Mufti of Egypt, Dr. Nasr

Farid Wasil in Cairo declared human cloning as a satanic act of disbelief and corruption that would change the nature with which God created human beings, thereby impacting negatively upon social order and practice. Accordingly, his juridical decision was that the technology had to be regulated and controlled by the government to protect Muslim society from such an inevitable harm.[10]

However, this position was disputed by another leading Egyptian legist Yusuf al-Qaradawi who, when asked if cloning was interference in the creation of God or an affront to God's will, asserted in no unclear terms:

> Oh no, no one can challenge or oppose God's will. Hence, if the matter is accomplished then it is certainly under the will of God. Nothing can be created without God's will facilitating its creation. As long as humans continue to do so, it is the will of God. Actually, we do not raise the question whether it is in accord with the will of God. Our question is whether the matter is licit or not.[11]

Although in these early rulings the issue of cloning technology was not given much serious consideration in Muslim ethical-religious discussions of cellular nuclear transplantation,[12] whether involving somatic or germ-line cells, there is much concern with anticipated biological and social effects of cloning on the underlying Islamic ethical framework and social fabric as discussed by al-Qaradawi. In brief, al-Qaradawi raises a fundamental question about the impact of this technology on the human life:

> Would such a process create disorder in human life when human beings with their subjective opinions and caprices interfere in God's created nature on which He has created people and has founded their life on it? It is only then that we can assess the gravity of the situation created by the possibility of cloning a human being, that is, to copy numerous faces of a person as if they were carbon copies of each other.[13]

The fundamental ethical question based on the laws of nature, as al-Qaradawi states, centers round a consideration whether this procedure interferes with growing up in a family that is founded upon fatherhood and motherhood. It is in a family that the child is nurtured to become a person. In addition, al-Qaradawi says, since God has placed in each man and woman an instinct to procreate this individual in the family, why would there be a need of marriage if an individual could be created by cloning? Such a procedure may even lead to a male not in need of a female companion, except for carrying the embryo to full gestation. Moreover, such an imbalance in the nature will lead to the corruption of human society, "leading to the illicit relationship between man and man and woman and woman, as it has happened in some Western countries." This reference to "Western" culture needs to be understood as the central issue in our search for cross-cultural communication about Islamic values of family life that would be affected by an invasive biotechnology.

In general, Muslim religious attitude has regarded Western European culture hegemonic in its imposition over the non-Western world. Traditional scholars have resisted this dominance in all areas of the modern culture in Muslim societies. The negative evaluation has been felt even in the area of International Law, which is regarded by many as the product of European cultural consensus without regard to the multi-cultural reality of the international community. Consequently, major moral problems confronting the world today are seen as the byproduct of Western materialist and anti-religious culture with little regard for the spiritual and moral well being of the people.

Cultural dislocations have evidently gripped many Muslim societies today. As a consequence of imported modernization programs without local cultural legitimacy Muslims have suffered "cultural homelessness" in their own societies since the early part of 20th century. The emerging oppositional discourse against Western encroachment on Muslim social values has symbolically led the militant Muslims to view anything and everything coming even in the form of scientific advancement as imposition of Western values on Muslim cultures. There is a fear of the further deterioration of social and familial values that are already affected by modern secular education and pervasive "CNNization" of mass media. Initial reactions to the news about cloning that were reported in the Muslim world expressed people's deep-seated fear about further erosion of family values through cloning. To be sure, science is not viewed amorally in the Muslim world. Any human action involves cognition and volition, the two processes that determine the moral course of an action. Hence, cloning of human beings was viewed with much suspicion in the beginning, and it was only gradually that more knowledgeable analyses took place in public.

The other issue taken up by al-Qaradawi against cloning is based on the Qur'anic notion about variations and cultural diversity among peoples as a sign from God who created human beings in different forms and colors, just as He created them distinct from other animals. This plurality reflects the richness of life. However, cloning might take away this diversity. A semblance through "copying" might even lead to the errors of marital relationship where spouses will not be able to recognize their partners, leading to serious social and ethical consequences. From the point of health also, as al-Qaradawi argues, one could presume that cloned persons, sharing the same DNA, will be afflicted by the same virus. However, he maintains that it is permissible to use the technology to cure certain hereditary diseases, such as infertility, as long as it does not lead to aggression in other areas pointed out earlier.[14]

It is interesting to note that among Muslim scholars there is almost no reference to eugenics in any of the opinions studied for this paper. In contrast, drawing from the modern European history, several Arab Christian scholars mention the danger of the abuse of cloning technology if used with the same intentions that eugenics was used for racial exclusion and generational improvement through extermination.[15]

When we examine the Shiite rulings we notice that their legists have endorsed the cloning technology as being part of the possibilities that are actually created within the natural forms of conception. The leading Lebanese jurist, Sayyid Muhammad Husayn Fadl Allah, in his judicial decision states:

> There are two points that deserve mention: first, does such a scientific advancement [in the field of cloning] mean that humans are interfering in God's act or that it is deviation from the religion? We do not regard the biomedical advancement as interfering in God's work and against religious thought in its doctrinal concern. In fact we have argued earlier in relation to test-tube baby that such a birth is not far from the God's law of creation. After all, the scientists have discovered this law. They know God's secret in the matter of procreation, by seeking to be guided by the laws that God has shaped for procreation. This is what we see in this new experiment with cloning, which is not proposing any new law of creation, nor is it formulating a new way of creation that would challenge God's power over creation. It has simply discovered some secrets of physiology and has cognized the dynamic of these secrets and its potentials in employing them to clone an animal or a human being.[16]

Hence, what the scientists are doing is not exactly "creation;" cloning is simply employing all that is potentially within the natural sphere to bring about the conception. However, such interventions are not without harm to the accepted social norms regarding marriage and parenthood. It requires information about its negative and positive implications based on the religious norms. These norms can define the ethical and legal parameters of a cloned person's relationship to his/her mother, with no reference to a father since no sexual intercourse was involved to enable semen to reach uterus. Moreover, since cloning is connected with artificial insemination in which a nucleus that carries the DNA is carried to an ovary and after its fertilization is returned to the uterus of another in order to acquire the fetus, it raises serious questions about the meaning of childhood that carries the DNA of the one whose nucleus is used without any concern for lineage and inheritance of the fetus. In addition, a number of subsidiary issues arise out of the social relations like marriage through proxy.

Obviously, Shiite scholars treat the term 'clone' more in its scientific sense of making identical copies of molecules, cells, tissues, and even animals involving somatic cell nuclear transplant cloning. Hence, it takes the position of endorsing the applications of the technology as long as it provides practical benefit in terms of improved human life.[17]

6. The Ethical Dimension of the Issue

At the center of the ethical debate about cloning in Islam, as pointed out by Muslim scholars, is the question of the ways in which cloning might affect familial relationships and responsibilities. In large measures, Muslim concerns in this connection resonate the concerns voiced by Paul Ramsey about the social role of parenting and nurturing interpersonal relations.[18] Islam regards interpersonal relationships as fundamental to human religious life. The Prophet is reported to have said that religion is made up of ten parts of which nine-tenths constitute interpersonal relationships, whereas only one-tenth forms human-God relationship. Since the fundamental institution to further these relationships is a family and since human cloning interferes with the workings of the male-female relations, the Muslim scholars have advised their governments to exercise extreme caution in endorsing the technology without necessary caution.

Some Muslim thinkers have raised questions about maneuvering of human embryos in IVF implantation in terms of their impact upon the fundamental relationship between man and woman and the life-giving aspects of spousal relations that culminate in parental love and concern for their offspring. Islam regards spousal relationship through marriage to be the cornerstone of the prime social institution for the creation of a divinely ordained order. Consequently, Muslim focus of the debate on genetic replication is concerned with moral issues related to the possibility of technologically created incidental relationships without requiring spiritual and moral connection between a man and a woman in such embryonic manipulation. Biotechnological intervention can jeopardize the very foundation of human community, namely, a religiously and morally regulated spousal and parent-child relationship under the laws of God. It is for this reason that among Muslim scholars the more intricate issues associated with embryo preservation and experimentation with stem cell have received less emphasis in these ethical deliberations. To be sure, since the therapeutic uses of cloning in IVF appears as an aid to fertility strictly within the bounds of marriage, both monogamous and polygamous as recognized in the Shari'a, Muslims have little problem in endorsing the technology. The opinions from the Sunni and Shiite scholars indicate that there is unanimity in Islamic rulings on therapeutic uses of cloning, as long as the lineage of the child remains religiously unblemished. In other words, to preserve the integrity of the lineage of a child reproduction must take place within the religiously specified boundaries of spousal relationship.[19]

Besides the significance attached to the spousal relationship for bearing and nurturing of children, another issue in Muslim bioethics is the problem of determining the moral status of the technology itself. In the world dominated by the multi-national corporations Muslims, like other peoples around the globe, do not treat technology as morally neutral. No human action is possible without intention and will. In light of the manipulation of genetic engineering for eugenics in the recent history, it is reasonable for the Muslims, like the Christians and the Jews, to fear political abuse of the reproduction technology through cloning. With its emphasis on egalitarian spirituality, Islam has refused to accord validity to any claims of superiority of one people over the other. The only valid claim to nobility in the Qur'an stems from being godfearing.

It is obvious that ethically cloning for purposes other than therapeutic lays enormously grave responsibility on humans in terms of genetic improvement of quality of human life, the authority that can make these decisions with necessary foresight and wisdom, and the criteria that can be used in evaluating the risks and benefits of such interventions.

7. The Principles of 'Equity' and 'Public Interest' As They Bear Upon Cloning

In Islam although religious, ethical and legal dimensions are interrelated, it is important to underline the legal doctrines that bear upon the decisions made by Muslim legal scholars in endorsing or prohibiting cloning. Without adequate legal reasoning based upon careful interpretation of the Qur'an and the traditions, in addition to certain rationally derived principles and rules, no Muslim legist can issue his judicial decisions on the subject. In connection with the cloning the legists evoked the two fundamental principles of 'juristic preference' (*istihsan*), allowing the legists to weigh the pros and cons of the technology and choose that which is in the interest of the people; and 'public interest' (*maslaha*), which requires to reject harm and promote the common good. These two principles have provided a religious basis for their legal decisions about the limited use of cloning for therapeutic purposes. Since the subject of technologically-assisted reproduction has no precedent in the classical juridical tradition, Muslim legists depend heavily on the scientific information supplied by researchers to deduce their judicial decisions.

Closely related to the two above-mentioned principles, there are three subsidiary principles or rules that are evoked to resolve ethical dilemma and derive judgments related to cloning: (1) 'protection against distress and constriction' (*'usr wa haraj*); (2) the principle of 'No harm, no harassment' (*la darar wa la dirar*), and (3) the rule about 'averting causes of corruption has precedence over bringing about benefit' (*dar'u al-mafasid muqaddam 'ala jalb al-masalih*).

8. Postscript on Responses to Inquiries about Cloning

Following the early opinions that were expressed immediately in the aftermath of the animal cloning in 1997, in the last four years more meaningful discussions, and legal rulings based on comprehensive understanding of the case, have emerged among Muslim jurists.[20] In fact, most of these rulings have introduction dealing with the scientific information needed by the religious scholars to understand the exact nature of the problem. For example, careful analysis of the kind of embryo splitting is part of the *responsa* literature. It is not uncommon to read the following introduction on human cloning:

There are two ways of acquiring animal or human embryo:

1. Natural: This procedure enables the male semen to reach the female egg through sexual intercourse. In the uterus semen encounters ovary, and enables one of them to fertilize the ovary, thereby coagulating the semen and implanting itself in the uterus. It then goes through the stages of clot, blurb, until it develops into a complete being.
2. Artificial: this is a new method developed some years ago to cure infertility. The method takes male semen and female ovary and fertilizes it outside the womb in a test tube in a

> special type of cytoplasm. After it fertilizes the embryo is returned to the womb in order to complete the biological process of becoming a complete being.

What is new in the juridical discussions is the biological science and its relationship to the religion - something that provides Muslims a fresh venue of communicating their rulings on the basis of culturally mutual issues confronting modern societies. The sophistication with which Muslim legists are differentiating and collating information on cloning to formulate their responses is unprecedented in the history of Islamic jurisprudence. The citations from the scriptural sources are kept to the minimum, but the selected passages from the Qur'an are relevant for providing new interpretation of the situation created by scientific breakthroughs. Following is another example of precise scientific information included in deriving fresh juridical decisions:

> After much experimentation and application of new techniques scientists have discovered a new method of producing a living being. With the success of the experiments of cloning animals and plants, Muslim scholars have declared that this is among possible techniques that could be applied to human beings. This technique is known as "cloning" of a fetus *(al-istinsakh al-jini).*
>
> The technique employs human female egg, splitting it to separate the nucleus, and implanting it in a cell from which the nucleus is removed. Then the nucleus is cultivated in the cytoplasm of a regular cell inside the female egg. By using electrode pitting the cell begins to divide in order to create embryo. Then this embryo is placed in the uterus of a woman in order to begin its biological development to another being.
> Among the features of this new being is its being completely a clone of the person to whom the nucleus belongs. Moreover, the reproduction has occurred without the natural procedure that requires a male and a female to engage in a sexual intercourse for the sperm and ovum to meet and fertilize. This new procedure, which has occurred through nuclear transplantation, needs the female only to carry it to its complete term. In

> fact, this creation of an individual in this manner takes place outside the framework of a family. This technique is known as cloning because it is not possible to distinguish the new creature from the original at all. It is said that this procedure will engender lot of ethical problems, especially when the experiments will use condemned criminals. In these situations the two persons will look alike and it will be impossible to ascribe the crime to the right person. It is important to keep in mind that scientists were able to clone Dolly after 337 failed attempts!

According to these rulings, the main objection to cloning, which has caused hue and cry and confusion among people, especially those who believe in and limit the power of creation to God, is that the embryo that is cloned is not in reality a result of a male impregnating a female in a legitimate conjugal relationship. Rather, it is the result of the splitting of a nucleus of one animal and implanting it in the cytoplasm of another. Hence the embryo does not carry the DNA except of one whose nucleus is used to clone. More importantly, as these scholars assert, this way of creating clones, disproves the conventional meaning of childbearing for all the parents.

It is in this light that the rulings become decisive precedents in resolving the problems that arise out of concern for lineage and inheritance of the fetus, and subsidiary issues related, for example, to the concern for regulating social relations that emerge because of the birth occurring outside the conventional marital relationship. Despite the fact that presently this technique does not apply to human beings except in its future application, yet because of this possibility, Muslims have resorted to their religious scholars to seek their opinions about the religious and legal basis for technologically assisted reproduction in general. They are also concerned to understand a stance the religious law takes regarding the relation of the child to the owner of the original nucleus used in cloning.

9. The Problems Raised in the Context of the Shari'a

A variety of questions have arisen in the context of familial and social relationships in Muslim cultures. These questions reveal both communitarian as well as universal ethical/legal concerns regarding the status and social placement of the cloned human being. More pertinently, the raised issues are borderless, they are human, and, as such they have far-reaching implications beyond Muslim cultures. Consequently, there is a need to examine these questions and their responses, in order to reach a consensus in matters that have cross-cultural application. Let us read these questions:

> Q.1: It remains to be established whether there is permission to actually conduct such experiments in the manner described above. If it is permissible, what are the conditions that must be met?
> Q. 2: If a child were born in this manner, what would his/her relationship be to the person whose nucleus was transferred to the cell, whether man or woman? Following are possible scenarios:
>
> - Is the child legally an offspring in the conventional sense in the light of the fact that he/she was created through an extracted cell instead of a natural process of being born through the coming together of a sperm and an egg?
>
> - Is the child to be regarded outside the legal relationship? How should he be related to the biological owner of the cell or to the DNA carrying nucleus? In other words, how should he/she be related in accordance with the type of creation that brought it into being?
>
> - How should the child be classified in terms of his religious affiliation? Is he/she to be regarded as Muslim or non-Muslim? Or should he/she be connected to the religion of the one from whose cell he was produced?
>
> - What is his/her lineage?
>
> Q.3: What is the ruling about his religious affiliation while still young: Is he to be considered a Muslim or a non-Muslim? Or, should his religious affiliation be the same as the donor of the cell? [This is in light of the ruling that a child's religious identity is connected with the father.]
> Q.4: What is the ruling about the responsibility of the blood relationship and blood wit that must be paid in case of homicide, and the responsibility for the crime? [This is related the Islamic

> penal code, where an unintentional homicide has to be compensated by the blood relatives.]
> Q.5: Are there any rights and responsibilities between the cloned offspring and the owner of the cell?
> Q.6: What is the ruling about marriage with other naturally born children of the owner of the cell, if the cloned child is regarded as an outsider? Can he marry, for instance, his daughter?
> Q.7: There is a possibility of cloning human organs in the laboratory and preserve them for that person or for someone else for transplant. Is this permissible? Does this permission include cloning organs of reproduction, since these belong to a person whose privacy must be guarded, according to the rules of decency in the Shari'a? Also, does this permission include cloning of the brain?

Complexity of issues related to emerging relationship between the cloned child and the donor of the cell and the egg is self evident in the above questions. They also reveal the cultural sensitivities specific to Muslim societies, like the rules about privacy and exposure of private parts. However, they also underscore universal legal problems that might arise across nations in settling disputes about ownership as well as accruing responsibilities for the child's welfare.

The following responses show the way some legists understand the technology and its ramifications for society. It is important to keep in mind that there is no unanimity among these scholars. The responses have been selected to provide rulings that would be commonly accepted as representing Islamic values:

> A1: As for the permissibility of undertaking to create another being by means of cloning technology, there is permission to produce another living being by means of this technique or any other means, by resorting to discover and apply the laws of nature that God has placed at human disposal. Hence, this procedure is not forbidden unless it involves morally objectionable acts. Moreover, a precautionary measure is necessary to avoid fertilization of the sperm of a stranger with the egg of anyone other than a legal spouse. This way the offspring can be legitimately ascribed to the two parents who are legally married. In principle, then, the experimentation with cloning is not forbid-

den, except when it leads to other forbidden acts that might adversely impact upon man-woman relationship.

However, there are some issues that require precaution and may as well lead to the prohibition of this mode of reproduction:

a. The argument that cloning is creation of a child outside the framework of a family.

   There is no ground to prohibit cloning because of that when there is no evidence in the Shari'a to restrict human scientific activity, and human ability to create by following his potential to discover the laws of nature. Rather, his development is tied to his ability to break new paths and employ the laws of nature entrusted to humanity by God through investigation and intuitive reasoning, within the framework of a family.

b. The argument that this technique will cause major ethical problems, because of the possibility that the criminals running away from justice might use it.

   As mentioned in the context of an earlier response, such a possibility does not necessitate its prohibition. The offenses, even if they happen to be forbidden acts, when a criminal person performs them in order to derive an advantage from them, cannot be prohibited. The criminals use a number of products and find these items more beneficial than this experiment for their criminal ends; and yet, no one has resorted to prohibit them. It is quite likely that cloning technology for cosmetic enhancement might provide great benefit to criminals. Yet, has anyone prohibited cosmetic use of technology because of this abuse?

c. The success of this technique is preceded by a number of failed experiments in which embryos, before they can develop into a full pregnancy, have been destroyed.

That which is prohibited in any such experiment is destroying a living being whose blood it is forbidden to shed. Also, it is prohibited to kill a fertilized ovum that is on its way to life. This is similar to abortion. It is not forbidden on a person to conduct an experiment in which a living being might die before the conditions for life are completed, without his having intentionally desired so. Hence, it is permitted for a man to approach his wife sexually when she is ready for pregnancy, even though the pregnancy might encounter miscarriage as a consequence of lack of conditions for life being completed. This could be because of weakness in the sperm or lack of other necessary conditions for the embryo to develop and grow into a child. At any rate, we do not see any objection for the technique, as long as it does not lead to any other forbidden act, like looking at the private parts and touching them, as required in the law.

A.2: The relationship of a cloned child to a man or a woman from whom the cell was extracted for nuclear transfer:

If the child was created the way described above, then he/she does not have a father in the conventional sense at all. The reason is that ascription of fatherhood is connected with the fertilization of the sperm and the egg to create a living being, as pointed out by God in the Qur'an: "Then He fashioned his progeny of an extraction of mean water." (K. 32:4). In this experiment there is no role for the sperm; rather it is the separated cell from the body. More particularly, when the cell is extracted from a woman it is inappropriate to attribute to her fatherhood for the cloned child. It has been narrated in several traditions that God created Eve from the rib of Adam. Regardless of the unreliability of these traditions, and depending on the apparent sense of these reports, no one can doubt that these reports necessitate that we must regard Eve as Adam's daughter! This clearly reveals that the standard used to determine the child-father relationship does not include that the off-spring should be created from a part of his body; it simply states that he should be created from his sperm, as mentioned earlier. As for the child-mother relationship, this follows the creation of a new life from her egg. It is clear that not all of her eggs can be the source of creation. Rather, only some fertilize from her nucleus. It is only then that the ascription of relationship to her materializes.

Nevertheless, it is difficult to rule out any relationship between the clone and the donors of cell and egg, just as it is not possible to rule out that the child is the cell or egg donor's brother, especially when it is the brother who shares with his brother one of the two parents. More importantly, the criterion for this ascription cannot be derived from the fact that the clone is the carrier of specific hereditary traits, because conventionally these factors are not critical for the ascription of relationship between the child and the parent.
It is important to keep in mind that in the final analysis it is the custom and convention that determines the criterion for ascription of relationship. The Sacred Lawgiver has depended on the custom to promulgate the ordinances related to social relationships. It cannot be assumed that the relationship between the clone and the donor is automatic, regardless of the normal, agreed upon social conventions regarding such relationship.

A. 3: The ruling about the child's religious affiliation while still young:

As long as the child remains incompetent to distinguish for himself about his religious affiliation, the rule that applies to the child who is under the custody of another person applies to this child. Similar is the case of a child prisoner in the care of his captor. When he attains maturity to know the good from bad, then he is under the religion to which he converts. Assuming that he adopts a religion other than Islam, he cannot be regarded as an apostate, even if the cell donor happens to be a Muslim. The reason is that the cell donor is not his father in the conventional sense.

A.4. The ruling regarding the child's lineage: (a) In terms of his responsibility to pay the blood-wit in case of a murder committed by his family member or his liability in the case of a crime:

Since the family connection is dependent upon relationship to the father, the clone lacks that connection to the donor of the cell in terms of being a son and the donor's being his father, as discussed earlier, it necessitates negating any relation to the family of the donor. Accordingly, he is not required to have his family to pay his blood-wit. In fact, his blood-wit is restricted to the one who is liable for the crime.

However, inasmuch as there is doubt in his relationship to the owner of the egg, there is also hesitation in connecting him to those related to the donor. In this situation his status appears to be like a grand child to her parents and her sisters are his aunts. In any case, there is no evidence to prove or disprove the relationship. Additionally, there is no proof to establish his vestiges or to deny them. Hence, the case requires caution in specifying definite legal rulings.

A. 5: As for the rights specified by the Shari'a between the cloned individual and the donor of the cell, since the conventionally acknowledged norms to establish relationship are absent, there are no rights.

A. 6: The rulings regarding permission for him to marry children related to the donor:

Since that which determines such close blood relationship between the donor and the clone does not exist, it is out of question to regard the donor's children as the clone's siblings. Nevertheless, some traditions suggest that in the beginning of the creation there was a proscription against marriage between Adam and Eve's children. This text, even when it cannot serve as an incontrovertible evidence to deduce a prohibitory ordinance, confirms the legal precedent for prohibiting marriage between the clone and other children of the donors. Hence, it is necessary to apply a precaution in permitting the marriage because there is a possibility to establish sonship to the donor of the egg. Actually, this precaution extends between the clone and all those who are connected to him through the donor of the egg, such as her sister, her son, her daughter, and so on.

A. 7: Permissibility of using the cloned parts of the body in the laboratory and preserving them for the future use for that person or for others when needed:

It is permissible to clone the body organs, including the sex organs. It is also permissible to look at them because of the lack of its attribution to a specific person, which is the criterion for its prohibition. Since attribution with specificity is the criterion in its prohibition, it is prohibited to transplant a male organ to a woman and vice versa. As for separating them from the body, it is problematic to regard it as prohibition.

In conclusion, we need to be cautious in overly utilizing these advancements of the modern civilization without putting in place proper restrictions to forestall harm and calamity to humanity. Indeed God created this universe to serve humanity and to advance it towards its own betterment, just as God says in the Qur'an: "It is God who has created everything on earth for you [human beings]" (K. 2:29). In another place, God says: "Haven't you seen that God has made serviceable to you all that is in the heavens and the earth, and has showered on you His external as well as internal blessings" (K. 22:63-65). Hence, one should not depart from God's purposes, otherwise we will deserve God's abandoning and punishment, as God says: "Haven't you seen those who exchanged God's bounty with ingratitude, and caused their people to dwell in the abode of ruin?" (K. 14:28).

10. Conclusion

The recent opinions expressed by Muslim legists around the world confirm my assessment of the ethical issues associated with cloning, namely, that in providing religious guidance in matters connected with the future of humanity, Islamic norms have been studied in the context of the social and cultural conditions of Muslim societies. A unanimity has now emerged among Muslim scholars of different legal rites that whereas in Islamic tradition therapeutic uses of cloning and any research to further that goal will receive the endorsement of the major legal schools, the idea of human cloning has been viewed negatively and almost, to use the language of the Mufti of Egypt, "Satanic." A further recommendation among

Muslims seems to be discouraging even the research aspects geared towards improvement of human health through the genetic manipulation because of the rule of prioritization based on the principle of distributive justice. In view of limited resources in the Islamic world and the expensive technology that is needed for research related to cloning, Muslim legists have asked their governments to ban research on cloning at this time.

Since technologically-assisted reproduction in Islamic tradition is legitimized only within the lawful male-female relationship to help infertility, somatic cell nuclear transplant cloning from adult cells for therapeutic purposes will have to abide by the general criterion set for this technology. In the case of cloning specifically for the purposes of relieving human disease, there is no ethical impediment to stop such research, which on the scale of probable benefit outweighs possible harm. I believe that research in human cloning from adult cells in the course of reproduction treatment should be allowed with necessary regulatory clauses to restrict abuse under penalty. My opinion is based on the principle of 'averting (and not interdicting) causes of corruption has precedence over bringing about that which has benefit.'

In our religiously and ethically pluralistic societies where there is a search for a universal ethical language that can speak to the adherents of different religious and cultural traditions, Islamic tradition with its experience in dealing with matters central to human interpersonal relations in diverse cultural settings can become an important source for our ethical deliberations dealing with the ideals and realities of human existence. For instance, I am deeply concerned the way we shy away from considering the subjective dimensions pertaining to human spiritual and moral awareness in setting our goals for research in human embryo. Our policies in the matter of cloning should be seriously informed from the perspective of corrective as well as distributive justice. On hearing my Christian and Jewish colleagues on human cloning I feel that there is a consensus to look into prioritization of national resources to achieve fair distribution of health care resources both nationally and internationally. From a standpoint of common moral commitment to the principle of distributive justice, it will be hard to justify a heavy investment in embryonic research related to animal cloning without addressing some immediate and serious problems of poverty around the globe. Moreover, the wealthy countries have a responsibility to share their material as well as scientific resources with other underprivileged nations whose immediate needs do not go beyond treating common diseases like malaria and tuberculosis.

Notes

1. Pellegrino, Mazzarella, Corsi, 1992, 13.
2. By 'scriptural' I mean not only that which is regarded by Muslims to have been revealed to Muhammad, the Prophet, by God; but also the pattern of conduct left by Muhammad himself, usually known as the sunna. In other words, 'scriptural' also denotes the normative in Islam. Throughout this work I have rendered sunna with capitalized 'T' in the translation of this technical term ('Tradition') meaning, all that is reported having been said, done, and sliently confirmed by the Prophet. The translation of *hadith* (the vehicle of the sunna, through which it is reported) is rendered with lower case 't' ('tradition') or simply *hadith*-report. The Sunna (=the Tradition) in religious sciences is comprised of major compilations of the *hadith*-reports which include the six officially recognized collections of the *sahih* ('sound' traditions) among the Sunni Muslims, and the four books among the Shiites.
3. In the recently published essays in the volume DuBose et al. 1994, various authors have critically assessed the relevance of principle-oriented bioethics in the context of growing consciousness about the need to meet the demands of ethical pluralism in the multi-cultural society of North America. There is no doubt that bioethicists need both principles and rules to determine why some moral judgments lead us to classify action as prohibited, required, or permitted, in certain circumstances.
4. To speak about such a possibility in the highly politicized "theology" of international relations is not without problems. Like development language for which modern Western society provides the model that all peoples in the world must follow, any suggestion to create 'meta-cultural' language of bioethics runs the risk of being suspected as another hegemonic ploy from the Western nations. However, there is a fundamental difference in the way development language is employed to connote Western scientific, technological, and social advancement, and biomedical vocabulary that essentially captures universal ends of medicine as they relate to human conditions and human happiness and fulfillment across nations. It is not difficult to legitimize bioethical language cross-culturally if we keep in mind the cultural presuppositions of a given region in assessing the generalizability of moral principles and rules.
5. Contemporary moral discourse has been aptly described as "a minefield of incommensurable disagreements." Such disagreements are be-

lieved to be the result of secularization marked by a retreat of religion from the public arena. Privatization of religion has been regarded as a necessary condition for ethical pluralism. The essentially liberal vision of community founded on radical autonomy of the individual moral agent runs contrary to other-regarding communitarian values of shared ideas of justice and of public good. There is a sense that modern, secluar, individualistic society is no longer a community founded on commonly held beliefs of social good and its relation to responsibilities and freedoms in a pluralistic society (see Heyd, 1996).

6. Mayer, 1991 in the Chapter 1 on "Comparisons of Rights Across Countries," has endeavored to analyze charges of cultural relativism against the Universal Declaration of Human Rights made by Muslim governments guilty of violating human rights of their peoples. However, in the process of arguing for the universal application of the UDHR document, she has paradoxically led to the relativization of the same by ignoring the historical context that actually produced the UDHR in the first place. See my review of her book in the *Journal of Church and State*, Fall 1992.
7. Deontological ethical norm determines the rightness (or wrongness) of actions without regard to the consequences produced by performing such actions. By contrast, teleological norm determines the rightness (or wrongness), of actions on the basis of the consequences produced by performing these actions. Deontological norms can further be subdivided into objectivist and subjectivist norms: objectivist because the ethical value is intrinsic to the action independently of anyone's decision or opinion; subjectivist because the action derives value in relation to the view of a judge who decides its rightness (or wrongness). See Hourani, 1985, 17 who introduces the latter distinction in deontological norms.
8. Ghazali, 1971, 174. Ghazali's "Five Purposes" (*al-maqasid al-khams*) has become an accepted phrase in the Sunni works on legal theory and often quoted in discussions about the principle of *maslaha*. See Shawkani, 1930, 216.
9. For various Muslim opinions collected from around the world see Campbell, 1997. In addition, for specifically Sunni opinions expressed by their leading religious authorities, see: *al-Majalla*, 1997 and *Sayyidati*, 1997. For the Shiite opinions besides *al-Istinsakh bayna*, 1999, see also Hasan, 1997.
10. *al-Istinsakh bayna*, 305; *f*or discussion, see a*l-Majalla*, 1997, 6.
11. *Sayyidati*, 1997, 64.

12. Dr. Hasan al-Turabi in his opinion on the subject has pointed out that many scholars have not paid attention to the various scientific facets of the issue, which they need to examine before formulating their responses. See *al-Istinsakh bayna*, 1997, 307.
13. *al-Istinsakh bayna*, 1997, 63.
14. *Sayyidati*, 1997, 62-63.
15. for instance, see the Catholic opinion expressed by Bishop Habib Pasha in *al-Istinsakh bayna*, 1997, 19-21.
16. *al-Istinsakh bayna*, 1997, 289.
17. The opinions regarding cloning coming out of Lebanon and Iran indicate more openness is accepting the technology and even the adult somatic cell transplant. See Ayatollah Khamenehi, 1996, 111-122 dealing with technologically-assisted reproduction.
18. Ramsey, 1970.
19. Among the Shi`ite jurists, Ayatollah Khamenehi, 1996, 117-122 seems to have sanctioned both surrogacy and sperm and egg donation without requiring the donor of the sperm to be the husband as required by the senior jurists like the late Ayatollah Khomeini. See Khomeini et al. (n.d.), 74-81.
20. Much of the information in this section has been collated and compiled from number of new studies dealing directly or indirectly with the ethics of cloning. The main text with which the rulings of other jurists, both Shiite and Sunni, have been compared is Hakim, 1999. The other compilations include Khitab, 1997; Hasan, 1997; Salama, 1996 and Jabiri, 1994.

References

Campbell, Courtney. "Religious Perspectives on Human Cloning." (Ph. D., Oregon State University) paper commissioned by National Bioethics Advisory Commission, 1997.

DuBose, Edwin R./ Ronald P. Hamel/ Laurence J. O'Connell (eds.). *A Matter of Principles? Ferment in U.S. Bioethics*. Valley Forge: Trinity Press International, 1994.

Ghazali, Abu Hamid Muhammad al-. *al-Mustasfa min 'ilm al-usul*, Muhammad Mustafa Abu l-'Alla (ed.). Cairo: Maktabat al-Jundi, 1971.

Hakim, Muhammad Sa'id al-Tabataba'i al-. *Fiqh al-istinsakh al-bashari wa fatawa tibbiya.* n.p., 1999.

Hasan, Sadiq Ja'far al-. *al-Istinsakh al-bashari fi ra'y al-Imam al-Shirazi.* n.p., 1997.

Heyd, David (ed.). *Toleration: An Elusive Virtue.* Princeton: Princeton University Press, 1996.

Hourani, George. *Reason and Tradition in Islamic Ethics.* New York: Cambridge University Press, 1985.

(al)-Istinsakh bayna al-islam wa al-masihiyya, The Center for Islamic-Christian Studies (ed.). Beirut (Dar al-fikr al-Lubnani), 1999.

Jabiri, Ahmad 'Amr al-. *al-Jadid fi al-fatawa al-shar'iyya li-al-amrad al-nisa'iyya wa al-'aqm.* Amman: Dar al-Furqan, 1994.

Khamenehi, Ayatollah. *Pizishki dar a'ineh ijtihad* (Medicine through the Process of Independent Reasoning). Qumm, 1996.

Khitab, 'Abd al-Mu'iz Khitab. *al-Istinsakh al-bashari: hal huwa didd al-mashiyyat al-ilahiyya?* Cairo: Dar al-Nasr, 1997.

Khomeini, Ayatollah, et al. *Pasukh bi-su'alha.* n.p., n.d.

(al-)Majalla: The International News Magazine of the Arabs, No.894, 30 March-5 April, 1997.

Mayer, Ann Elizabeth. *Islam & Human Rights: Tradition and Politics.* Boulder et al.: Westview Press, 1991.

Ramsey, Paul. *Fabricated Man: The Ethics of Genetic Control.* New Haven: Yale University Press, 1970.

Salama, Ziyad Ahmad. *Atfal al-anabib bayna al-'ilm wa al-shari'a.* Beirut: Dar al-Bayariq, 1996.

Sayyidati, No.843, 3-9 May, 1997, 62-64.

Shawkani, Muhammad b. Ali al-. *Irshad al-fuhul ila tahqiq al-haqq min ʻilm al-usul*. Cairo: Maktabat Muhammad Ali Subaih, 1930.

The Debate about Human Cloning Among Muslim Religious Scholars since 1997

Thomas Eich

Abstract: For most Muslim religious scholars, cloning is a new technique of procreation within the framework of Godís will rather than an infringement of Godís prerogative of creation. Consequently, their discussion puts the focus on the effects of cloning rather than on the act itself. Those who wish to forbid cloning argue that it would lead to a blurring of the lines of genealogy (nasab). Therefore, a considerable number of sunni statements have emerged that reproductive cloning can only be forbidden outside marriage. Several shi'i scholars also argue for a redefinition of blood-relationship terms. The argument that cloning is in contradiction with the concept of human dignity is by and large absent from these discussions. However, an analysis of the statements about therapeutic cloning shows that many scholars who consider it permissible only have a limited understanding of the techniques involved. Recently, several scholars have argued in favour of prohibiting it, because it would imply the destruction of an embryo.
This article gives an overview of the arguments exchanged among Muslim religious scholars in the debate on human cloning that has evolved since the birth of Dolly in 1997.

Key Words: Islam, Religion, Shari'a, Cloning, Blood relations (nasab), Argument of social consequences, Argument of human dignity.

Muslim religious scholars do not consider cloning to be an infringement of God's prerogative of creation. They argue that it is nothing but the discovery of a new technique of procreation, which is still within the framework of God's will. Consequently, their discussion puts the focus on the effects of cloning rather than on the act itself. Those who want to forbid cloning argue that it would contradict the principle of variety in God's creation and lead to a blurring of the lines of genealogy (*nasab*). The latter term is only to a limited extent defined biologically. Of equal importance is the legal framework of the procreative act, i.e., the existence of a marriage. Therefore, a considerable number of sunni statements have emerged that reproductive cloning can only be forbidden outside marriage. Several shi'i scholars also argue for a redefinition of blood-relationship terms. The argument that cloning is in contradiction to the concept of human dignity is by and large absent from these discussions. Analysis of the statements about therapeutic cloning shows that many of those Muslim religious scholars who consider it to be allowed only have a limited understanding of the techniques involved. Recently, several scholars have argued in favor of forbidding it, because it would imply the destruction of an embryo.

This article gives an overview of the arguments that are exchanged among Muslim religious scholars in the context of the debate on human cloning that has evolved since 1997, the year in which the birth of Dolly the sheep was announced.

1. Is There an *Islamic World*?

In the following I will not speak of the Islamic World but of countries with predominantly Muslim populations. The difference between the two terms is that the former suggests a given unity, which would allow easy generalizations. This unity would be based on the particular characteristic of Islam. This depiction would eliminate a plethora of other characteristics of certain regions within the so-called Islamic World such as political, social or economic situation implying that Islam is the most important of them. However, accepting for the sake of the argument that "Islam" would be the unifying characteristic of a certain region, we would have to consider what "Islam" actually would mean in this context. Would it be the religion of Islam? Or a shared historical heritage? Or a common social structure resting for example on the family unit than on the individual?

The alternative term "countries with predominately Muslim populations" has several advantages. *Countries* refers to modern states which have given boundaries, structures and representatives. *Muslim populations* describes a simple sociological fact, namely that people inhabiting these countries term themselves Muslims. This definition also implies the possibility of variation over time and place: by putting the focus on the self-description - a certain action - of people, it seems reasonable to assume that this very action might have different implications and meanings depending on the given context. Finally, the word *predominantly* reminds us of the important fact that a substantial number of non-Muslims also live in these countries.

This also implies that I do not refer to *all* countries with predominantly Muslim populations to the same extent. Therefore, in the following overview I will always try to put the sources to which I refer into their national or regional context. For example, I will analyse statements by the former Muftis of Egypt and Tunisia. The statements of these representatives of religious scholarship in the respective countries are primarily of importance in those specific countries. Only to a limited extent is it possible to generalize from these individual opinions about prevailing opinions in countries with predominantly Muslim populations.

2. Islamic Law (*sharî'a*)

Islamic law is based on two textual sources. The Coran is according to Muslim understanding the verbally inspired word of God that was conveyed by the archangel Gabriel to the prophet Muhammad on the Arabian peninsula between 612 and 632, when Muhammad died. The second source is the *sunna*, the traditions about the sayings and doings of the prophet and his companions, which are seen as ideal. These texts contain fundamental guidelines about the proper way of living as well as a number of explicit, sometimes very concrete and particular, rulings. Two additional derivative sources on which an Islamic jurist can rely in order to reach a judgement about a certain behaviour are the *ijmâ'* (*consensus* of the religious scholars) and *qiyâs* (analogical reasoning).

In their attempts to arrive at an assessment of modern medical technologies such as cloning, Muslim religious scholars (*'ulamâ'*) face some difficulties since the problems created by these technologies are without precedent. Therefore, they cannot rely on existing judgements (*Coran, Sunna, Ijmâ'*), and only to a limited extent is it possible to use analogies (*qiyâs*). This is why these topics are termed *min al-masâ'il al-ijtihâdiya*, which means that the *'ulamâ'* have to rely largely on their own reasoning without basing their final opinion on any precedent. In doing so, they are guided by the fundamental principles as well as by the basic, abstract rules of the *sharî'a*. One of these rules is called *sadd al-dharâ'i'*, for example, which means that something has to be considered *harâm* (taboo or forbidden) if the results of a certain act might inflict harm on an individual or on society, even if the act itself is not viewed as intrinsically bad.

The *sharî'a* does not distinguish between law and ethics. Therefore, the two are usually not dealt with separately. Additionally, many of the texts by contemporary Muslim religious scholars dealing with medical issues within the framework of the *sharî'a* are so-called *fatwâs* (legal opinions). These *fatwâs* are always in the form of an answer to a certain question. It is important to keep these two points in mind, in order to avoid reading something into these texts which is actually neither intended nor expressed by their author. For example, if a certain scholar is pondering the question who would be the legal father of a chimera, he is just doing so because he has been confronted with the question in that way. His words do not imply any statement about the permissibility of or a moral assessment of the creation of a chimera. This would have been a different question, not "who would be the legal father of a chimera?" but "does the *sharî'a* allow the creation of a chimera?."[1]

In most countries with predominantly Muslim populations the *sharî'a* is no longer applied in its entirety. But this does not mean that the judgements of the *'ulamâ'* are irrelevant. First, some of these countries, such as Iran and Saudi-Arabia, actually apply the *sharî'a*. Particularly Saudi-Arabia has the financial means and the infrastructure to carry out bio-medical research. This research is framed by laws that are deeply influenced by the opinions of the *'ulamâ'*. Second, countries such as Egypt have declared Islam the religion of the state. Therefore, laws can be questioned on the grounds that they would contradict fundamental tenets of Islamic Law. Consequently, laws are cross-checked by *'ulamâ'* informally before they are submitted to the secular political institutions for enactment. Third, by using the mass media such as newspapers, TV, radio, tapes and - more recently - the internet, certain prominent *'ulamâ'* e.g. Yûsuf al-Qaradâwî for example can - and do - reach big audiences all over the world, thus shaping public discussion about certain topics quite substantially.

3. Mixing Biological and Moral Assessments: The Example of Lineage (*nasab*)

For the *sharî'a* the question of the legitimacy of an act influences the assessment of its effects. For example, the illegitimacy of sexual intercourse leading to pregnancy and the giving birth to a child influences the assessment of the social position of that child. The *sharî'a* is a product of history. It is important to keep this in mind when considering the discussions about reproductive technologies. Before the 1970s it could usually be assumed that the woman who carried a child and gave birth to it was identical with the donor of the egg cell involved in creating that child. Therefore, the classical discussions of the *'ulamâ'* about lineage focussed much more on the question who was the *father* not the mother of the child. *Nasab,* for example, the Arabic term for genealogy/lineage, refers exclusively to the relation between the father and the child. But this relation is only to a limited extent merely biological for the *sharî'a*. At least of equal importance is the legitimacy of the act, in other words whether or not the sexual intercourse occurred between husband and wife. Adultery (*zinâ)* has immediate consequences for the relationship between the man and the child, the result of his illegitimate sexual intercourse, since the term *nasab* refers primarily to the relationship of a father to his legitimate child. It is important to understand that there is no term in Arabic for the relationship of an illegitimate child to his father. The *sharî'a* term *walad al-zinâ* (child of adultery) refers to a person, not to a relationship. The only way to create a *nasab* relationship between the father and the child would be a legal procedure called *iqrâr,* in which the father has to accept the child as his own.[2]

This intermingling of biological and moral aspects is beautifully captured in the title of a *fatwâ* of the *fatwâ* committee of the Ministry of Religious Foundations of Kuwait, which is translated here as closely to the original as possible: "Where a man has committed adultery, the son he has from this adultery is not ascribed with a *nasab* to him."[3] The *'ulamâ'* recognize the biological side of the story of course by stating "the *son* he has from this adultery," but they refrain from ascribing *nasab* to the *walad al-zinâ.* Since there is no relationship between the man and the child, the child is not considered his child at all in a legal sense. The *walad al-zinâ* has no rights of inheritance or financial support from his biological father, for example. Some jurists even allow a man to marry his biological daughter, who is the result of *zinâ.*[4]

This concept of several elements constituting *nasab* also becomes clear in the discussion about IVF. Muslim religious scholars have, almost unanimously, agreed that sperm donation should be forbidden, because it would lead to *ikhtilât al-nasab*, which could be translated as *the mixing of lineage*: since the two elements that constitute *nasab* - the biological element of sperm and the legal element of a valid marriage - would be provided by two distinct parties, it would become impossible to identify the legitimate holder of *nasab.*[5]

To sum up, it is important to highlight three things: first, the *sharî'a does* take into account the biological aspects of a given problem, but - second - it does not do so exclusively, which means, thirdly, that the biological and legal aspects are not dealt with separately.

This phenomenon has some relevance for culture-transcending bioethics. It is important to note that when people from different religious or cultural backgrounds seem to arrive at the same opinion about a question, the way these opinions were reached might nevertheless be of some importance. As one participant at the Bochum conference suggested in the discussion following his presentation: "If we can arrive at a universal ban on cloning, it does not really matter *why* the different parties want to forbid it." Yet, it seems reasonable to spend some time looking at the arguments and trying to understand them. For example, on Tuesday, 4.2.2003, the Arabic newspaper *al-Hayât* published an article on page 6 entitled "The Shaikh al-Azhar rejects human cloning" (*Shaikh al-Azhar yarfud al-istinsâkh al-basharî*). It is interesting to read in the article itself what the Shaikh al-Azhar Muhammad Sayyid Tantâwî really said: "cloning, by which it is intended to create a human out of nothing in a way other than marriage, is forbidden by the *sharî'a*."[6] What Tantâwî said is neither surprising nor sensational at all. He simply stated that according to the *sharî'a* procreation outside of a legal marriage is forbidden.

4. Is Cloning an Alteration of God's Creation?

There are a variety of issues in the context of cloning (Arabic *istinsâkh*) upon which the *'ulamâ'* agree, regardless of whether they will ultimately oppose it or approve of it in the end. It is commonly held, for example, that the cloned person and the cell donor are certainly two different persons. This is because the influence of differing times and circumstances under which the two humans grow up are taken into account. In addition, ensoulment does not take place at the moment of inception in Islamic teachings but at a later stage of pregnancy. Therefore, the concept of the singularity of every human is not challenged by cloning for the *'ulamâ.'*[7]

In the early phase of the debates among Muslim religious scholars about cloning after 1997 there was some discussion whether *istinsâkh* would constitute an alteration of God's creation and consequently infringe on his prerogative of creation and his almightiness. Those who expressed this view usually referred to a Coranic passage in which the devil says to God: "Most certainly I will take of Thy servants an appointed portion: (...) and most certainly I will bid them so that they shall alter God's creation [*khalq Allâh*]." (Coran 4:118-119)[8]

This argument was countered in two ways. The first counter-argument referred to the Coranic story of the creation of Jesus. According to the Coran, Jesus was born without having a father (Coran 19:20-22 and 21:91). In Islam Jesus is recognized as one of the prophets preceding Muhammad. Accordingly, he is viewed as a mere human being, and the concept of Jesus having had no biological father is void of any ideas that have evolved around the same concept in Christian christology. Reference to the example of Jesus served to show that in fact sexual intercourse between members of the two sexes might be the usual but not necessarily the only way to achieve procreation. Therefore, it was said that cloning could not be forbidden on the grounds that it would be procreation without any mixing of the genetic materials of two persons and consequently be in conflict with the only way God wanted man to procreate.[9] This argument in turn has been countered by stating that the creation of Jesus without a father was a *mu'jiza*, a miracle, to show the almightiness of God and that it therefore could not serve as a precedent in any way.[10] To my knowledge, the "Jesus-argument" has now vanished from the discussions about cloning among Muslim religious scholars.

The second counter-argument stated that cloning was nothing but the discovery of ways of procreation that simply had not previously been known to mankind. Accordingly, it was argued that it did not question the almightiness of God, since it was part of the rules God had given to his creation. In addition it was said that *istinsâkh* only made use of materials that already exist and therefore cannot be seen as creation (*khalq*), which is defined as creating something new out of nothing.[11]

To sum up, it can be stated that Muslim religious scholars agree on two basic points. First, a cloned human being would clearly be a different person from the cell donor. Second, cloning does not constitute an alteration of God's creation, it is only the discovery of new techniques within it. Therefore, the discussions about the permissibility of cloning do not focus on the act itself but on its effects. Differing assessments of these effects lead to a variety of opinions on *istinsâkh*.[12]

5. Arguments Against Human Cloning

First, it was argued that cloning would contradict the principle of variety in God's creation. This was based on several passages of the Coran such as "we made you into different peoples."[13] Interestingly, it is sometimes linked to another argument from the natural sciences: if cloning were to become common, mankind/nature could be affected much more easily by diseases. It should be noted here that all those who wish to forbid cloning on the strength of the plurality-argument restrict this entirely to humans: it is commonly agreed that the cloning of animals and plants has to be allowed.

The second argument is directed at the *nasab*-issue: the genealogical relatedness of the clone to his/her social environment would be unclear. Would somebody who carries exactly the same DNA as another person be that other person's son or brother? Would he have a mother at all, if he did not carry any of the DNA of the person who gave birth to him?.[14] This line of argument is sometimes linked to the question of the well-being of the child, who would suffer mental harm because society could not view him, like all other children, as its integral part. The avoidance of harm in turn is a guiding principle of the *sharî'a* as was mentioned above.

6. "Counter-Arguments"

For what follows now it is important to note that for many authors the point is not so much that cloning should be allowed because it would be a good act. They simply say that the arguments summarized so far do not suffice to forbid cloning according to the *sharî'a*.

A fundamental argument is that cloning is something entirely new, therefore, it is not covered in the texts of Coran and *sunna* and the existing rulings of the *sharî'a*. In such cases something can only be declared *harâm* because of the harmful effects it would have on individuals or society. But - it is argued - the effects are not yet known. Therefore, cloning would have to be considered as allowed (*mubâh*) in the first place.[15]

The second argument is directed at the plurality issue. For example, Ayatollah Taskhiri, the Iranian representative at the Islamic Fiqh Academy of the OIC (Jedda), stated in 1997 that the influences of the environment on the individual are underestimated.[16] Another shi'i scholar argued that the traditional way of procreating would most probably not vanish, since people would continue to have sexual desires and would consequently continue to procreate in this way, therefore securing the biological diversity of mankind.[17] Muhammad Taufîq Moqdâd, the *sharî'a* representative of Iranian Supreme Leader Khamenei in Lebanon, argues that reproductive cloning would only constitute a threat to society if clones of a particular person were to be produced in large numbers.[18]

The third argument is directed at the *nasab* issue. Some religious scholars simply state that the genealogical position of the clone has to be defined. For example, 'Abd al-Karîm Fadlallâh from Lebanon argues that the clone is clearly the brother or the sister of the cell donor, because they would carry the same DNA. It would, therefore, be clear that they have the same father. Only the question of who is the mother needs to be discussed.[19] Muhammad Taufîq Moqdâd says that if the cell donor is a man, he has to be considered the father of the clone, the egg donor as his/her mother.[20] The same argument is put forward by Ayatollah Hasan al-Jawâhirî from Qom/Iran, who is also a member of the Islamic Fiqh Academy of the OIC.[21] Other authors argue that in the Coran the biological elements that constitute fatherhood and motherhood are not clearly defined and can therefore be defined by positive law (*qânûn*).[22] It might be noted that all these statements aiming at a redefinition of concepts of blood relationship are made by shi'i scholars.[23]

Sunni scholars seem to argue more for an integration of cloning into the existing legal framework of reproductive medicine.[24] To start with, Nâsir b. Zaid ad-Dawûd, a judge from Ahsa‘, an eastern province of Saudi-Arabia, argued in 1997 that cloning could be allowed if it used the DNA of a man and the egg of his wife. Similarly, the prominent expert on modern *sharî'a,* Wahba Zuhailî from Damascus, considered cloning within the bonds of a legally valid marriage to be unproblematic.[25] Ra'fat ‘Uthmân, former dean of the *sharî'a* Faculty at al-Azhar University and member of the Egyptian National Committee for Bioethics, is hesitant to prohibit this particular constellation of cloning (DNA from a man, egg from his wife within a marriage). It could only be forbidden if the clone were to suffer any harm. He therefore recommends waiting for the results of experiments in the Western world.[26] In a recent MA thesis from Azhar University a slightly different argument is put forward. If, for example, you cloned the son of a legally married couple by implanting his DNA into a denucleated egg of his mother, who would give birth to the clone, there would be no blurring of the lines of genealogy since all the elements that constitute *nasab* are there: the legally valid marriage, the mixing of the DNA of the two spouses (if this is considered to be necessary) and the pregnancy of the wife.[27]

But it is of utmost importance to stress that none of the mentioned religious scholars wishes to allow reproductive cloning indiscriminately. It is explicitly stated that reproductive cloning has to be subject to at least two conditions, even if it is carried out within a legally valid marriage. The first is that the ends of cloning cannot be achieved in any other way in every single case, i.e. all other techniques of reproductive medicine have failed. Second, cloning has first to become a safe technique which creates healthy children.[28]

7. The Argument of Human Dignity?

It might also be worthwhile to point out the virtual absence of a certain argument from the debate among Muslim scholars about cloning, namely the argument that it would infringe on human dignity. It only features prominently in the contributions by Muhammad al-Mukhtâr al-Salâmî and Nasr Farîd Wâsil, the former Muftis of Tunisia and Egypt respectively.[29]

They base the concept of human dignity (*karâmat al-insân*) on several arguments. First, they mention the Coranic statement that God made man his viceregent on earth (*khalîfa fî l-ard).* (2:30) Second, God has enabled man to arrive at the truth by himself and make his choices. Additionally - referring to the Coran again - the angels are reported to have bowed in front of man after he had been created by God. (2:34)

Nasr Farîd Wâsil goes on to argue that there is an obvious link between man's elevated position in God's creation and the fact that the Coran is very explicit and concrete about the way in which man should procreate, namely by the mixing of the sexes, because on the other hand the Coran speaks about the procreation of animals only in very general terms. Therefore, it would be no problem at all to apply the technique of cloning to animals, which leads to the conclusion that doing likewise with man would put him on the same level as animals. Consequently, interfering with man's customary way of procreation would immediately infringe on his dignity.[30]

Salâmî concludes from the concept of human dignity that "man is blessed, he is an end not a means" (*al-insân mukarram - wa ghâya lâ wasîla*). The fertilized ovum already had a given amount of this *karâma*, which would increase with the biological process of growing. Therefore, it would not be possible to use it "as a means for something else than itself" (*wasîla li-ghairihâ*).[31]

He closes his argument presented at the annual meeting of the Islamic Fiqh Academy of the OIC in 1997 as follows:

"God has bestowed dignity on man in the individual's life. Man has no value if he loses his dignity. To consider man as a thing like any other thing, science can deal with; to give one copy the right to life and to grow to maturity and to freeze another copy just to get it going for purposes of substitution or to destroy it - in this lies the destruction of his dignity."[32]

8. Therapeutic Cloning

Obviously Salâmî's statement was also directed at therapeutic cloning. This issue has received much less attention from Muslim religious scholars. In several statements evidence can be found that therapeutic cloning would be allowed by the *sharî'a* because it would be for the benefit of mankind (*sharî'a*-principle of *maslaha*). For example, the shi'i scholar Muhammad Sa'îd al-Hakîm argued that way,[33] as did the sunni Hasan 'Alî al-Shâdhilî in his contribution to the session of the Islamic Fiqh Academy of the OIC in 1997.[34] Nasr Farîd Wâsil also used the *maslaha* argument in a similar way.[35] In January 2003 Ahmad al-Tayyib, who became the Mufti of Egypt in March 2003 only to lose his post in fall of the same year, issued a *fatwâ* on behalf of the Egyptian Ministry of Justice,[36] allowing therapeutic cloning. This legal statement gained some importance in the UN debates about a universal ban on all forms of cloning: Iran's suggestion to postpone the decision, which eventually won a majority, was based on the argument that for the *sharî'a* therapeutic cloning did not constitute a problem at all. In this statement Iran relied among other things on the *fatwâ* of Ahmad al-Tayyib.[37]

Yet it seems clear that these scholars have only vague ideas of the practical implications of the issues at hand, namely that therapeutic cloning necessarily implies the destruction of embryos. This interpretation is substantiated by an interesting passage in an article by 'Abd al-Fattâh Idrîs from Azhar University, a specialist in medical jurisprudence. Summing up several statements which allow therapeutic cloning, he gives the reasons for this stand: First, the *maslaha* argument is used. Second, it does not infringe on human dignity:

"The cloning of organs does not affect human dignity, since these organs are taken from genetically manipulated cloned animals or from human body cells. These cells are not growing to become embryos or human beings, but to become human organs *by way of multiplying these cells*."

On the other hand, several statements have recently emerged that are more critical of therapeutic cloning.[38] Yûsuf al-Qaradâwî, for example, declares that therapeutic cloning should be forbidden because it would imply the destruction of an embryo. Only if techniques could be developed in which cells are implanted directly, and without destroying embryos, into animals to create the needed organs or tissues - only then could therapeutic cloning be allowed.[39] Ra'fat 'Uthmân from al-Azhar University issued a *fatwâ* to the same effect.[40] Similarly, Shaikh Muhammad al-Moqdâd stated in an interview with me in Beirut in October 2003 that therapeutic cloning should be forbidden if it involved the destruction of an embryo. Iranian scholar and member of the Islamic Fiqh Academy (Jeddah), Hasan al-Jawâhirî, distinguishes between three forms of "therapeutic cloning": first, the direct copying of an organ, second, therapeutic cloning via the creation of an embryo, and third, using the organs of a clone after his/her birth. Jawâhirî considers only the first technique, the "direct cloning of organs" to be allowed.[41]

Analysing the statements of Muslim religious scholars about cloning, it becomes obvious that there are several linguistic problems and inconsistencies at the bottom of these opposing assessments. First, scholars such as Qaradâwî and Jawâhirî obviously do not distinguish between techniques that have to be termed *therapeutic cloning,* because they necessarily imply the production of a cloned embryo in the first step, and other techniques that do not. It is possible to create artificial noses or ears from human body cells, which are implanted into host organisms and can be subsequently transplanted into human patients. This would be what

Jawâhirî refers to as the first of his three techniques. For this technique it is not necessary to first produce a cloned embryo first. Therefore, it cannot be considered as cloning at all and should be dealt with separately. Secondly, the term *therapeutic cloning* is in itself problematic. In bioethical discussions in the so-called West the term is a highly contested one, since it suggests a fundamental difference between therapeutic and reproductive cloning. But therapeutic cloning implies the creation of a cloned embryo in order to subsequently destroy it. Therefore, the two acts only differ by their aims and consequently their results. Critics argue that the term *therapeutic cloning* is used to veil the real nature of the procedure. In the case of Arabic this aspect is even more pronounced. The Arabic equivalent for "cloning" is *istinsâkh*. Literally translated, this term could be rendered as "the wish to bring about a copy." Reproductive cloning is translated as *istinsâkh basharî,* "human copying," therapeutic cloning as *istinsâkh al-a'dâ'*, which means literally "the copying of organs."[42] It comes as no surprise that Muslim religious scholars, when asked whether "the copying of organs or tissues" should be allowed, immediately answer in the affirmative because this would obviously be to the benefit of mankind and would not contravene any *sharî'a*-rule.[43] This is also true of the *fatwâ* of Ahmad al-Tayyib.

Most of the statements against so-called therapeutic cloning referred to above are by and large based on existing assessments of the *sharî'a on* the abortion issue. Of course it has to be asked if these rulings also apply to the extra-corporal embryo. Ayatollah al-Hakîm from Iraq argues that embryos left over from an IVF procedure can be destroyed.[44] 'Umar Sulaimân al-Ashqar from Jordan takes the same line, adding that the production of a surplus of fertilized eggs should be avoided.[45] This was also the line of argument in a decision by the Islamic Fiqh Academy of the OIC in 1990 and of Muzammil as-Siddiqi, former president of the Islamic Society of North America.[46,47]

It would be beyond the scope of this paper to delve into the statements of Muslim religious scholars about the status of the extra-corporal embryo. But it might be pointed out that all these statements have in mind embryos that are left over from IVF procedures, i.e. embryos that had been created with the intention of implanting them so they might eventually grow into a human being. Therapeutic cloning is a different matter of course, since it aims at the creation of an embryo for the sole purpose of destroying it in the interests of an assumed scientific gain. As has hopefully become clear in this paper, the purpose of an act is essential for its assessment in the *sharî'a.*

Notes

1. The answer to the latter question would clearly be "no," because this would be an alteration of God's creation (see, for example, Sabzawârî, 2003, 181 and 'Uthmân, 2003a, 6. For a discussion of the former question see the statement by the Lebanese shi'i scholar Muhammad Taufîq Moqdâd 11 Jumâdâ II 1422 (1.9.2001), who quotes the Supreme Leader of the Islamic Republic of Iran Ayatollah Khamenei.
2. Note that the status of a *walad al-zinâ* cannot be reversed. *Iqrâr* can only be made with children whose *nasab* is only disputed and not refuted yet. Therefore the *iqrâr* procedure basically aims at proving that at the time of sexual intercourse a marriage existed, which for formal reasons was legally invalid. Kohler, 1976, 59f, 80-128.
3. *al-zânî lâ yunsab ilaihi ibnuhu min al-zinâ.* (http://alwaei.awkaf.net/fatwÂa/article.php?ID=35).
4. Salâma, 1996, 175; Kohler, 1979, 59.
5. It might be noted in this context that the *'ulamâ'* are very critical of genetic tests. See the *Qarâr al-majma' al-fiqhî al-islâmî* Nr.16/7 (January 2003) published by the Muslim World League (Mecca) (www.Muslimwordleague.org).
6. *al-Istinsâkh alladhî yuqsad bihi îjâd basharin bi-ghair tarîq al-zawâj muharram shar'an.* The Azhar is a university in Egypt. Its *sharî'a* department is one of the most important in countries with predominantly Muslim populations. A session of the Islamic Fiqh Academy of the Muslim World League (Mecca) from 21.-25.10.1422/ 7.-11.1.2002 condemned human cloning. Among the eight reasons given for this stand, the necessity of the legal framework for procreation is referred to twice (reason 1 and 6). See Turki, 2003, 32f.
7. see, for example, Qaradâwî, 2002; 'Uthmân, 2003b, 46f; Moqdâd, 2001; Sabzawârî, 2003, 87.
8. See, for example, 'Âmilî, 1999; Shams ad-Din, 1999; Wâsil, 1999.
9. Hakîm, n.d., 19; Salâmî, 1997, 149f.
10. Khadmî, 2001, 86f (Khadmî is a Tunisian sunni religious scholar and member of the COMSTECH International Committee on Bioethics (CICB) of the Organization of Islamic Congress (this committee has been formed in January 2003)); Sabzawârî, 2003, 105.
11. see, for example, Fadlallâh, 1997; Moqdâd, 2001; Qaradâwî, 2002; Khadmî, 2001, 71f.
12. In this context see also the Statement of Abulaziz Sachedina under http://www.people.virginia.edu/~aas/issues/cloning.htm.

13. see, for example, Tayyâr, 1997; Qaradâwî, 2002; Salâmî, 1997, 156; Turki, 2003, 33; Tabrîzî, 1997.
14. Shâdhilî, 1997, 191-194; Khadmî, 2001, 97-101; Qaradâwî, 2002; Salâmî, 2001, 164f; Hakîm, n.d, 21-23.
15. Moqdâd, 1999 and 2003; Sabzawârî, 2003, 142-145.
16. Ayatollah Taskhiri, 1997, 225.
17. Sabzawârî, 2003, 130.
18. Muhammad Taufîq Moqdâd, 2001 and 2003; see also Jawâhirî, 1998, 297f, where he explicitly counters the argument put forward by Tabrîzî, 1997, and the sunni scholar 'Uthmân, 2003, 48.
19. Abd al-Karîm Fadlallâh, 1999, 273.
20. Muhammad Taufîq Moqdâd, 2001.
21. Ayatollah Hasan al-Jawâhirî, 1998, 297, quoting the Iranian scholar Ayatollah al-Sayyid Kâzim al-Hâ'irî.
22. Sabzawârî, 2003, 126f, 137f, 170f; Rispler Chaim, 1998, 569 referring to the Iranian scholar Ayatollah Mustafâ al-Harandî from Qom.
23. Shi'i thought traditionally gives more room to historical development in its hermeneutical approach to the textual sources of the *sharî'a.* It could be speculated that this might be the reason why shi'i scholars are more likely to admit, that concepts of blood relationship are socially and historically defined and might consequently change over time.
24. But see also the statement by the shi'i scholar Abdulaziz Sachedina, 1998, 241 to the same effect.
25. Wahba Zuhailî, 1997, 124f.
26. Ra'fat 'Uthmân, 2003, 42-45. For a similar statement by a shi'i scholar see Jawâhirî, 1998, 300.
27. 'Uqûbî, 2002, 153. An MA thesis does not of course represent any kind of official statement by the University where it was submitted.
28. Dawûd, 1997; 'Uqûbî, 2002, 153f, 156; 'Uthmân, 2003, 45; Sabzawârî, 2003, 132, 135.
29. The aspect of human dignity is touched upon in many contributions to the debate among Muslim religious scholars of course, but is usually not the key argument. For example, in the statement issued by the IFA (Mekka) in 2002 which aims at forbidding human cloning it is first referred to the argument that procreation has to be brought about within the framework of a marriage. Only secondly can it be read "[God] has honoured [man] by creating him in the best of moulds. Therefore, it is prohibited to manipulate it at any stage of the human being's creation." (Turki, 2003, 32)
30. Wâsil, 1999, 420-429.

31. Salâmî, 2001, 162f.
32. Salâmî, 2001, 171. On the concept of human dignity in the writings of modern Muslim thinkers see Wielandt, 1993, 179-209.
33. Muhammad Sa'îd al-Hakîm, n.d., 27f, 31f.
34. Hasan 'Alî al-Shâdhilî, 1997, 209-213. As mentioned before, Salâmî's contribution to the same session is obviously opposed to therapeutic cloning. It is interesting to note that the decision of the committee dated July 1997 simply omitted the issue.
35. Nasr Farîd Wâsil, 1998, 126f.
36. The right to issue *fatwâs* is not necessarily linked to an official position as Mufti.
37. Iran was the UN representative of the OIC at that time. For a description of the UN debate see the contribution of Alexander Capron in this volume. I am greatly indebted to LeRoy Walters from Georgetown University for referring me to this document and providing me with the relevant political background information.
38. A decision of the Islamic Fiqh Academy of the Muslim World League (Mecca) from December 2003 apparently viewed therapeutic cloning critically, too. Until the date of submitting this paper only a rough draft translation of this document was at my disposal, not the original. This translation raised more questions than it provided answers. For this reason it is only referred to in passing.
39. Yûsuf al-Qaradâwî, 2002 and 2003.
40. Ra'fat 'Uthmân, 2004.
41. Jawâhirî, 2002, 34. In a statement to the same effect Fâdil al-Mîlânî explicitly refers to the famous picture of the mouse with an implanted human ear on its back, which shows the enormous effect of this particular picture in the whole debate (www.alkhoeifoundation.com, Arabic site, accessed 22.1.2004).
42. Wâsil, 1997, 458 uses the term *al-Istinsâkh al-juz'î,* which translates as "partial cloning."
43. Organ transplantation is widely permitted. For a groundbreaking study about the relation of the bioethical discussions about the issue and the establishment of organ transplantation programmes in countries with predominantly Muslim populations see Grundmann, 2004, 27-46.
44. Ayatollah al-Hakîm, n.d., 39f.
45. Umar Sulaimân al-Ashqar, 2001, 307f.
46. *Qarârât*, 1998, 117f.

47. Siddiqi's argument is put forward by Anwar Nasim in several presentations in exactly the same wording; Nasim is the head of a bioethics committee of the OIC which was created in 2003.

References

'Âmilî, Muhammad Jamîl. "al-Istinsâkh…bauq Iblîs." *al-Istinsâkh baina l-Islâm wa l-masîhiya*, 1999, 95-101.

Ashqar, 'Umar Sulaimân al-. "al-Istifâda min al-ajinna al-mujhada au al-zâ'ida 'an al-hâjja fî l-tajârib al-'ilmîya wa zirâ'at al-a'dâ'." *Dirâsât Fiqhiya fî qadâyâ tibbiya mu'âsira,* vol. I., Idem et al. (eds.). Ammân: Dâr al-nafâ'is li-l-tashrî' wa l-tauzî' al-Urdun, 2001, 305-310.

Dawûd, Nâsir b. Zaid al-. "al-Istinsâkh ... bi-l-shurût al-khamsa!" *al-Muslimûn*, 644 (6.6.1997).

Fadlallâh, 'Abd al-Karîm. "al-Mustansakh … yakûn akhan wa shaqîqan li-man ukhidhat minhu al-khilya." *al-Istinsâkh baina l-Islâm wa l-masîhiya,* 1999, 271-278.

Fadlallâh, Husain. "al-Istinsâkh wa l-dîn." *al-Istinsâkh. Jadal al-'ilm wa l-dîn wa l-akhlâq*, 'Abd al-Wahîd 'Alawânî (ed.). Damascus, 1997, 97-102.

Grundmann, Johannes. "Scharia, Hirntod und Organtransplantation: Kontext und Wirkung zweier islamischer Rechtsentscheidungen im Nahen und Mittleren Osten." *Orient,* 45 (1/2004), 27-46.

Hakîm, Muhammad Sa'îd al-. *Fiqh al-istinsâkh al-basharî.* n.p., n.d. (for an English translation of this text see the contribution of Abdulaziz Sachedina to this volume).

(al)-Istinsâkh baina l-Islâm wa l-masîhiya, Center for Islamic-Christian Studies (ed.). Beirut: Dâr al-fikr al-Lubnânî.

Jawâhirî, Hasan al-. *Buhûth al-fiqh al-mu'asir*, II. n.p., 1998. (available under www.al-shia.com).

Jawâhirî, Hasan al-. "al-Istit'âm wa l-Istinsâkh." *Qirâ'ât fiqhiya mu'âsira fî mu'tiyât al-tibb al-hadîth*. Beirut: al-Ghadîr, 2002, 11-36.

Khadmî, Nûr ad-Dîn. *al-Istinsâkh fî dau' al-usûl wa l-qawâ'id wa l-maqâsid al-shar'îya*. Riyâd: Dâr al-Zâhim li-l-nashr wa l-tauzî', 2001.

Kohler, Christian. *Das Vaterschaftsanerkenntnis im Islamrecht und seine Bedeutung für das deutsche internationale Privatrecht.* Paderborn: Schoeningh, 1976.

Moqdâd, Muhammad Taufîq. "al-Istinsâkh al-basharî ... baina l-ilm wa l-akhlâq." *al-Istinsâkh baina l-Islâm wa l-masîhiya*, 1999, 195-200.

Moqdâd, Muhammad Taufîq. *Handwritten statements about cloning dated 11 Jumâdâ II 1422* (1.9.2001) and *4 Dhû l-Qa'da 1423* (8.1.2003).

Qaradâwî, Yûsuf al-. *Two statements about cloning dated 30.12.2002 and 6.1.2003* (available on his homepage www.qaradawi.net, accessed 2.1.2004).

Qarârât wa tausiyât majma' al-fiqh al-islâmî min al-daura al-thâniya hattâ l-'âshira 1406-1418 (h)/ 1985-1997. Damascus: Dâr al-Qalam, 1998.

Rispler-Chaim, Vardit. "Genetic Engineering in Contemporary Islamic Thought." *Science in Context*, 11, 1998, 567-573.

Sabzawârî, 'Alî l-Mûsâwî al-. *al-Istinsâkh baina l-taqniya wa l-tashrî'*. Beirut: Mu'assasa al-A'lamî li-l-ma'bu'ât, 2003.

Sachedina, Abdulaziz. "Human Clones: an islamic View." In *The human cloning debate*, Glenn McGee (ed.). Berkeley, 1998, 230-244.

Salâma, Ziyâd Ahmad. *Atfâl al-anâbîb baina l-'ilm wa l-sharî'a.* Beirut: al-Dâr al-'arabiya li-l-'ulûm, 1996.

Salâmî, Muhammad al-Mukhtâr al-. "Istinsâkh." *Majallat majma' al-fiqh al-islâmî,* 10.3, 1997, 142-160.

Salâmî, Muhammad al-Mukhtâr al-. *al-Tibb fî dau' al-îmân*. Beirut: Dâr al-Gharb al-islâmî, 2001.

Shâdhilî, Hasan ʻAlî al-. "al-Istinsâkh. haqîqatuhu, anwâ'uhu, hukm kull nau' fî l-fiqh al-islâmiya." *Majallat majma' al-fiqh al-islâmî.* 10.3, 1997, 165-213.

Shams ad-Dîn, Muhammad. "Istinsâkh al-bashar … amr ghair mashrû' qat'an wa yaqînan." *Istinsâkh baina l-Islâm wa l-masîhiya*, 1999, 131-136.

Siddiqi, Muzammil. *Statement about surplus embryos.* (http://www.islamicity.com/articles/Articles.asp?ref=IC0202-404, accessed 12.1.2004).

Tabrîzî, Ayatollah al-Mirzâ Jawâd al-. *Fatwâ about cloning* dated 20 Dhû l-Qa'da 1417 (30.3.1997) (www.tabrizi.org/html/bo/sirat/3/34.htm#94, accessed 30.4.2003).

Taskhiri, Muhammad Ali al-. "Nazara fî l-Istinsâkh wa hukmihi al-Shar'î." In *Majallat Majma' al-fiqh al-islâmî*, vol. 10.3, 1997, 215-234.

Tayyâr, ʻAbd Allâh b. Muhammad al-. "al-Istinsâkh." *al-Muslimûn*, 638 (25.4.1997), 8.

Turki, Abdullah al-. "MWL Condemns Human Cloning Operations." *The Muslim World League Journal*, 31, 2003, 31-33.

ʻUthmân, Ra'fat. "al-Istinsâkh fî l-nabât wa l-hayawân wa l-insân." *Qadâyâ fiqhiya mu'âsira*, I, edited by a group of Professors from the Sharî'a Faculty of Azhar University, Cairo: Azhar (2003a), 23-50.

ʻUthmân, Ra'fat. "al-Istinsâkh fî dau' al-qawâ'id al-sharî'a." *al-Akhlâqiyât al-hayyawiya* (a magazine published regularly by the National Bioethics Committee of Egypt), III (2003b), 6.

ʻUthmân, Ra'fat. *Fatwâ about therapeutic cloning* (available on http://www.islamonline.net, accessed 15.2.2004).

ʻUqûbî, Muhammad ʻAbd al-Fattâh al-. *al-Istinsâkh.* unpubl. MA-thesis Azhar University, 2002.

Wâsil, Nasr Farîd. "Ra'y al-Duktûr Nasr Farîd Wâsil Muftî Misr." *al-Istinsâkh baina l-ʻilm wa l-falsafa wa l-dîn*, Husâm ad-Dîn al-Shahâda (ed.). Damascus: Markaz al-'ilm wa l-salâm li-l-dirâsât wa l-nashr, 1998, 124-127.

Wâsil, Nasr Farid. "al-Istinsâkh al-basharî wa ahkâmuhu al-tibbiya wa l-'ilmiya fî l-sharî'a al-islâmiya." *Ru'ya islâmiya li-ba'd al-mushkilât al-tibbîya al-mu'âsira*, vol. II, 'Abd ar-Rahmân 'Abd Allâh al-'Awadî and Ahmad Rajâ'î al-Jundî (eds.). Kuwait: IOMS, 1999, 411-464.

Wielandt, Rotraud. "Menschenwürde und Freiheit in der Reflexion zeitgenössischer Muslimischer Denker." In *Freiheit der Religion. Christentum und Islam unter dem Anspruch der Menschenrechte*, Johannes Schwardtländer (ed.) Mainz: Matthias Grünewaldverlag, 1993, 179-209.

Zuhailî, Wahbâ al-. "al-Istinsâkh, al-jawânib al-insâniya wa l-akhlâqiya wa l-dîniya." *al-Istinsâkh. Jadal al-'ilm wa l-dîn wa l-akhlâq*. 'Abd al-Wahîd 'Alawânî (ed.). Damascus: Dâr al-Fikr, 1997, 115-131.

The Debate on Human Cloning: Some Contributions from the Jewish Tradition

Y. Michael Barilan

Abstract: Contemporary rabbinic discourse on cloning is limited to non-procreative cloning and to the use of cloning technology to overcome an otherwise insurmountable infertility problem between a husband and wife. Without exception Jewish sources do not grant moral status to extracorporeal pre-embryos which are not destined for future implantation and which have not reached the stage of formation. Therefore, the creation of such pre-embryos, even by means of cloning, is permissible and even necessary for the sake of finding a cure for diseases. Some rabbis prefer cloning to other modes of infertility treatment, since cloning does not require the donation of genetic material from a third party. In this paper I analysis the bioethical problem of cloning in the light of the rabbinic literature. I argue that the most problematic aspect of cloning is the indirect, but potentially serious, pressure on women to accept technologies of infertility which are far from innocuous. I also argue from the Jewish experience that overt stigmatisation of cloning might bring about discrimination and even abuse of cloned humans, should this practice become prevalent. Although Jewish law is quite liberal with regard to cloning, Judaism does not try to "preach to the nations". Rabbis and Israeli legislators respect communities who object to cloning, but they also expect tolerance towards Jewish receptivity to this technology.

Key Words: Judaism, Halakha, Human cloning, Infertility treatment, Family relationships, Human dignity, Gender

1. Cloning: the Problems in Hand

Human cloning typically comprises a few basic components:

i. Artificial steps (imitating natural processes) and manipulation (of natural processes) in human procreation.
ii. Creation of human beings that are genetically (chromosomally) identical to other humans.
iii. Alteration of the human genome.
iv. Technology, research and development.
v. Socio-cultural responses to clones and cloning: does he/she have parents or siblings? What is the nature of this parenthood or siblingship?

vi. Partial (non-reproductive) cloning.

These are *different and contingently related* elements. Assisted reproductive technologies (ART) have already introduced the first component, which cloning will only accentuate.

A good model for deliberating the artificiality aspect separately from the genomic one would be that of an artificial uterus that incubates an IVF baby who will never have lived inside a woman's body and which will be parented by the donors of two intact gametes. The thought experiment of the artificial uterus distinguishes genetic manipulation from non-genetic artificial intervention.

Assessing artificiality in fertility is not a simple task. Drugs that induce ovulation may be considered more natural than IVF. On the other hand, these drugs are associated with the unnatural and undesirable outcome of multifoetal pregnancies (triplets, quadruplets etc.). Ironically, it is the greater "artificiality" of IVF which substantially reduces the risk of multifoetal pregnancies, since the number of embryos implanted is under control. In the same vein, the currently high rates of miscarriage and medical problems among cloned animals throw into relief the perception of distorted artificiality inherent in this technique. Another example would be that of contraception. Both Judaism and Christianity articulate a commitment to "natural" sex. *Halakha*, which is more empiricist than rationalist with regard to natural order, and which is more sensitive to phenomenology than Natural Law, regards the pill as more natural than artificial barriers such as the condom.[1] The rabbis are aware of the physiological impacts of the pill. However, since the pill does not interfere with erotic sensibilities and with physical intimacies, it is more natural. The emphasis on phenomenology will resurface in the discussion on cloning.

If cloning technology becomes safe, it may enable us to bypass defective genes and reduce the incidence of genetic disease. Whenever we clone a healthy person, we do not have to worry about such conditions.

Since prehistory humans have mastered many techniques of genetic manipulation (see Genesis ch. 30) in grafting plants and in breeding animals. Therefore, it is not the lack of biotechnology that has prevented humans from breeding a race of "tall warriors" or "beauty queens," but a complicated mesh of social facts and moral norms. Even pro-slavery and totalitarian societies have not undertaken such endeavours.

In contrast to germ-line therapy and genetic engineering, cloning only replicates a genetic array that has been created naturally. Cloning is not about *tampering* with natural genomes, but about *copying* them. This distinction sheds light on two different kinds of artificial intervention: mimicking nature and altering it.

These four modes of intervention, non-genomic vs. genomic and copying nature vs. changing it, may combine in ways that could bear on moral status ("human like you and me" or not) and personal identity ("who I am and how I am related to other people: 'parents,' 'siblings,' other 'clones' - i.e. all those who have been cloned, although not of the same genetic profile").

The genetic overlap between a clone and his or her "original" is equal to that between identical twins. Hence, being genetically identical to another person should not be morally problematic in itself. If something is wrong in human reproductive cloning, it must be sought within either the *process* or within the *intentions*. The genetic manipulation associated with cloning might be of significance only in the context of mass cloning.

Cloning on a large scale may be manifested in two ways: mass cloning of a single genome and a high prevalence of cloned people, each originating from a different genetic source. In the first case one person "sires" numerous genetic copies. In the second case, many people resort to cloning as a means to reproduce, each "siring" only a few clones. Both options need to be distinguished from a third mode of reproductive cloning, i.e., cloning as the last means of begetting genetically related progeny. Because severe and refractory infertility is quite rare, cloning as a means of treating it is not likely to bring about demographically significant mass cloning in either of its manifestations.

Many arguments against cloning appeal to socio-cultural (component 5 or some combination of 4 and 5) misgivings such as the "enslavement" of clones or some other derogatory and exploitative relationship between "cloners" and their "products." These reservations relate neither to the technology nor to the essence of cloning, but to projected socio-cultural reactions to it. Possibly, none of these reservations would apply in the context of cloning as a last means of begetting genetically related children.

Unfortunately, the current debate on the ethics of cloning pays little regard to these and other distinctions, which might be found essential for formulating practical, non-dogmatic and morally sensitive solutions to the debate.

Besides, there is a widespread fear that the enterprise of cloning might go astray, thus bringing about unexpected mutations, maladies and social problems. But this concern relates to the reliability and efficiency of the *technology* involved, not to the *concept* of cloning.

Research, development and commercialization of human cloning necessitate extensive experimentation on human beings, on women who would carry cloned babies within their wombs and on the clones as well. The problem of experimenting on humans applies to all fertility technologies, not specifically to cloning.

Partial cloning is the use of cloning-related technology in order to create, at least in the eyes of the creators, something less than a human being endowed with human rights.

Therapeutic cloning is the only kind of partial cloning currently under discussion, but other purposes might surface as well, basic science for example. Partial cloning deserves discussion within the context of research ethics.

2. Human Dignity

The principle of human dignity is pivotal in Western ethics as both the humanistic and the Judeo-Christian traditions regard it as a *fundamental* value. They link human dignity with membership in the human species, with human rationality, and with the principle of neighbourly love. In this context, a discussion of the human dignity inherent in human pre-embryos makes sense. They are human.

In the Jewish tradition,[2] however, ethics is very loosely linked with questions of metaphysics. Human dignity is associated with the possession of a unique human body,[3] a personal human frame, which is or has been alive. Its dignity stems from the capacity of the human body to procreate and to talk. Jewish law, *Halakha*, regards the duty of neighbourly love as separate from the and indeed superior to duty to respect human dignity. This is possibly why the *Halakhaic* discourse invokes the term "human dignity" mainly in the context of treating *dead* human bodies. When a person is alive, his or her values, feelings and wishes, shape the content of our neighbourly love towards him or her. According to the Talmud, neighbourly love applies only to persons after birth, whereas human dignity is accorded to formed foetuses.

According to the tradition of Natural Law, neighbourly love is shaped by the concept of the common good.[4] In Jewish law and ethics, neighbourly love is determined by the subjective state of mind of the other person, his or her feelings and thoughts, no matter how bizarre or immature.[5]

Although many Talmudic sources refer to formed foetuses as moral human beings the rabbis simply do not find this relevant to determining their moral standing. The sources never explain why.[6]

According to the Talmud, neighbourly love applies only to persons after birth. The rabbinic sources persistently refrain from rational arguments in defence of this position. It is attributed to a divine decree. For the Israelites, God decreed that legal personhood begins at birth (Talmud *Niddah* 44a).

In this light we can see why contemporary rabbis and theologians sanction the utilisation of in vitro pre-embryos. They are not endowed with human dignity and even if they were, the commitment to the lives and health of the living takes precedence over the vital interests of fetuses.

Moreover, cloning expands human procreation and creativity. It also serves the interests of those who wish to pursue it. Their investment, aspirations and liberties should not be obstructed without good reason. From a Jewish point of view, restrictions on science, medicine and entrepreneurship *frustrate* the promotion of human dignity, which is based on the covenant between God and humankind, "fill the earth and subdue it" (Genesis I:28).

These differences as well as the similarities in construing the value of "human dignity" within Judaism and in the West challenge us to look beyond the words "human dignity" and face the beliefs, motives, ethoses and other factors operating behind them.

3. Naturalism and Positivism

Elsewhere I dwell at length on Jewish law being mainly a positive legislation whose scope is limited to the Jewish people.[7] Not only does naturalism not characterize Jewish law, but the principle of non-interference with nature does not even exist in Judaism. The sources harshly rebuke those who wantonly destroy natural objects. But there is no objection to transforming nature in order to create, to improve upon or even to convert reality into a better one. "Better" in this respect is explicitly defined from a human point of view.

The Torah forbids the cross-breeding of plants and animals. The rabbis usually interpret these laws narrowly. The ban only applies to some animals mentioned in the Torah (e.g. ox and donkey) and plant species (edible roots and plants), while all other modes of biotechnology are permitted.

However, some rabbis distinguish between exploiting nature and creating new species. Only the latter is offensive to God and to His Creation.[8] A prominent Israeli Rabbi and bioethicist, A. Steinberg concludes that reproductive cloning is not objectionable since it does not create hybrids or new species.[9]

The absence of natural, holistic and systematic analysis of moral problems is liable to bring about dry formalism that is insensitive to the complexities of human life and to the richness of morality. For example, the rabbis, who grant no legal status to in vitro pre-embryos, do not realise that their standing would be seriously problematised should biotechnology find ways to facilitate the extra-uterine development of in vitro pre-embryos into formed foetuses. This is not a fanciful proposition, but the next step ahead. Would it be permissible to sacrifice an artificially bred human foetus for the benefit of a born human? Is it moral to farm foetuses in incubators for the sake of dismemberment, experimentation and other forms of medical use? Can the supremacy of neighbourly love over human dignity and the valuation of born people over foetuses be extended in this way?

The Rabbinic sources and their methods of deliberation are open to different strategies of handling these and similar questions. In the absence of full and comprehensive discussion, the current Rabbinic approval of cloning - this is, for example, the official position of the Chief Rabbinate of Israel - cannot be regarded as a mature position.

Ethicists operating within the framework of natural law have opposed many practices which are now widely accepted, for example, vaccination, corneal transplantation and in vitro fertilization. Encyclical Catholic objection to contraception on the grounds of natural law thinking also raises concerns about the true meaning and applicability of this notion. Even Catholic theologians and clergy are less than pleased with these decrees. It is very difficult to see how such rulings conform to natural law, being eternal, universal and rational.

Critics such as Adorno and Foucault have repeatedly warned against the seeds of tyranny inherent in enlightenment ethics and universal naturalism. Jewish positivism, which does not appeal to absolute moral laws or to the eternal laws of nature, but to a covenant between God and a small nation, is more receptive to multiculturalism than Catholicism, Kantianism, Utilitarianism and similar schools of ethics, which speak in the name of universal rationality.

According to the Jewish tradition, only a few and very basic laws pertain to all people and peoples, the prohibition of murder, for example. Jewish disregard of metaphysics and Jewish strong emphasis on casuistic discourse places the destruction of pre-embryos in territories far removed from those of killing people. While Jewish thinkers may harshly condemn the abuse of human rights anywhere in the world, the treatment of pre-embryos is regarded as a matter to be decided by each and every human community. The division of ethics into a fundamental, straightforward, universal and unbending set of precepts on the one hand and elaborate, refined and particularized systems of conduct on the other is a stimulating model for multiculturalism that does not drift into mere relativism.

My interlocutor might insist on knowing the *principles, the systematic laws* which draw the line between universalism and pluralism or which *define* the beginning of the person. My response would be that such pursuit of rules and definitions characterises *doctrinal* religions and ethics. Resisting strict doctrinal formulations is not necessarily an avoidance of ethical commitment and discourse, rather it may strengthen it. This is borne out by a Talmudic vignette.

According to Jewish law, a chick found fifty cubits away from a nest, belongs to the finder, not to the nest owner. When a rabbi inquired into the status of a chick that straddles the fifty-cubit mark, the rabbi was expelled from the *Beth Midrash* (Rabbinic academy) (Talmud, *Babba Bathra* 23b). This story conveys a Wittgensteinian lesson: pushing moral discourse beyond its scale of exactitude and thoroughness brings about nihilism, rather than clarity or perfection.

4. Community Values vs. Individual Liberties

Jewish law and theology have no objection to *beneficial* actions that do not violate the laws of the Torah. Neither the rabbis nor the civil authorities can add a single prohibition to the Torah. They can only introduce *temporary* and *local* prohibitions and only in order to promote or to prevent harm to either religion or people. Deliberations focus on the nature of anticipated harm, the evidence in support of its occurring etc. General terms such as "offense to human dignity," "being against nature," or "playing God" are avoided. The Jewish tradition recognises the right and duty of each community to enact laws that promote the interests of the public and the values of religion. Sometimes rabbis overrule community regulations and laws on the grounds of disproportionate harm to a person or a group of people. Generally speaking, they are quite tolerant with many restrictive regulations. Rabbinic law distinguishes between principal legislation and rules (*takanoth*) enacted by leading rabbis and community magistrates in order to address specific circumstances. Regulation of genetic technologies may only fall within the ambit of *takanoth*, which, by definition, are always temporary, and whose authority does not stretch beyond the local community from which they derived.

As I see it, this could serve as a prudent and moral strategy. We know too little about cloning to justify its excommunication as if it were a cardinal sin. The rabbis teach us that everything is permitted until we find a positive reason to forbid it. Yet we have to keep in mind that these rabbinic rules were created long before the age of technology. Close monitoring of social and scientific developments, monitoring which is highly sensitive to all aspects of the problem, seems to be a more appropriate and productive framework for shaping public policy.

On the other hand, Rabbi Shafran (a lecturer in medical ethics at the Hebrew University in Jerusalem) points out that *Halakha* cannot cover all aspects of public policy.[10] The sages had no legal pretext to prevent certain harm. For example, King Hezekiah, and not the sages of the time, put away "a book of medicines" that had misled the people (Talmud *Pesachim* 56a). Shafran calls for a temporary prohibition on cloning coming from the community, not from the narrow circles of *Halakhaic* sages (*Poskim*). This is the essence of the *takanoth*, a civil legislation enacted by religiously committed Jews with the participation of the local rabbis.

Today, many ultra-orthodox Jews (*Charedim*) refuse to recognise the authority of the state of Israel, whereas secular Jews tend to ignore theology and *Halakha*. This situation stymies genuine efforts at explicating a viable "Jewish" or "Israeli" attitude to communal issues such as cloning.

Historically, many *takanoth* have been introduced so as to meet legal or moral standards prevalent in the non-Jewish environment. The rabbis and the Jewish leadership have been highly sensitive to the image of Judaism in the eyes of the non-Jew. Jewish communities may observe their own laws and ways of life, but they must never create an image of moral laxity. Rabbis have made the Jewish laws on abortion stricter in order to better conform with Christian morality on sensitive subjects such as human life and dignity. Jewish authorities have annulled local laws which clashed with the law of the state.[11] Elsewhere I have conjectured that leading rabbis might enact *takanoth* against stem cell research and cloning if "slippery-slope" arguments are found cogent or if these practices become near-universally unacceptable.[12]

The Jewish discourse within Israel is much less sensitive to non-Jewish morality. Rabbis and ethicists stress Jewish self-determination as well as a sense of alienation from "Christian ethics" that failed to stop the Nazis.

Even if some rabbis try to introduce *takanoth* that restrict cloning as an infertility treatment, others might overrule them on the grounds of causing disproportionate suffering to vulnerable and needy individuals.

Indeed, contemporary Rabbinic discourse is disinclined to restrict the liberties of individuals whose conduct does not collide with *Halakha*. One notable example is a recent ruling by Rabbi Dr. Halperin, an expert on fertility and the bioethical advisor to the Israeli Minister of Health.[13] Discussing the ethics of IVF and surrogate uteri, he expresses discontent with many aspects of ART and articulates concern about the degradation of "family values." However, acknowledging the significance of child-bearing to Jewish life and the fact that IVF does not violate any religious law, he thinks society cannot interfere with women's access to IVF. The Torah forbids extra-marital sexual intimacy. But IVF involves neither adulterous eroticism nor illicit sexual contact. Halperin explains that although rabbis cannot forbid what the Torah does not forbid, they often resort to subtle modes of discouragement. He interprets the silence of a famous rabbi with regard to IVF as an indirect way of expressing disapproval. He cannot and would not restrict the liberty of people who otherwise abide by the laws of the Torah, but neither would he endorse their actions.

This exemplifies how relative liberalism combines with non-naturalism to allow individual choices of reproduction that are somewhat incompatible with the values of Jewish society. From a strictly Rabbinic point of view, only the laws of the Torah, not social values thus formulated, can restrict the liberty of individuals.

However, leading rabbis (*Poskim*) sometimes forbid things which they find inconsistent with the "spirit" of the Torah, even if not with its literal laws. Such rulings are usually issued in the context of *intimate* and *personal* consultation, a weaker context than that of *takanoth.* Ultimately, once they have been made public, they diffuse into Halakhaic discourse.

Halperin cites the same Rabbi, Auerbach (d. 1995), who opposed the preservation of the semen of an unmarried man who was about to undergo chemotherapy that would render him sterile. The Rabbi did not explain this ruling.[14] Possibly he sided with the opinion that bachelors were not obliged by law to procreate, so the man in question had no *Halakhaic* motivation to preserve semen. This line of thinking does not explain why Rabbi Auerbach *forbade* the act. It also fails to explain why the Rabbi did not instruct the man, who was already betrothed to a woman, to get married right away and then to freeze his semen. The lessons we may learn from this case are that even a powerful and fundamental ethos like procreation does not bend *Halakhaic* modes of legal reasoning, and that, on the other hand, rabbis may invoke "the spirit of the Torah," which is possibly a euphemism for some form of naturalism, in order to restrict human conduct, at least when this conduct falls within a "grey area" of the law. I assume that Auerbach wishes to convey that the value of procreation is not merely about achieving an end - having natural progeny, but *also* about the means harnessed to that end, and about the chain of intentions and events that link an immediate action to that end. Auerbach permits the emission of semen by a husband in the framework of a couple's infertility treatment.[15] Emission of semen by an unmarried man who might soon die is probably too many steps removed from paradigmatic family life. Besides, if he dies, the frozen semen might loom over the future of his partner and her ability to make a fresh start. Auerbach's avoidance of systematic discussions and of explicating whether this was only a isolated ruling tailored to the personal situation of one specific man is a ploy of avoiding doctrinal statements and of setting a precedent.

Rabbis might object to cloning due to considerations of *degree* rather than of *principles*. They may also seek *Halakhaic* ways, even legal tricks, to prevent acts that entail too much deviation from acceptable parenting, or to pre-empt acts that invite interpersonal conflict.

My conjecture is born out by a short *Responsum* on cloning by Auerbach's son-in-law, Rabbi Goldberg.[16] He rules that if registration of all the people involved in the process is properly carried out in a way that prevents disputes over identity, cloning is permissible. In the same *Responsum* he permits the freezing of semen taken from an unmarried man who is about to undergo chemotherapy. Auerbach, who objected to premarital preservation of semen, would probably have objected to cloning as well.

Opinions vary over time. My impression is that contemporary rabbinic rulings are much more receptive to ART than those issued twenty or thirty years ago. Contemporary *Halakha* appears to accept the ethos of fertility rather than be engaged with it in a dialectical process of inspiration and restraint.

Goldberg's *Responsum* on cloning is short. It contains one word: "permissible," without offering explanations or entering a discussion. This extreme brevity, as well as Auerbach's silence, illustrates the difficulty inherent in analytical attempts at understanding *Halakha* or to behold it through a narrow historical window. Even if such modes of discourse are functional in certain communities, they are far from being a corpus of knowledge ready for translation into official guidelines at the national or international level.

5. The Problem of "Designer Babies," the "Right to an Open Future" and the "Right to a Unique Genetic Profile"

As far as I know, Jewish rabbis and theologians have not discussed these topics. Of course, philosophers and rabbis all wish for a humane society. The differences lie in the methods and language used. This situation creates an opportunity to re-examine the motivations and values behind the notions of open future, genetic identity and genetic heritage.

The idea of an open future reflects a prevailing wish for humans not be controlled or programmed by external agents. Chance in child bearing is not a value in itself; promoting it merely expresses the rejection of engineering and design. However, nobody denies that responsible parenting requires prudence and discipline. Parents who exert no influence on the future of their children are considered negligent rather than responsible. Does a child who was conceived promiscuously and hence accidentally have a future that is more "open" than a child begotten by a married couple and within the context of family planning?!

Let us consider a child whose parents have "genetically designed" two of his traits: eye colour and an athletic disposition. The child's upbringing was mild, being neither too permissive, nor too restrictive. Compare this child to his neighbour whose parents, whilst not interfering with his genes at all, made decisions that had a fateful effect on his future: they sent him to a very authoritarian religious boarding school and denied him the means to pay his way through college. Does the latter child have more of an "open future" than the former? Does untoward "design" of people apply only to somatic intervention and not to cultural and psychological influence?

Many parents strive to direct and control the future of their children. Some over-ambitious parents push their young into competitive sports or beauty pageants; others encourage total dedication to the cultivation of a particular talent. A tennis player since the age of six who breaks her leg at twelve is in the same predicament as a cloned "Elvis" who is found to be hoarse.

I believe that this thought-experiment illustrates that the real issue at stake is *moderation* and *proportionate judgement*, very old virtues which have nothing to do with the language of rights and with genetics.

Shifting the discourse from the question of the *kind* of control, an issue that is always relative to particular socio-historical circumstances, to the question of the *degree of control*, may help us construe a cross-cultural, pragmatic and morally sensitive deliberation of the ethics of parenting.

6. Women's Health and Other Feminine Issues

So far we have focused on the *concept* of cloning. However, cloning is deeply implicated in IVF, which, to date, necessitates the exposure of women to supra-physiological doses of potent hormones and then subjecting their bodies to the invasive retrieval of eggs (usually done under general anaesthetic) and the implantation of embryos.

The overlap of fertility medicine and research is a source of serious bias and confusion. One solution is to separate the two realms of practice. The UK allows ovulation to be induced and eggs to be harvested solely for the sake of scientific research. A ban on redirecting eggs from procreative to scientific goals will complete the process of separation, but it might also create a disagreeable situation in which supernumerary pre-embryos are destroyed while women are subjected to the travails of egg harvesting.[17]

Principally, fertility medicine should aim only at achieving successful pregnancies, with zero supernumerary embryos. Research on cloning is dependent on supernumerary eggs and embryos, wherever the non-procreative extraction of eggs is prohibited, as in Israel. The conflict of interests is obvious and in need of regulation.

Besides, the Israeli National Health Insurance generously supports fertility treatment. Indeed Israel ranks high in terms of IVF per capita. Many Israeli clinics prescribe large doses of hormones and tend to implant more than one embryo in each cycle of treatment. I find these practices highly problematic. Although all the women treated sign a declaration of consent, most of them are unaware of alternative regimens, such as longer periods of natural attempts at conception, lower doses of hormones, and implantation of single embryos. The centrality of procreation both in Islam and in Judaism puts considerable pressure on women to submit themselves to the risks and pain associated with fertility treatment. Possibly some of them would defer treatment in favour of longer attempts at natural conception if it were not for the pressure exerted on them by husband, family and culture. Possibly the eagerness of doctors to succeed and to "help" their "patients" intensifies at pressure, and the unchecked pursuit of conception results in a vicious circle.

Procreation is a fundamental value in Judaism, and many women take grave medical risks in order to have their own children. On the other hand, Rabbinic Judaism has always been mindful to protect women from *legal* pressures to beget children.[18] The duty of procreation is binding only for the male, and this is attributed to the unequal distribution of suffering and risk between the genders. Moreover, many rabbis permit abortion whenever the woman suffers too much from the pregnancy. Rabbis have given permission to volunteer for medical experiments that involve some risk - but not mortal risk - whenever those experiments are likely to contribute to saving the lives of others.[19] Procreative cloning is far from being life-saving, and the risks associated with IVF and pregnancy are not trivial.

In my view, until clear and efficient regulations of fertility medicine exist in Israel (as well as in other countries), talking about "supernumerary" embryos is no more than a red herring, particularly as Israeli fertility clinics suffer from a *shortage* of eggs and pre-embryos. Can supernumeracy truly coexist with a shortage?!

I do not find aggressive fertility regimens ethically acceptable. In my view this should be the most important ethical consideration in the whole evaluation of assisted reproduction in general and of cloning in particular.

Besides, I personally suspect that the bioethical emphasis on the human dignity of the cloned person in the discourse on cloning reflects an innate cultural disregard of women's health and their human rights. I fear also that the trendy focus on regulating cloning and allied techniques signalises an implicit, but unjustified, despair of successful regulation of fertility medicine in ways that protect and promote women's health, their moral status and social standing.

7. Maternity, Paternity and the Family

The issue of women's health invites us to take a close look at cultural concepts of maternity and family. The risks of pregnancy and labour are widely accepted in all human societies, even in times and places in which those risks are quite considerable. Many women willingly *sacrifice* themselves for the sake of *their own* children.

What constitutes the essence of maternity, what makes a child one's own?

There are a number of components,[20]

1. Sexual: begetting through the sexual fertilization of one's ovum.
2. Phenomenological: intimate relationship, conception, pregnancy and labour.
3. Genetic: from a purely genetic point of view, maternity (or paternity) through cloning is the most exact mode of parenthood: there is complete chromosomal overlap between parents and offspring.
4. Social: acceptance of a child as one's own. Adoption is an example of social parenting. Although it is not a method of pro-

> creation, but a solution for parentless children, many parents regard adoption as a solution to their own childlessness.

These paradigms are not complete. Babies conceived by means of IVF from the gametes of their social parents are unanimously considered the "natural children of their parents." Similarly maternity is not compromised when vaginal delivery is substituted by Cesarean section. As long as the deviations from the typical paradigm of parenting are minor, parenthood does not become problematic.

The above components might also overlap. Usually options 1,2 and 3 overlie. Surrogate motherhood divides maternity between a woman who assumes role 1,3, and 4 and a woman who plays role 2.

Halakha recognises only biological parenthood, thus excluding option number 4. Adoption does not exist in Jewish law. A child may find a home in a foster family, but *legal* parenthood cannot be separated from biological parenthood. Rabbi Halperin thinks that anonymous sperm donation and certain forms of adoption violate a person's "right to know the identity of his or her biological parents."[21]

Rabbis debate the essence of biological parenthood. They consider potential conflicts between the different components to be a serious problem. Some see this as a reason to ban surrogate motherhood; others believe that legal arrangements can anticipate and resolve potential conflicts.

When parenthood is already divided, some rabbis argue that maternity is determined by birth, not genes. Others insist that both women, the "genetic" and the parturient mother, be *Halakhaically* regarded as the child's mothers.

Rabbi Sheilat makes the case that the *Halakhaic* duty to procreate is consequentialist:[22] one must have a living and fertile child, no matter how unusual its conception might have been. Children who died in the lifetime of their father play no role in this fulfilment of the duty of procreation.[23]

He also points out that according to *Halakha*, a child conceived through illicit sex between a slave and a Jewess is considered a father-less person.[24] The semen of the slave is believed to rot in the body of the free Jewess, leaving no impact on the foetus beyond stimulating the egg to divide *The Ethics of Human Cloning* and the *Sprout of Human Life*.[25]

Shafran suggests that in the absence of a father, the only parent, namely the biological and social mother, takes over the role of the legal father in addition to being the legal mother.[26]

Following these discussions, Sheilat approves of procreative cloning. He realizes that cloning may facilitate single motherhood in a way that does not transgress *Halakha*. Yet, he thinks that cloning might be opposed to the true spirit of neighbourly love. He believes that bringing a child into a parental unit of less than two people is unfair. I find in this ruling an opening to what I later refer to as “new eugenics.”

Following Rabbi Auerbach, Sheilat suggests that using a donated somatic cell for cloning is problematic even when its DNA is substituted by that coming from the “parents” of the clone. The Talmud (*Yebamoth* 32b) traces the etymology of the word bastard (*mam-zer*) to “foreign defect” (*mum zar*). Even a non-genomic trace of a “foreigner” may provoke a conflict in the family and cause confusion over identity. On the other hand, Sheilat finds the use of donated eggs less of a problem than the use of somatic cells. His reasoning is based on an analogy from a Talmudic discussion on a hermaphrodite who sires a child.

Virtually all rabbis support cloning as means of treating infertility in a married couple. The genetic make-up of the child will thus be quite unusual, but the rabbis, I believe, are concerned with genetics and sexual fertilization only where conflicts over paternity may arise. An adopted child will always live with the psychological and cultural ghosts of his biological parents. Many adopted children and adopting parents manage to completely cast aside the shadow of the “other” parent from their daily lives, but thinking of absent parents and trying to find them is always a *possibility*.[27] Where a child is constantly searching, consciously or subconsciously, for his or her “true” parent, and men and woman are similarly embark on a search for their “lost children,” serious interpersonal and intrapersonal conflicts may ensue. The “foreign defect” is not a biological taint, but a psycho-cultural presence, the “ghost” of another parent,[28] and the “absent presence” of a child, a sibling or even that of an uncle or a cousin. One advantage of cloning over other fertility techniques lies in the fact that it does not compound the division of parental roles; rather it reduces it.

This way of thinking neutralizes the problem of overt exploitation: harnessing one’s body and health to the fertility agenda of another person. Following this school of Rabbinic teaching, I wish to infer that as long as a child has only two parents (or claimants to parenthood), the precise mode of biological continuity is of little significance (provided that the child’s health is not affected).

If my conjecture with regard to forestalling interpersonal conflicts is cogent, there should be no difference between donating a somatic cell and donating an egg. Each donation involves the presence of a foreign element, and in the case of an egg donation, this element - the genome - is much more significant, biologically and culturally. However, we have to bear in mind that my conjecture is an attempt at a philosophical generalization whereas *Halakha* is, at least to some extent, resistant to such endeavours.

Sheilat himself probably senses this difficulty as well. He finds no "foreign defect" in a cloned pre-embryo that contains *no genetic material of another woman* and which is implanted inside a surrogate mother, but he points to people's awareness of the *social* presence of the other mother as a serious moral issue.

Possibly here *Halakha* is not interested in a theory of motherhood and in analyses of the kind I have addressed. The rabbis are merely concerned with claimants to parenthood or to siblingship.

I find the Rabbinic position wanting in the sense that it focuses only on potential conflicts between the family and other people, thus ignoring implicit conflicts between husband and wife and between the values of society and the well being and human rights of women.

8. The New Bastardy

Broadly speaking, a bastard is a person whose social status is compromised because of the way his or her parents brought him or her into the world. Bastardy is not merely an abstract legal status; it is often a powerful personal stigma and a considerable social disability. The Hebrew Bible and virtually every pre-modern society has its own system of bastardy. Jewish law is relatively severe because even legitimate descendants of bastards are not allowed to marry into the main community, no matter how remote he or she is from the original bastard (see Deut. 23:3 and *Shulkhan Arch, Yore De'a* 265).

Not only do the rabbis acknowledge bastards as innocent, but they also describe them as inconsolable victims of [Halakhaic] injustice (*ashuk*) (*Leviticus Rabbah,* 32). The problem of bastardy also casts a shadow on Jewish messianic visions. One Talmudic authority claims that the Prophet Elijah, who according to tradition will herald the coming of the Messiah by resolving all *Halakhaic* and theological disputes, will also expose all bastards. Another Talmudic authority disagrees, pointing out that the second coming of Elijah is meant to bring peace on earth, not to exacerbate social tensions (*Mishna Edduyoth*, 8:7).

The Midrash offers only a futurist solution: in the World to Come, God will cleanse bastards from the alloy (*pessoleth*) within them in the same way as He will heal the cripples and the sick. Rabbis of later generations ruled that suspicions of bastardy should not be pursued (*Tosephoth* on Talmud *Zebahim* 45a). Most rabbis tend to prefer peace and reconciliation to the exposure of "true identities." In other places the sages indicate that the value of peace justifies lying and even erasure of the Holy Name of God (Talmud *Sottah* 7a).[29] The fundamental Rabbinic code of law, *Shulkhan Aruch*, (*Eben Ha'Ezer,* 2:5) rules against inquiries into allegations of bastardy and against the exposure of bastards. This was a landmark ruling, since it enabled bastards to emigrate under assumed identities and to assimilate with the Jewish community of their destination. It is notable that no Orthdox rabbi calls for declaring the old law of bastardy obsolete.

The problem of bastardy divided Jewish society in late antiquity. Religious zealots, dedicated to the idea of social "purity," persecuted bastards (and alleged bastards) in order to brand them, physically as well as socially.[30] The lenient and revolutionary Rabbinic ruling cited here developed as a result of the terrible internecine bloodshed? brought about by the laws on bastardy.

I am concerned that the taboo on cloning might create a modern bastardy which will stigmatize clones, or alleged clones. They will live in a society that promotes the belief that their very creation violated "human dignity." The police will monitor fertility labs, and agents of the law might force alleged clones and their purported "parents" to submit DNA samples in order to verify their status.[31] Children will grow up while, in the background, legal proceedings take place in order to verify the extent of the genetic manipulation perpetrated on their own selves and while experts testify to the degree of similarity between their genes and the genes of some other person or persons. The net cast to fish clones will catch many more disturbing findings such as illegitimate children whose prevalence in the West is estimated in the range of a few percent. Such findings will certainly disrupt many happy families. Genetic privacy will be jeopardized.

My interlocutor might argue that fierce condemnation of child abuse does not harm - rather it protects - the victims of abuse and the dignity of children. Indeed so. But child abuse is an offense against a child. Being a clone is what the child *is*. Policing fertility is different from monitoring suspected molesters. It will also discriminate against infertile people who will find themselves under scrutiny, whereas the sexual life of fertile people is protected by the right to privacy.

Besides, if we do grant cloned persons full human and legal rights and if we treat them like any other human being, as everybody agrees that we should do, and if they assimilate successfully with human society at large (e.g. "normal" men will marry "cloned" women etc.), there will come a day when we will wonder what is wrong with cloning people in the first place.

I believe that these are the very mechanisms responsible for the disappearance from modern society of bastardy as an important factor. We may condemn certain forms of sex, but unless we have a *personal interest* in it, we do not pry too much into the circumstances of the begetting of our friends, colleagues and family.

We usually pride our society for not policing irresponsible conception, resulting, e.g. from un-protected sex between a penniless schizophrenic and her alcoholic and abusive friend, or an illicit relationship between a married woman and her lover. Is cloning the first mode of procreation that deserves control by the state?

Those who are mindful of "slippery slopes" with regard to cloning have to be equally concerned with the "slippery slopes" associated with policing sexuality and family life.

9. The Interests of the Child/Clone and the New Eugenics

Animal models of cloning show very low rates of success and significant rates of morbidity. These observations cause uneasiness about applying current techniques of cloning to humans. Until we can guarantee clones a reasonable level of health, we must not bring them into the world. The *Shulkhan Aruch* (*Eneb HaEzaer* 2:7) forbids a marriage that is certain to produce offspring with serious disabilities such as epilepsy. It is bound to object to cloning that is based on current technologies.

When animal models mature, developers of ART might proceed with human cloning. I can see no circumstances that will justify this kind of experimentation on humans. Sterile couples might wish to take the risk, but they cannot do so on behalf of the future child/clone.[32]

Somebody, somewhere, nevertheless, may develop a safe, evidence-based method of cloning humans which will not put the future child/clone at risk.

The international declarations on the human right to a unique personal genome seem to rise to this challenge. In my opinion they are also the harbinger of a new wave of eugenics. If a person has a right not to have a genome that is identical to someone else's, why not grant him or her a right not to suffer from a genetic disease, or a right not to be born to incompetent parents? The chances of a man and a woman who are both afflicted with a psychiatric disease begetting offspring with a psychiatric disease is in the region of 50%. Are we going to interfere with such marriages or to condemn them?

Besides, why focus on genetic identity and not on complete HLA matching? Indeed, immunological (HLA) identity is a case in point. Many envision an international data bank of HLA profiles which would allow bone marrow donors to be matched with needy patients. Accidental sharing of HLA identity is not found to undermine human dignity, rather, it enhances human solidarity and it saves lives. Complete genetic identity, if properly construed, may also serve as a source for organ donations and other forms of altruistic gifts.

However, if the clone is as similar in appearance to his or her parent as are identical twins to one another, this might cause a problem of human dignity.[33] The Jewish ethos of human dignity is based on morphological diversity. Strong resemblance to one's social parent is not an issue,[34] but the possibility that a few socially unrelated but morphologically identical people walk the streets might inadvertently erode our deep conviction about the equality and dignity of each and every human being. However, this is more of a concern with regard to mass-cloning than with regard to sporadic cloning. Do I care if somewhere in Indonesia there is a person whose genome is identical to mine?

10. Mass Cloning, the Mad Scientist and the Infertile Couple

Thanks to novels and films such as *The Boys from Brazil* (1977), the concept of cloning is almost irrevocably associated with the idea of mass-cloning. In that book, a fugitive Nazi doctor clones a gang of young "Hitlers," using DNA taken from the original. Even within less fantastic contexts, the word *cloning* invokes powerful associations: totemism, tribalism, branding, and mass reproduction of commodified people.

Our discussion leads us to a crucial distinction between two paradigmatic "cloners." One is a person who cannot procreate children by using his or her own gametes. For this person, cloning is the only way of having biologically related child. So far, I have found no reason to object to this way of coping with sterility, provided proper protection of the woman and the future child is guaranteed.

The other is the "mad scientist" who pursues cloning to replicate specific individuals or a specific "genetic recipe." His agenda is exploitative and manipulative. Both the mad scientist and those who fear him tend to over-estimate the role of genes in human life. Even worse, this form of geneticisation undermines our belief in the freedom of choice and in moral life (can a German dictator grow in the Jungles of Brazil and during a time of peace?). If we really think that a cloned Hitler will become a human monster, why blame Hitler himself for his crimes? After all Hitler was not responsible for his own genetic profile.

Mad scientists must be stopped simply because they exploit women for the sake of unsound and possibly immoral schemes. Megalomaniac agendas also seem incompatible with the moral and normal parenting of the cloned "tribe." Mass cloning cannot be carried out in democratic countries. Mad scientists always need brutal dictators to supply them with women to experiment on and to exploit as well as with an authoritarian system to "educate" the cloned tribe. As a matter of fact, this madness is so absurd that I doubt we will ever face the need to stop a single mad scientist. On the other hand, fear of mad scientists should not stigmatise infertile parents who suffer from "gametic incompatibility."

So far, there has been no need of legislation to deter men who conspire to emulate King Solomon by siring children with a thousand women, thus forming a clan of their own. Enforcement of paternal responsibilities will make the procreation of children by cloning as much a burden (and a blessing) as having them naturally. A mad scientist can operate only when backed by stupendous amounts of money or by a regime that will exempt him from conjugal and paternal liabilities.

Obviously, responsible parenthood entails much more than paying alimony. My point is that if moral insight does not stop people from begetting children they cannot bring up properly and decently, the financial burden will restrain them.

Fathers are accountable for their children because they are accountable for making a personal and moral choice, having sex. People are expected to control their sexual behaviour, their gametes, but they cannot control the whereabouts of their genes. Cells slough away unnoticeably. The Talmud (*Haggiga* 15a) discusses the case of a woman who is impregnated in a public bath by semen that has been accidentally emitted by a bather. This is really an exceptional situation. Cloning might be less of a challenge since *complete* genetic identity is evidence *against* sexual conception for which the participants may be held responsible. This fact also shows that the degree of genetic overlap does not necessarily correlate with the degree of affiliation.

Who, therefore, will be responsible for cloned children? The women who give birth to them. If cloning is sanctioned only when women's health and rights are fully respected, the birth of a clone is a sure sign of someone's free choice to become a mother. The mothers of clones will have the same status as other women who choose to become single mothers. In contrast to many "natural" children, a clone can never be a product of an "accident."

Obviously, men may commit themselves to being the fathers of those clones, before or after conception. However, my proposed emphasis on the mother of the clone might relieve women from the pressure of providing a solution to male sterility.

Before I close this section, I would like to call attention to the sad fact that the evasion of paternal responsibilities is a painful social, moral and legal problem world-wide. Men need not wait for genetic technology to mature in order to exploit women and to disown children. The grave problem in hand is a social and an ethical one, and it is as old as humankind itself.

11. Futurism in Philosophy

In contrast to hotly debated topics such as euthanasia, pornography and the like, the discourse on cloning is anchored in extrapolations into the future. We can research the reality behind prostitution, but not the sociology and psychology of cloning.

Prof.Kasher, a leading Israeli authority on applied ethics,[35] repeatedly warns against the "slippery slope" of "slippery-slope arguments." Rabbi Tendler, a prominent American rabbi specializing in medical ethics, warns against the invocation of "slippery-slope arguments" when life-saving research and practice is at stake.[36] As Prof. Rabbi Steinberg observes the Jewish tradition is usually wary of "slippery-slope arguments," particularly those which are not strongly supported by solid experience.[37] Since *Halakhaic* discourse is casuistry-bound, it is dependent on real-life cases to flourish. Rabbi Steinberg also recognises this as a potential weakness when facing the need to formulate *general guidelines and principles* regulating *technologies and practices of the future.*[38]

12. Social Ethoses of Sexuality and Fertility

An Israeli doctor and rabbi writes that one reason why Judaism looks favourably on cloning is the Biblical ethos of the purity of the species and the Biblical disapproval of exogamous marriage.[39]

I have argued earlier that the rabbis worry about potential conflicts of paternity, not about the "purity" of origins as such. Indeed, the Bible does not base its prohibition of exogamy on essentialism (=racist arguments) but on consequentialist considerations such as the corruptive influence of mixing with pagans or conflicts relating to the division of land (Exodus 34:15, Talmud *Ta'anith* 30:2).

However, this singular Rabbinic observation on cloning calls attention to an interesting paradox in the history of ideas. Judaism, which cares so much for lineage, is nowadays receptive to the idea of cloning, whereas Christianity, which rejected the question of origins - baptism being a fresh start for every Christian - is currently concerened with the initiative to clone people. I conjecture that the Catholic spirit of humanism is one source of objection to a method of procreation that is not a union of two *different* people.[40]

The rabbis may not necessarily object. The Hebrew Bible already cast marriage as a separation from one's parental family (Genesis 2:24). However, since procreation is a fundamental religious duty of every man, the rabbis endorse cloning only as an alternative to not having biological progeny at all.

Another paradox is the observation that "pro-life" activists, who object to abortion even when the child is expected to suffer medically or socially, are usually opposed to bringing healthy and happy children into the world by means of cloning. These activists would not hear of abortion, but they would preach to sterile couples to accept their barrenness. We may object to genetic manipulation, immoral sex, exploitation of people etc., and I can see how cloning might become illegal as a by-product of other restrictions, such as those related to research ethics. However, I do not see how a genuine "pro-life" approach can criminalise the very creation of new human life which is otherwise healthy and possibly happy.

13. Gender - Sensitive Perspectives

Often we forget that we owe our very existence to unknown and remote ancestors who struggled to continue the chain of humanity against all odds. There were times when most children died young and obstetric mortality a high. Life expectancy at birth was at times less than five years (sic).[41] As late as the nineteenth century urban mortality rates in London exceeded birth rates. Even today many people struggle to have children in the face of horrendous epidemics such as HIV and in the face of oppression and famine. Countries which were encouraged by international agencies to restrict family size, now face a demographic catastrophe of diminishing population due to the AIDS epidemic.

In such circumstances, no doubt, many people have questioned the reasonableness of having children at all. Procreation and devotion to rearing children were heroic activities.

The Talmud (*Sottah* 11b) attributes the deliverance of Israel from Egypt to women who resorted to erotic tricks in order to conceive by husbands who had decided not to bring children into slavery and persecution. These stories have been told and retold throughout Jewish history as well as during the holocaust.

Any attempt to check risky techniques of fertility will have to reckon with an archaic and powerful ethos of sacrificial drives to procreate, an ethos that is possibly a viable and indispensable strategy of survival as well.

I contend that the moral challenge is located exactly at the division between procreation and non-procreative exploitation of our procreative drives, eroticism, parental sentiments, moral values, and human biology.

Moreover, in Western culture procreation has been regarded as a somewhat strange, almost unnatural, event within a stable natural order.[42] Begetting children has not been considered an essential event in personal life. The Jewish tradition underlines generation as the main blessing within creation, and refers to childless people as personally and spiritually deficient. Even within Western philosophy some influential voices have seen the capacity to reproduce as an indispensable element in personal self identity.[43]

Following feminist thinker O'Brien,[44] I construe two gender-sensitive polarities of understanding procreative cloning. From a feminine point of view that emphasises biological continuity cloning is one more aspect of personal fulfilment and growth?. From a masculine point of view that is anxious about social order and that is alienated from the *abject* bodily experience of procreation, cloning is transgressive and futile.

In Jewish thought both points of view are accentuated. While male-dominated rabbinic discourse is highly and openly "effeminated," anxieties about the preservation of a patriarchal social order also dominate social life.[45]

Mass cloning, "designer babies" and gender selection as well as scientific and therapeutic cloning all take advantage of our faculties of procreation and of our willingness to put ourselves at risk for the sake of our own children and their future. Mass cloning and the self-serving "design" of children seem to originate from the masculine drive for power and control rather than from attitudes of care and concepts of the self that transcend the boundaries of one's own body.

Is procreative cloning a propitious and natural development of feminine notions of self and rationality or merely an elaborate step in the subjugation of female bodies and persons to technologies of alienation? Is the suppression of cloning one more manifestation of dominance over women's bodies and over their potential for independence from men? The answers to such questions are not inscribed in a given set of moral laws or a metaphysical order. They depend on the way we actually construe the technology of cloning along with its regulation and social acceptance.

14. Summary: An Outline of Guidelines on Cloning Which Israeli and Jewish Thinkers Might Adopt.

a. Although procreation is a fundamental value of human society, it does not justify extreme risks to and sacrifices on behalf of individuals. It is usually less important to procreate than to save existing human life.
b. Fertility technologies should be strictly regulated so as to protect the health, dignity and other interests of the participants. These regulations must reflect awareness of sources of bias, such as commercial incentives to achieve quick results, the interest of science in supernumerary eggs and embryos, and untoward pressures on women to submit themselves to fertility technologies.
c. Since human dignity is conferred on *formed* embryos, artificial creation of such embryos, i.e. live embryos that have recognisable human characteristics such as head, face, limbs etc, will not be permitted, even for the sake of life-saving medical procedures.
d. In vitro pre-embryos that have been ethically procured and which are not needed for procreation may be utilized for the purpose of promising scientific research.
e. Whether "parents" have the right to destroy their pre-embryos rather than to donate them anonymously for the sake of procreation might be an issue of contention. I tend to believe that the rabbis will object to the donation of embryos.
f. Since the Jewish tradition is highly sensitive to biological parenthood, safe cloning is preferable to treating infertility by means of donated gametes or pre-embryos. Jewish and Christian ethics probably diverge here. The former seeks phenomenological diversity (different body shapes and different personalities), whereas Christianity also stresses the diversity of origins. Judaism holds a consequentialist view of origins, whereas Christianity stresses its naturalism.
g. Invasive fertility technologies should be used only in exceptions, e.g.

    i. The last means of begetting biological progeny.
    ii. Prevention of disease (unless a safer way of doing so is feasible).

iii. It is preferable to experiment on pre-embryos than on sentient animals and on corpses. Secular and Christian ethics prefer experiments on the dead to experiments on sentient creatures.

h. Commercialisation of the human body, e.g. the sale of organs and eggs or the hiring of uteri will not be permitted. (Some rabbis and philosophers approve of a properly regulated market for such purposes).
i. Rabbis and theologians might wish to conform to international law, should it condemn cloning. Judaism does not wish to be regarded as morally lax in the eyes of the non-Jews.
j. From a legal point of view, some rabbis recognise dual maternity (=two mothers to one person) and "hermaphrodite parenthood" (a mother who serves both as a father and a mother). I find these models inspirational with regard to expanding our concept of the "family" in ways that accommodate gay couples and other less traditional set-ups. The Orthodox rabbis certainly would not endorse such a move.

Notes

1. http://www.moreshet.co.il/shut/shut2.asp?id=34718. A responsum by Rabbi Sharlo, January 22, 2004.
2. For a comprehensive discussion see, Barilan Y.M. and Seigal G., 2004 or Barilan Y.M. "Abortion in Jewish religious law and Israeli law: Neighborly Love vs. Imago Dei." (In press). Statements about Judaism and the Talmud which have no separate references in this paper are to be found in these sources.
3. According to Maimonides, human individuality is achieved by virtue of having a unique [moral] personality. Barilan Y.M. "Refusing medical care in the Jewish sources." (In press).
4. Simon, 1967, 97-109; Tierney, 1997, 22-8, 63ff.
5. Barilan, 2003a, 82, 92-4.
6. I have tried to do so in another paper. See Barilan Y.M. "Abortion in Jewish religious law and in Israeli law." (Forthcoming).
7. Barilan, 2003b, 143-70.
8. Zuriel, 1988, 901.
9. Steinberg, 1998, 30.
10. Halperin, 1996, 155.
11. Karpi, 1974, 843.
12. Barilan and Seigal, 2004.
13. Halperin, 1996, 13-48.

14. It is not likely to be related to the prohibition of wasting semen. See Auerbach in *Noam* 1958; 1:145-166.
15. *Noam* op. cit., 157; Abraham, 1990, 8. See p. 9-11 for a debate on whether the man is considered the legal father of the child. Rabbi Sofer keenly observes that in situations such as public bath impregnation and artificial insemination, there is a gap in the *human* relationship, during which the semen is suspended in a non-human environment. Due to this disruption, relationships of parenthood begin *only* when the foetus is formed (40 days). This naturalistic distinction is based on the Talmudic and Biblical assertion that moral status belongs to foetuses only in their capacity as "a human within a human." It also explains Auerbach's opposition to pre-marital preservation of semen. If the husband dies before the embryo is formed, the child has no father in the full sense of the word.
16. Goldberg, 1989, 65-6, 45-9.
17. This is only a problem when a surplus, rather than a shortage, of pre-embryos prevails.
18. Barilan, 2003b.
19. Steinberg, 1994, 490. Steinberg also writes that infertility treatments are not justified when done for the sake of medical research. Steinberg, 2003, 253.
20. Strong, 1997, 13. Strong offers a somewhat simpler model of *begetting, gestating and rearing*.
21. Halperin, 1996, 74. *Halakha* does not speak the language or rights. Halperin probably wants to convey that hiding such information from a person harms him or her unjustly.
22. Sheilat, 1998, 137-49.
23. Unless they left grand-children to their father before they died.
24. Slavery has not existed in Judaism for more than a thousand years. The laws of slavery are part of the Torah and are still actively studied, often with practical implications like this one. Holistic dedication to every law and part of the Torah is a cardinal value of Orthodox Judaism.
25. Western medicine discovered the ovaries and the ovum only in the 19th century. A 17th century Jewish Rabbi Azaria of Pano (*Responsa*, #117) draws an analogy between a chicken that lays eggs without a rooster and a woman's capacity to be impregnated by a spoiled seed. Apparently, the rabbi did not know the difference between a fertilized and a non-fertilized egg.
26. Shafran, 1998.

27. A need for organ or marrow transplant could be a case in which biological parenthood is sought in spite of excellent psychological and social suppression of it?. These offerings of biomedicine weaken the notion of *total* and *irreversible* adoption, if it ever existed.
28. Rabbi Feinstein (USA d. 1985) represents a different way of *Halakhaic* thinking, casting aside the biologisation implied by Sheilat. According to Feinstein, the only defect is conception through illicit sex. Artificial insemination, therefore, can never create a bastard. *Iggroth Moshe, Eben Ha'Ezzer* part I, #71.
29. *Midrash Sechel Tov* on Genesis 31 and *Midrash Tanhuma*, Buber ed. On Leviticus, *tzav* #10. The messianic cleansing of bastards is compatible with the Jewish vision of the World to Come as a realm of cosmic peace. Positive knowledge and techonological utopianism are not part of Jewish messianic visions.
30. Barilan, 2000.
31. Policing AFR in order to protect women's health and rights does not require microbiological scrutiny into people's genes and tissues.
32. One might object to any limitation of people's right to find their own mate for procreation. In that case it will be quite difficult to restrict cloning as the last resort in fertility treatment even when the risk to the child is significant. However, I assume that banning *experimental* techniques is different from banning ordinary attempts at child bearing.
33. Shafran, 1998, 152.
34. According to the Midrash (*Rashi*'s gloss on Genesis 25:19), the resemblance between children and their fathers helps dissipate rumours doubting paternity. The Talmudic source that links human dignity with morphological diversity (*Sanhedrin* 37a) also emphasizes the endless family-tree of Adam, thus, possibly hinting at a concept of diversity which is based on the fanning out of lineage, body and character.
35. http://www.health.gov.il/download/docs/vahadut/obar_enushi/prot02.doc (in Hebrew).
36. Tendler, 2000.
37. Steinberg, 2000.
38. Steinberg, 1998, 640ff.
39. Levinger, 2003.
40. Pre-modern Canon law was much more restrictive than *Halakha* with regard to exogamy.
41. Alter, 1983, 23-41.
42. Jakob, 1989, 19.

43. Calguilhem, 1964, 24-43.
44. O'Brien, 1989.
45. Boyarin, 1993 and 1997.

References

Abraham S.A. *Nishmath Abraham*, vol. III. Jerusalem: Schlesinger Institute, 1990.

Alter, G. "Plague and the Amsterdam Annuitant: a new look at life annuities as a source for historical demography." *Population Studies* 37, 1983, 23-41.

Barilan, M. "The attitude towards Mamzerim in Jewish society in late antiquity." *Jewish History* 14, 2000, 125-70.

Barilan, Y.M. "Her sorrows come first: abortion in Jewish law and in Israeli law." *Refua U'mishpat* 23 (2003a). (in Hebrew).

Barilan, Y.M. "Revisiting the problem of Jewish Bioethics: The case of terminal care." *Kennedy Institute of Ethics Journal* (2003b), 143-70.

Barilan, Y.M. and Seigal G. "Stem cell research: Jewish and Israeli perspectives." In *Grenzüberschreitungen. Kulturelle, religiöse und politische Differenzen in der Stammzellforschung*, Bender W, Hauskeller C. and Manzei A. (eds.). Münster, Agenda Verlag, 2004.

Barilan, Y.M. "Abortion in Jewish religious law and in Israeli law." *Review of Rabbinic Judaism* (Forthcoming).

Barilan, Y.M. "Refusing medical care in the Jewish sources." (In press).

Boyarin, D. *Carnal Israel: reading sex in Talmudic culture*. Berkeley, University of California Press, 1993.

Boyarin, D. *A radical Jew*: *Paul and the politics of identity*. Berkeley, University of California Press, 1997.

Canguilhem, G. "Monstrosity and the monstrous." *Diogenes* 40, 1964, 27-43.

Dobnov, S. (ed.) *Pinqas medinath Litta (Lithuania communities' book).* Berlin, 1925.

Goldberg, Z.N. "On egg donation, surrogate uteri, freezing semen of a bachelor and retrieval of semen from the dead." *Assiah,* 1989, 65-6, 45-9. (in Hebrew).

Halperin, M. "*Kavim le'Dark.o*" In *Medicine, Ethics and Halakha - The Second International Congress.* Jerusalem: Schlesinger Institute, 1996.

Jacob, F. *The logic of life.* Harmondsworth: Penguin, 19, 1989.

Karpi, D. (ed.) *Pinkas va'ad Qehilath Kodesh Padua (Padua community book).* Jerusalem, vol. I., 1974.

Levinger, U. "Human cloning - the Jewish view." *Iddkun Bi'refua Penimith* 42, 2003, 10-15.

Noam, Abraham S.A. *Nishmath Abraham,* vol. III. Jerusalem: Schlesinger Institute, 1990.

O'Brien, M. *Reproducing the world.* Boulder: Westview Press, 1989.

Shafran Y. "Genetic cloning according to Halakha." *Tehumin* 18, 1998, 150-60. (in Hebrew).

Sheilat I. "Cloning according to Halakha." *Tehumin*? 18, 1998, 137-49. (in Hebrew).

Simon Y. *The tradition of Natural Law: a philosopher's reflection.* New York: Fordham University Press, 1967.

Steinberg A. *Medical Halakhaic Encyclopedia.* Jerusalem: Falk Institute, vol. 4., 1994.

Steinberg A. "Cloning humans: scientific, moral and Jewish aspects." *Assiah* 62-2, 27-42, 1998. (in Hebrew).

Steinberg A. "Stem cells: medical, ethical and Halakhaic aspects." *Thumin*? 23, 2003, 241-54. (in Hebrew).

Strong C. *Ethics in reproductive and perinatal medicine: a new framework*. New Haven: Yale University Press, 1997.

Tendler M. "Stem cell research and therapy: A Judeo-biblical perspective." *Ethical issues in human stem cell research*. Vol. III: religious perspectives. Rockville (MD), 2000.

Tierney B. *The idea of natural rights*. Grand Rapids (MI): Eerdmans, 1997.

Zuriel M.Y. (ed.) *Otzaroth HaRe'i'ya*. Tel Aviv: Segal; vol. II., 1988.

Bioethics from the Perspective of Universalisation

Christofer Frey

Abstract: This introduction to a presentation of Jewish bioethics follows the guideline of universalisation – a challenging question concerning the genesis of norms within a particular religious history. It is not the history of a simple ethics of goods, founded in an unchangeable minimum of human features, but the development of obligations frequently combined with monotheistic beliefs. The Jewish Torah does not provide an abstract Kantian approach; the Torah is closer to a vision of human life integrating presumed universal desires and interests. The main problem is teleology, especially in the first stages of the embryo. Some hints in the Halakhic discussion prove that the Rabbis referred to Aristotelian thought concerning the status of the embryo. Many of them seem to have accepted different stages of becoming a human being, the telos (teleology!) being reached at birth. A present new interest in the Torah and Halakhah, even outside Judaism, seems to be founded in a sceptical attitude towards an ethical deduction from a God's eye view and an interest in considering whether particular traditions hold potential solutions to contemporary problems.

Key Words: Anthropology, Philosophical, Dignity, Human, Embryo, Ethics, empiricist or principle-oriented, Judaism, Neo-Aristotelianism, Person, Reason, Practical, Teleology, Torah, Universalisation.

1. Different Types of Ethics

Many discussions in bioethics are centered on the idea of human dignity, which - according to Kant - is not a matter of value and evaluation.[1] Dignity relies exclusively on reason - pure practical reason (*Vernunft*), whereas value and evaluation appeal to the Verstand, to emotions and to *Urteilskraft*. Kant argued in contrast to the British ethics of his time, especially against the trend towards emotivism and intuitionism. This ethics is not restricted to pure reason, but takes an empiricist attitude towards the world. Today, the two contrasting types can be identified as (1) *empiricist ethics* and (2) *principle-orientated ethics*.[2] But we acknowledge that both presuppose certain differing ideas of reason; and unfortunately there is no utterly reasonable position from which to settle the dispute between these two suppositions of reason. However, a meta-argument is possible and can be pursued as an enquiry into the details and basic assumptions of rational perspectives. At the same time, however, this meta-argument must take into account certain material aspects of human life that are the result of particular histories and traditions. Whenever the idea of history and the notion of tradition arise and join the idea of reason, traditions and cultures that are different from Greek and subsequent Western thought can contribute to a universal idea of reason by enriching and even modifying Scholastic and Enlightenment viewpoints.

2. Empiricist Ethics and the Consequences

The first considerations of this essay concentrate on the empiricist type of moral reasoning. In a time when the natural sciences and technical disciplines are pre-eminent, an empiricist spirit invades even the ethical argument. What is called *empiricist ethics* in this paper comprises either a relativistic theory of goods or a *moral relativism* combined with a rather abstract idea of some very formal common goods.[3] A utilitarian view may assume that everybody controls his or her own scale of desires and the coordinated goods ranking from the highest to the lowest; but nobody has exactly the same scale.[4] Complete relativism, however, is an absurdity, because a common life of different individuals presupposes at least some *basic goods* for all (1.1), for instance, the freedom to strive for one's own optimum.[5]

If this empiricist theory of the good is transposed into a legal theory, it needs a special foundation: The natural law theory of H.L.A. Hart - a theory without principles, as Dworkin, his successor, criticises) - presupposes a set of *common goods* such as the protection of life, the fulfilment of basic desires or a minimum of respect for the other.[6] The great challenge of microbiological research is whether these basic rules also apply to the embryo.

What are the consequences of these theories in the field of bioethics, especially with regard to the beginning of life?[7] They recognise that the embryo does not yet have subjective preferences and does not claim for goods. None of the variants of these theories attributes an objective good to the embryo, because according to them there is none. The liberty to realise goods is generally attributed to persons revealing distinct features of consciousness.[8]

Ethics in the empiricist perspective tries to settle moral disputes by *weighing goods* and emphasising one side of a given alternative. Curing a person suffering from Parkinson's disease can be ranked as a higher good than the destiny of an early embryo, if the attribution of the status of personhood (to an adult or - more generally - to a person already born) and the mere fact of a stem cell make a difference. Or the life and the personal and moral self-affirmation of a hypothetical or real mother may carry greater weight than the life of an embryo. Moral philosophers know many arguments for and against such assumptions; in everyday life, however, conflicts of this kind are never resolved once for all.

A further example: In the late nineties famous microbiologists and geneticists met at the UCLA (University of California, Los Angeles) to debate the future of their research. They argued fervently in favour of germ-line therapy and asked the political authorities to permit the necessary procedures.[9]

If universal guidelines are proposed within this trend of moral or ethical arguments, they could support not only an individualistic relativism of goods, but also the standard that people must subjectively manifest their interests before they become serious candidates for the ascription of human dignity.

3. Anthropological Presuppositions

To summarise the first part: There seems to be a particular *anthropological idea* behind the arguments of empiricist ethics - an idea that is frequently hidden, but sometimes brought to the fore: Human life can either be seen as a biological fact (a) or as an organic phenomenon on a higher level with the quality to feel pain (b), or - yet again - as the life of a person who demonstrates consciousness and is capable of developing visions of a personal future and of plans to realise those visions.[10] From the empiricist perspective, however, it is hardly accepted that human beings are able to support each other without the promise of reciprocity and to ascribe the status of the person to another human being, because the empirically restricted type of ethics is founded on a methodological (and in some cases even ontological) individualism, which sometimes resembles social atomism.

This type of ethics relies on Locke and his theory of personhood.[11] The embarrassment produced by this theory is evident: Human beings without consciousness - not only sleeping persons, but also people in a coma - can hardly be integrated into the considerations or reflexions of moral thought. To avoid some of the fatal consequences of this theory of the constitution of personhood, individualistic atomism must be transcended by the idea of *vicarious coexistence*: Other human beings maintain the status of personhood vicariously, if a person is unable to do so for him or herself, and even the moral and the legal system corroborates the interrelatedness. If empiricists ever accepted this idea, they could continue the debate by asking when and in which cases a vicarious act is legitimate. But a humanist approach to practical reason transcends empiricist perspectives.

What *alternatives* might be suggested? A radically individualistic theory of the human good (or of human goods in general) includes a relativism and therefore an ambiguity. *Relativism* could be accompanied by ethnocentrism. If something is felt to be a good or even a foundational value in a certain culture, it does not have to be good or necessary for someone in another culture. The American philosopher Martha *Nussbaum* and other representatives of so-called Neo-Aristotelianism contradict relativism and ethnocentrism by giving emphasis to some human essentials which may vary in their outlook according to different cultural implementations, but which can never be totally relative.[12] By analogy, it is necessary to enquire into a culture-transcending bioethics. Transcending one's own culture, without denying it, could be one of the ways to come closer to the idea of *universalisation*. Universalisation is either a quasi-logical idea or it is the process of going beyond the family and the clan ethos; or it tries to secure minimal conditions of a dignified common human life. But could modern empiricism adapt to universal morals which hitherto appear to be rather unspecified?

4. Ethics and Universalisation

The great problem of universalism exists in the alternative of *inclusivism* or of moral *expansion*. The universal perspective of *Islam* seems to be inclusive. Although Mohammed sought to restore the moral laws proclaimed by God the creator, he maintained the special obligation of Muslim faith to obey the divine imperatives with a holy fervour and to fight (more or less spiritually) for the inclusion of all mankind into the 'umma.' The true religion is considered an integral part of this inclusive type of universalism.[13] *Christianity* favoured the hope of an *inclusive* salvation and at the same time an *extension* of natural morals founded on creation and therefore without a particular regard to religion.[14] Consequently, a distinction must be drawn between religion and society (as in non-Puritan Protestantism and in modern Catholicism).[15] The modern idea of a *civil society* presupposes this very important differentiation. Universality does not start on the highest neutral point of reason or with the neutral observer of morals and interests, but with a tradition - religious or secular - which promotes the idea of transcending its own background without denying or losing it.

Both Islam and Christianity claim to incorporate Abrahamitic religion, believing in a God who is regarded as the only one and more or less replacing the other gods with their particular capacity for political and social legitimation. And this is exactly the challenge of the first and genuine Abrahamitic religion, the religion of Israel and of Judaism. But how does this religion and its admirable tradition relate to universalism? The *universality* of its God, the *particularity* of its belief in a special election - as an event in the history of salvation - and the tradition of halakhic argument appear to be contradictory.

Ethics with a universalistic affinity can achieve different forms: Seemingly the moral argument could start on the highest and most abstract point of the moral view, for instance a variation of the 'Golden Rule.'[16] This approach results in a kind of deduction. By proceeding through several steps of moral and situational specification, the argument should come gradually closer to the given case; but long experience with this type of dispute suggests that the deductive approach does not develop any continuous line of moral reasoning. The universal never converges with the concrete by a consistent method or a specific logic of moral reasoning. This is the problematic aspect of practical or ethical reason.

5. Ethics as Reflection on Morals and the Concept of Human Dignity

This ostensible bottom-down approach must, therefore, be redefined by a complex fabric of rational procedures: A *concrete situation* always conveys a *reflective dimension*. For instance: Should embryos outside the womb (= the fact or the case) be seen as aspirants of human dignity (= a reflective evaluation)? Ascribing dignity to the earliest embryonic status of a germ cell is problematic. But the germ cell can no longer be a mere biological fact. To introduce "dignity" as a reflective evaluation connotes an *intuition*,[17] which signifies something like the following: This case is more than a mere fact; or: it is a form of life for which we are responsible. When intuitions initiate a critical moral reflection, they come within reach of the *categorical imperative*: Accept as rational the norm which other rational beings could adopt by their own rational nature![18] Any criticism of habitualised or even tacit norms relies on a meta-empirical basis of thought beyond the surface of reality, on a stratum of ideas which could be called "transcendental" (with Kant) - the basis of moral knowledge and rationally reconstructed reality.

The rationale of *universalisation* could be a very *abstract* point of view if each rational being is assumed to make the same decisions on the same topics. These rational beings are apparently bodiless and represent an anthropological problem.[19] Abstract reason is therefore remote from life. Any form of reason which concerns concrete situations of life is somehow connected with history, with traditions and the bodily and social existence of persons. Moreover, moral norms in concrete situations are usually not imperatives, but they are included - often implicitly - in inclinations of a person and in his or her values and perspectives, which reveal the direction in which that person lives his or her life.

When the effects of moral reason are reflected, they must comprise - at least implicitly - a *vision of human life*. This was one of the main arguments of *Hegel*. He went beyond Kantian morals with the assumption that the *subject* - the bearer of reason - should become *substance* - an idea that develops a condensed and institutionalised reason by transforming a given society and tradition towards final accomplishment.[20] Although Hegel included a philosophical and theological perspective - the all-embracing final union in the absolute -, history was not proceeding in the way he outlined, but rather moving in the reverse direction, leading mankind into abominable moral catastrophes. This commission, however, to embody reason in social institutions challenges mankind after Hegel, but in a non-absolute, relative sense. One of these basic institutions is the law, a system of norms monitored by external sanctions. We should, however, prefer internalised norms and sanctions guaranteed and institutionalised by the freedom which respects the actual or hypothetical freedom of fellow persons.

The application of this type of moral argument to bioethics has several consequences: First, there is a necessity to establish legal and even moral regulations which incorporate elements of universal reason; these norms are imperative in new fields of research. Second, these norms must be related to integrating perspectives of human life as they develop in different given cultures - either positively or critically. Third, they cannot operate on the basis of a methodological individualism, but must define personhood and human dignity in terms of correlation to and respect of others.[21] Before an individual develops a sense of his or her own personhood, he or she should be accepted as a person by others, especially by the parents. The ascription of the personal status precedes the social and individual realisation of the personal being. Personalisation adjusts to the preceding status.[22] This "transcendental" ascription could have a special relation to the idea of vicarious being, acting and even suffering, as *Isaiah* 53 demonstrates.

The ascription of the personal status - initiated by other persons - is one of the most interesting points in the bioethical debate. The rather odd discussion on the status of the embryo should relinquish the ontological terminology and adopt terms of non-arbitrary, but fundamental ascription. The embryo is not a simple assemblage of cells, but a being entrusted to social and individual responsibility - under certain conditions and within certain limits.[23]

All these considerations prove the need for a basic anthropology which reflects the human vision and elaborates what it means to be a human being: A person is either a citizen of *two worlds* - the empirical *and* the rational one - or of *one world*, as assumed, for instance, by sociologists.[24, 25] Whenever empirically oriented scientists want to explain the totality of being, they tend to define the embryo as a mere biological fact; in the eyes of more idealistic philosophers, on the contrary, the material aspects of life could be neglected.[26]

As regulating ideas or perspectives of human life (and its integration) are included in moral (or ethically reflected) arguments, each moral or ethical thought starts in a *given history* (and even in a particular life-story) and therefore inspires further thoughts starting with contingent views and not at the highest point of an absolute or *neutral observer*. Universal arguments are elaborated in historical circumstances, they rarely precede them. Even the natural sciences and especially their results are tied up with cultures and - unfortunately - with ideologies. Sciences cannot promise moral neutrality; if they help to analyse given situations, they must admit that there are implicit or explicit criteria already valid and - combined with them - an anthropological perspective which is probably working as a meta-criterion. Such meta-criteria could be the respect for life,[27] the idea of creation (as in the Biblical tradition), and the integration of a personal life.

6. Judaism and Teleological Thought

Judaism - in spite of its venerable tradition - is challenged by these thoughts, too. What is the Jewish sense of universalism? Since the reconstitution of Judaism under the guidance of Gamaliel and the development of a distinctive Rabbinic interpretation of the Hebrew Bible that concentrated on the Mosaic Law, the Torah was considered to be God's eternal law and therefore never subject to a transformation by historical developments. Judaism strengthened its spiritual force in a legitimate defence against the moral system of late antiquity and Greek philosophy and relied on its own particularity, perhaps in contrast to the universal position of the unique God it reveres. Jewish bioethics seems to be founded in this unique and special tradition. Is it then a particularistic type of scientific and moral reasoning without universal claim?

But the need for orientation in bioethics is a worldwide challenge, and, by analogy, the question of reason - theoretical, foundational, and pragmatic - is a theme concerning all mankind. Particular traditions may contribute to the discussion of these questions and the development of motivations towards universal solutions.

These emerging common norms are not the same as the anthropological meta-norms or meta-criteria which were suggested above. But they could help to identify the often hidden anthropological precepts of our present arguments. The most important precept in bioethical discussions today is not very frequently discussed, although it is imperceptibly effective in directing bioethical conclusions. It can be identified as what is traditionally called "*teleology*."[28] This concept was usually linked with the so-called "*natural law*," which developed from Stoic thought and on Christian biblical grounds. From a religious perspective it was interpreted as the law of the Ten Commandments, given with the creation to all men, and reaffirmed on Mt. Sinai in the presence of Moses. Mediated by this system of thought, a universal claim of reason is presumed to be at work; and this claim is centred on a special teleological idea which conflicts with biological presuppositions. It considers the totality of being in this world as a well-ordered cosmos relying on a special necessity which even includes contingence.

It is precisely this particular ontological precondition which renders bioethical debates difficult. The modern biological idea of teleology is different from the ancient philosophical one. Certain biologists use teleological ideas in a methodological way to explain the coherence of systems of life.[29] Aristotelian philosophy uses teleology to explain the innate tendency of an organism to develop its own destiny; but the Catholic version - although related to Aristotle - is a metaphysical one: Each single being has a final propensity as its very essence.

Aristotelian thought seems to have influenced even Rabbinic arguments on this special issue. According to a Rabbinic tradition, the embryo turns from mere water into a human being on the fortieth or eightieth day of pregnancy. But this is considered only as one step in the development towards the complete human being which takes its final form at birth.[30]

What is the status of a single cell embryo or a blastula against this background? A biologist applying the methodological teleological concept cannot verify a type of life in the early embryo which possesses a final destination of its own. The high rate of spontaneous abortion before nidation seems to support a biological restraint. Even the older organismic view could not exclude that an organism falls short of its destination;[31] there is no direct way from the organic telos to a moral consequence. But the essentialist way of thinking is forced to locate the ontological essence in the early single cell insofar as it contains the genetic programme of a human individual.[32] All European disputes concerning the status of the embryo - and maybe even the worldwide discussions - have to do with teleological assumptions of this kind.

We should return to the assumption of a natural law in human morality: The idea of a natural law has undergone certain changes in modern times - either being converted to a secular canon of reason in society or manifested in religious interpretations: Some Catholics pleaded for a historically evolving law, whereas many Protestants locate this law in a flexible principle of otherness and reciprocity, as apparently did Rabbi Hillel as early as the first century.[33] This assumption points to a pre-individualistic ascription. Whoever exposes himself to his fellow needs a vision of what is good for the life of the other. Therefore, some Protestants conclude that the predicate of human dignity is principally valid in all stages of human life, but can only be morally and pragmatically effective in situations where the presence of the other as a member of humanity is evident.[34] This idea can be compared to the divine law and its application: "You shall not murder" is meaningless if there are only a number of cellular tissues. Human dignity with its derivates, human rights, cannot be extended without limits. The protection of life remains an important moral principle, although a grading within the legal status of the embryo and the foetus is inevitable.

What are the consequences for stem cell research? Kryokon frozen germ cells seem to lose their biological teleological capacity; it is a risk to transplant them if they have been preserved for any length of time. If they are to be discarded, should they be ritually buried or could their life capacity - in an extraordinary procedure - be transformed into many generations of stem cells? This could paradoxically harmonise with the enthusiastic notion of life held by defenders of the primordial dignity of human life. Are the first cellular stages of growth individuals because of their unique genetic programme, or have they lost their individuality because they have been deprived of a part of their teleological capacity? Biological terms cannot establish moral reasons, but any moral discussion needs empirical indications of morally relevant cases. The insecurity concerning their capacity to develop to a complete human being might be an indication that under certain extraordinary conditions transformation into lines of stem cells could be tolerated.

7. Judaism and Universalism

These discussions call for a worldwide idea of (practical) reason; if there is none, rationality could be at risk of being restricted to the methods of discourse. Is the idea of practical reason - on the contrary - only a substantial one if certain important ideals, which developed in special cultural traditions, are incorporated into the features of reason? At the beginning of the 20th century, Hermann Cohen, a German-Jewish neo-Kantian philosopher, rediscovered his Jewish roots, because he taught that the reality of life is not a simple datum given from outside, but it should be discovered by a basic judgement including a cognitive and even a valuating approach. His famous book "Die Religion der Vernunft aus den Quellen des Judentums" was a milestone in the great period of German Reform Judaism,[35] when Leo *Baeck* claimed that Moses had discovered for Jews - and indeed for mankind - what Kant was to explain centuries later: The Torah, the Jewish Law, should be regarded as the first expression of universal reason and morality.[36] These theories of more or less liberal Jews have now faded, but two challenging moral questions still remain:

the seriousness to live one's own tradition and
the possible universal appeal of one's own moral belief.

The latter seems to be a particular challenge to Judaism today.

Transcultural bioethics relies either on the foundation of a universally valid practical *logic* that is tacitly included in the manifoldness of diverse moral systems (and even elaborated ethics) or on a natural *moral stratum* which serves as the common - and maybe universal - basis of elaborated moral systems.

Historically, an interest in the behaviour and orientation of *all* men and not only of a tribe or a clan seems to be a special feature of monotheistic religions. This interest sometimes transcends the limits of an indigenous religion. But the suggested solutions to the problem of universality initiated by certain particular ideas are different, as universalistic forms of belief demonstrate:

> Protestantism tends to transcend the stratum of plainly proposed commandments (the "statutes," as Kant called them) in order to find a basis of reflection concentrated on principles including some general orientation marks in the reality of social life (such as Law and Gospel, creation and salvation).[37]

- The Catholic version endorses two levels of moral rules, a system of natural (moral and legal) law and a system of divinely revealed commandments; both are assumed to be transcultural, albeit in different ways.[38]
- Both confessions embrace a certain form of cooperation with philosophy, especially in discourses about the natural law, which is either interpreted as a metaphysical or as a historically evolved order.[39]
- Natural law was ultimately converted into rational law.[40]

In the West the crisis of the idea of a universally valid reason resulted in a kind of religious and philosophical postmodernism, which - especially in Christian theology - tries to revitalise the Jewish Torah within Western thought. Postmodern thinkers find in the Torah and its tradition a manifoldness free from the coercion to develop a deductively organised system of norms that proceed from unity to diversity. But even Jewish ethicists admit the difficulties implied in this reaction, as the following quotation demonstrates:

"Jewish religious law is a central aspect of the Jewish normative tradition, but when Jewish bioethics focuses on this aspect, it can become a discourse of authority, taking the form of commandments and injunctions addressed to those bound by that law. Such teachings, it has been forcefully argued, cannot validly claim relevance to non-Jews or even to secular Jews."[41, 42]

What are the reasons for the anti-universalist turn? This essay can only put forward a number of suggestions: There may be parallels to communitarian thought which was developed mainly by Jewish intellectuals,[43] who favour the idea that norms develop inside particular communities and cannot be deduced from the point of view of a sovereign neutral observer or from the so-called God's eye point of view. How then might we approach contemporary Jewish moral thought?

- The theme of history is a very crucial one: Natural law tends to be a-historic. It was considered to be valid as long as creation lasts, whereas the unique perspective of Jewish faith is essentially historical. The Torah results from a particular history; but according to many Rabbis the Law - although given in history by a special revelation on Mt. Sinai - is eternal.
- The method of moral argument in the tradition of the Torah, especially in the Halakhah, is not reflective; nor is it deductive or inductive, both of which are features of classical Western casuistry. Halakhic discussions reveal a casuistry by analogy, which resembles Aristotelian topics and rhetorics. Casuistry seems to have some abductive tendencies and to permit a considerable degree of flexibility.[44]
- However, the problem of universality remains unresolved. What is the moral relevance of inner-Jewish arguments to non-Jews? Some argue that the so-called Noahide laws could be considered to be principles for all people, whereas the Sinaitic Law is for Jews only.[45]

The article already quoted elaborates: “Certain writers, however, find in halakhic discourse a more universal aspect, whose norms do not draw their authority from the particular commitments of the Jewish covenant. Some locate this aspect of halakhah in the seven ‘Noahide Laws,’ which are seen to represent a sort of natural law. Others find the universal aspects in specific broadly humanistic themes. But in either case, the challenge of imbuing Jewish bioethics with general relevance consists in identifying the appropriate strands within halakhah. Since halakhah is not only a positive legal system but an articulated code of values, the significance of its teachings need not depend on accepting the legal system’s authority en bloc. The fact is that halakhic discourse often appeals to reasons that are intelligible independently of religious authority. It can thus be seen as constituting a language of moral discourse that is widely accessible and relevant and that bears its own distinctive emphases and valuations.”

- But there is no explicit bioethical commandment in the Torah and the Halakhah (as it is not in the Christian or the Muslim tradition).[46] When new problems arise and their pressure on public morality increases, especially under the impact of scientific and technical development, moral traditions have to be redefined, or they become irrelevant.

- Either new rules could be derived from old ones by analogy; or morals become bifurcated: erudite Jewish people, interpreting the traditional law, can admit that traditional law in modern societies regulates marriage, whereas economic law is adapted from countries like Britain or Germany.[47] And bioethical standards? Whenever such division is alleged, a moral system bound to a particular identity and a pragmatic system of social regulations operate side by side.
- Although rules and especially legal standards in Israel are frequently taken over from Western countries, the situation in bioethics seems to be inverted: The former prime minister of Northrhine-Westphalia, now minister of economic affairs in Berlin, chose to secure certain interests of people engaged in stem cell research in an Israeli scientific Institute at Haifa. He seemed to believe that Jewish ethics provides strong moral support for stem cell research even in Germany to overcome a certain one-sided principle-orientated tendency prevailing there.

8. Some Questions about Cloning

Many people believe that there is no moral difference between therapeutic and reproductive cloning - except in the subjective intention of the researcher. This moral opinion depends on a specific ontology which combines an essentialist view of the human germ with a strong form of teleology and a universalistic perspective: The argument should be valid in *all* cases of a totipotential stem cell.

But the controversies result from the ontological foundations in the background of the discussion. Traditionally the argument of unique human dignity is expressed by the idea that only human beings - although they are animals (Latin: animalia) - have a rational soul. This implies a strong teleology from the beginning. But does a human being have this soul from the moment of the fusion of oocyte and sperma? The Aristotelian teaching concerning the soul supposes that the soul enters the body on the fortieth (male) or the eightieth (female) day. This strange hypothesis seems to have been taken over by some rabbis.[48] Older Christianity is split by the contradicting opinions either of a transfer of the soul from the parents to the embryo at the moment of procreation or of the creation of the soul in an already existing embryo. Certain natural law traditions regard individualization as the moment from which human dignity is valid. (Individualization presupposes - according to this opinion - nidation and the exclusion of the possibility of identical twins developing.) But what does it mean to be an individual in the strong sense? When Jews frequently emphasise the birth, they recognise the personal history of human existence in the midst of a given community. The radical application of the notion of human dignity to the human germ is a modern idea that runs counter to the threatening trends of an instrumentalisation of human life (and life in general). In each case there are difficulties: Human dignity does not develop in stages; nor does the general application of human dignity to a germ cell make sense: Should a cell be buried if it is not transferred to the womb?

Moral ambiguities like this do not permit an absolute position; the universalistic perspective is not uniform, but conforms only to certain exclusions: No manifestation of human life can be entirely reduced to the status of an object; and research - if legal - must respect certain limits. Therefore, research with germ cells should be a strictly regulated exception.

Notes

1. According to Kant there is a strict difference between value ('Preis') and dignity: "Im Reich der Zwecke hat alles entweder einen Preis, oder eine Würde. Was einen Preis hat, an dessen Stelle kann auch etwas anderes, als Äquivalent, gesetzt werden; was dagegen über allen Preis erhaben ist, mithin kein Äquivalent verstattet, das hat eine Würde." (Kant, 1786). Unfortunately this frequently quoted proposition is hardly ever interpreted consistently. An ethics of life cannot exclusively rely on pure reason, but has to integrate bodily life and emotions; therefore the moral discussion of issues of life must overcome the Kantian separation of reason and sensuality, but cannot be restricted to a Humean type of ethics; an ethics of life, however, would - according to Kant's definition - be an ethics of value.
2. William A. Frankena explained the main difference in ethics as the contrast between teleology and deontology, an ethics of comparison of goods and an ethics of strict obligation. The background of this discrimination is a different mental construction of the social world of human beings - the first one from the empiricist and psychological point of view with goods relative to human desires, the second one relying on the obliging power of reason, but not separated from real life. Cf. Frankena, 1993. - The first to divide ethical theories into the categories 'teleological' and 'deontological' was Muirhead, 1932. The

term 'deontology' asserts that certain duties are intrinsically right or wrong, but without reflexion of the 'good' or 'bad' of the consequences combined to an act resulting from an explicit duty.

3. See footnote 2, implying the differences between a Humean and a Kantian type of ethics in modern terms.
4. J. Bentham favoured a relativistic ethics of goods founded in a democratic belief and in a kind of an anti-Aristotelian view: not the descendent of the educated class, but everybody should be able to determine her or his goods: Bentham, 1970.
5. Thoroughly relativistic theories were never successful, as the example of Westermarck proves: Westermarck, 1906-8.
6. Cf. Hart, 1994. - He discusses primary and secondary rules, relying on the nature of social rules in general. - Dworkin, 1977, criticises the lack of foundational reflexion in Hart's oeuvre.
7. The italic passages in this essay try to furnish examples of bioethical applications of moral rules.
8. cf. Singer, Kuhse. Singer, 1979. Singer proposes his arguments in the context of a Lockean idea of personhood. To be a person means to be conscious of oneself and one's own personality in social settings. Embryos cannot be 'persons.'
9. Engineering the Human Germline: The Immediate Prospects for Pre-Implantation Genetic Therapy, a special one-day symposium co-sponsored by The UCLA Center for the Study of Evolution and the Origin of Life. This consultation was held on March 20, 1998, cf. http://www.ess.ucla.edu/huge.
10. Cf. footnote 7.
11. The context of the modern doctrines of personhood is the history of a merger of philosophical and Christian (theological) ideas: Cf. Brasser, 1999.
12. Nussbaum, 1993, 323-363.
13. Cf. Leaman: "Most theologians were Ash'arites [...], which meant that they were opposed to the idea that ethical and religious ideas could be objectively true. What makes such ideas true, the Ash'arites argued, is that God says that they are true, and there are no other grounds for accepting them than this. This had a particularly strong influence on ethics [...] , where there was much debate between objectivists and subjectivists, with the latter arguing that an action is just if and only if God says that it is just. Many thinkers wrote about how to reconcile the social virtues, which involve being part of a community

and following the rules of religion, with the intellectual virtues, which tend to involve a more solitary lifestyle."

14. Ilting, 1978, 245-314, Finnis.
15. Protestantism (except Puritanism) used the doctrine of the 'two kingdoms' to differentiate between the revealed commandements of God (divine law) and the civil rules given in creation or in the regimen of conservation of the world fallen into sin and chaos.
16. Hare, 1963, proposes a Golden-Rule Argument for testing moral maxims: Whoever wonders whether he should do this or that, should reflect whether he can prescribe that every human being should act in the same way, whatever the world and his role might be. Hare, 1981, distinguishes two levels of moral reasoning - the one as 'critical' and the other as 'intuitive.'
17. Moore, 1903, especially chapters 5 and 6, argues in favour of the following propositions: (a) intuitions are general, (b) the truth of utilitarianism is self-evident, (c) we can experience by intuition the goodness of some properties. - Ross, 1930, especially chapter 2, proposes a pluralist version of intuitionism. He continues his arguments in 'The Foundations of Ethics,' Oxford 1939, especially 79-86 and chapter 8: Persons intuit general principles about the moral relevance of types of actions.
18. The historical roots of most of the contemporary arguments in favour of respect for persons are found in the moral philosophy of Kant, esp. in the 'Grundlegung zur Metaphysik der Sitten' (see footnote 1!) [transl. with notes by Paton, 1948]; he determines humanity as an end in itself.
19. Schopenhauer, n.d., 345-655, criticised the purely rational and moral beings of Kant as 'lovely little angels' and not as human beings. - Therefore the modern transcedental philosophy of Karl-Otto Apel postulates a 'Leib-Apriori,' a bodily and social Apriori of human existence (Apel, 1973).
20. Hegel, 1821, proposes a theory of law and morality and unfolds a social and political philosophy. The famous but often obscurely presented theory of objective spirit includes (1) a special conception of freedom, (2) a theory of the whole of social life as 'ethical life' and (3) regards it as the ultimate 'reality of reason.'
21. Cf. Donagan, 1977 tries to overcome the monomial principle of abstract reason and therefore to derive moral duties from a basic and comprehensive principle of respect for persons. Outka, 1978, demon-

strates the principal character of agape-love in contrast to the abstraction of the Golden Rule reasoning.

22. Berger, 1976.
23. Anselm, 2003; Bünker, 2001; Schockenhoff, 1993.
24. Gadamer, 1986, 173-199.
25. Ruse, 1986, 173-192.
26. This inclination could even be a temptation of many religions.
27. Schweitzer, 1993, 121-132, 133-141; Dworkin, 1993 and others.
28. Darwin, 1859. - An anthropological and ethical reaction: Taylor, 1964.
29. Engels, 1982.
30. Cf. Wick, 2004, presenting strict Jewish positions of embryonal life and Oeming, 2004, presenting liberal Jewish positions.
31. see Aristotle.
32. the purpose is simultaneously the imperative of being - see Hans Jonas (Schweitzer, 1974, 375-402; Jonas, 1973, 1979, 1984).
33. Hillel, living at the time of king Herod, was the main figure among the moderate pharisees, reducing the commandments of the Torah to the Golden Rule (bŠab 31a).
34. Frey, 2003, 161-78.
35. Cohen, 1966.
36. Baeck, 1988.
37. 'Law and Gospel' is a major theological conception of the Reformers, inspired by an inner-Jewish dispute of Paul and the Pharisees in the New Testament and emphasising the difference between morals and salvation. It helps to differentiate between religion and social policy.
38. The Roman-Catholic version of the difference between religion and social policy goes back to the doctrine of natural law, the latter being dependent however upon the interpretation of religious authorities who know about the foundations of the natural law in creation.
39. Cf. footnote 13!
40. Hobbes, 1889.
41. see Halakhah.
42. Zohar.
43. Walzer, 1983; Etzioni, 1993.
44. 'Abduction' goes back to the American philosopher Pierce. Pierce maintained that in the long run the only one method of reasoning could form a consensus by an agreement. This process he called 'abduction' (an 'inference to the best explanation'). He thought that propositions were not true because they were arrived at by abduction; but that they were true simply because they were universally accepted.

45. Gen 9, 4-7. - Years before the Sinai events - according to the historical fiction of the Hebrew Bible - the surviving mankind got the commandment not to shed the blood of human persons.
46. Zohar, Steinberg, 1988.
47. Cf. Falk, 1991, 1: "<Surrendering our logic to the Almighty and embracing the logic of Sinai> [quotation from an American rabbi] would render our moral faculties superfluous and constitute an act of ingratitude vis-a-vis their Creator. A meaningful evaluation of the Torah cannot be obtained only <from within> and cannot exclude rational, historical, psychological and utilitarian considerations."
48. Cf. Zohar: "Rabbinic traditions regarding abortion are far from consistent. Two often-quoted passages (Mishnah Niddah 3: 7 and Babylonian Talmud Yevamot 69b) refer to pregnancy up to forty days as 'mere fluid.' Beyond this point, some mandate abortion for the sake of a moderate interest of the woman, arguing that the embryo is simply 'her body' ('Arakhin 7a)."

References

Agape, Gene Outka. *An Ethical Analysis*, New Haven/Ct. and London, 1978.

Anselm, Reiner und Ulrich Körtner (eds.). *Streitfall Biomedizin*. Göttingen, 2003.

Apel, Karl-Otto. *Transformation der Philosophie*, vol. 2. Frankfurt a.M., 1973.

Baeck, Leo. *Das Wesen des Judentums*, 8th ed. Wiesbaden, 1988.

Bentham, J. *An Introduction to the Principles of Morals and Legislation* (1789). London, 1970.

Berger, Peter und Brigitte Berger. *Sociology. A Biographical Approach*. Harmondsworth/Middlesex, 1976.

Brasser, Martin (ed.). *Person. Philosophische Texte von der Antike bis zur Gegenwart*. Stuttgart, 1999.

Bünker, Michael und Ulrich Körtner (eds.). *Verantwortung für das Leben. Eine evangelische Denkschrift zu Fragen der Biomedizin*. Wien, 2001.

Cohen, Hermann. *Religion der Vernunft aus den Quellen des Judentums (1919)*. Darmstadt, 1966.

Darwin, Charles. *The Origin of Species*. London, 1859.

Donagan. *The Theory of Morality*. Chicago/Ill. and London, 1977.

Dworkin, Ronald. *Taking Rights Seriously*. London, 1977.

Dworkin, Ronald. *Life's Dominion*. London, 1993.

Engels, Eve-Marie. *Die Teleologie des Lebendigen*. (Erfahrung und Denken 63). Berlin, 1982.

Etzioni, Amitai. *The Spirit of Community. Rights, Responsibilities and the Communitarian Agenda*. New York, 1993.

Falk, Ze'ev. *Religious Law and Ethics*. Jerusalem, 1991.

Finnis, John. "Natural law." In *Routledge Encyclopedia of Philosophy* (digital version).

Frankena, W. *Ethics*. Englewood Cliffs/NJ, 1993.
Gadamer, Hans-Georg. "Bürger zweier Welten." In *Der Mensch in den modernen Wissenschaften*, Krzystof Michalski (ed.). Stuttgart, 1983.

Hare, Richard M. *Freedom and Reason*. Oxford, 1963.

Hare, Richard M. *Moral Thinking: Its Levels, Method and Point*. Oxford, 1981.

Hart, H. L. A. *The Concept of Law*, 2nd ed. with postscript. Oxford, 1994.

Hegel, G.W.F. *Naturrecht und Staatswissenschaft im Grundrisse. Grundlinien der Philosophie des Rechts*. Berlin, 1821.

Hobbes, Thomas. *The Elements of Law Natural and Politic (1640)*, F. Tönnies (ed.). London, 1889.

Ilting, Karl Heinz. "Naturrecht." In *Geschichtliche Grundbegriffe, vol. 4*, Otto Brunner a.o. (eds.). Stuttgart, 1978.

Jonas, Hans. *Organismus und Freiheit*. Göttingen, 1973.

Jonas, Hans. *Das Prinzip Verantwortung. Versuch einer Ethik für die technologische Zivilisation*. Frankfurt a.M, 1979.

Jonas, Hans. *The Imperative of Responsibility: In Search of an Ethics for the Technological Age*. Chicago (Ill.), 1984.

Kant, Immanuel. *Grundlegung zur Metaphysik der Sitten*. Riga, 1785, 2nd ed. 1786.

Leaman, O. „Islamic philosophy." In *Routledge Enyclopedia of Philosophy* (digital version).

Moore, Gerald E. *Principia Ethica*. Cambridge, 1903.

Muirhead, J.H. *Rule and End in Morals*, Oxford, 1932.

Neuenschwander, Ulrich (ed.). *Straßburger Predigten*, 3rd ed. München, 1993.

Nussbaum, Martha. "Menschliches Tun und soziale Gerechtigkeit. Zur Verteidigung des aristotelischen Essentialismus." In *Gemeinschaft und Gerechtigkeit*, Micha Brumlik/ Hauke Brunkhorst (ed.). Frankfurt a.M., 1993.

Oeming, Manfred. "Bioethische Probleme des Klonens in alttestamentlicher und jüdischer Perspektive." In *GlLern* [Glaube und Lernen] 19, 2004.

Paton, H.J. *Groundwork of the Metaphysics of Morals*, London, 1948; repr. New York 1964.

Ross, William D. *The Right and the Good*. Oxford, 1930.

Ross, William D. *The Foundations of Ethics.* Oxford, 1939.

Ruse, Michael and Edward O. Wilson. "Moral Philosophy as Applied Science." *Philosophy* 61, 1986, 173-192; repr. in E. Sober (ed.). *Conceptual Issues in Evolutionary Biology*, 2nd ed., Cambridge (MA), 1994, 421-438.

Schockenhoff, Eberhard. *Ethik des Lebens. Ein theologischer Grundriß.* Mainz, 1993.

Schopenhauer, Arthur. „Die beiden Grundprobleme der Ethik." In *Sämtliche Werke*, Bd. 3, Eduard Grisebach (ed.). Leipzig, o. J.

Schweitzer, Albert. *Kultur und Ethik*, 3. ed. München, 1960.

Schweitzer, Albert. "Die Ethik der Ehrfurcht vor dem Leben." In *Ges. Werke II.* München, 1974, 375-402.

Singer, P. *Practical Ethics*. Cambridge, 1979, 2nd ed. 1993.

Steinberg, A. (ed.) *Encyclopedia of Medicine and Jewish Law.* Jerusalem, 1988.

Taylor, Charles. *The Explanation of Behaviour.* London, 1964.

Walzer, Michael. *Spheres of Justice*. Oxford, 1983.

Westermarck, E. A. *The Origin and Development of the Moral Ideas*. London and New York, 1906-8.

Wick, Peter. "Nicht ein Gott der Toten, sondern der Lebenden." In *GlLern* [Glaube und Lernen] 19, 2004.

Wilson, E.O. *Sociobiology*. Cambridge (MA), 1975.

Zohar, J. "Bioethics, Jewish." In *Routledge Encyclopedia of Philosophy* (digital version).

The American Debate on Human Cloning

Nigel M. de S. Cameron

Abstract: The US debate over embryo stem cell research and human cloning has been a dominant theme of public and political focus during the Bush administration. President Bush's first televised broadcast to the American people was on embryo stem cell research, and while efforts at federal legislation on cloning have been frustrated, US influence helped secure a prohibition on all forms of cloning in the United Nations Declaration on Human Cloning.

Keywords: Cloning, Bioethics, United States, Congress, Biotechnology, George W. Bush, Embryo, Somatic cell nuclear transfer, Embryo stem cell research

1. The Significance of This Debate

The announcement in February 1997 of the cloning of Dolly the sheep took the media by storm and offered cover copy to every news magazine as the uncomprehending mammal stared out at scarcely more comprehending publics around the globe. Yet despite the news hype that accompanied the announcement and has followed its sequelae, one can argue that the final significance of Dolly was, and continues to be, underestimated.

For the first mammalian cloning bifurcates the history of the world. It demonstrates to humankind as has nothing before that we are in line to become the creatures, the products, of our own inventive selves: *Homo sapiens* in the hands of *Homo faber*. As the newspapers and the television talk shows jumped straight from the Scottish sheep to the meaning of this story for the human future, Dolly proved the catalyst of a vast debate about biotechnology and its significance for human nature. On trial, alongside human nature itself, is the capacity of our public institutions to respond to what some commentators have begun to recognise as their greatest challenge.

This point was well-made by President George W. Bush in the April 2002 White House speech in which he called for a comprehensive ban on human cloning in the United States:

> Science has set before us decisions of immense consequence. We can pursue medical research with a clear

> sense of moral purpose or we can travel without an ethical compass into a world we could live to regret. Science now presses forward the issue of human cloning. How we answer the question of human cloning will place us on one path or the other.

This is the benchmark debate of the biotech century. For in the prospect of human cloning we begin to see the bizarre ironies, and the tragedy, of the extension of our human powers over our own selves and our own kind. In our attempt to serve as our own creator, we are revealed as usurpers capable only of manufacture, supplanting the inter-personal mystery of human sexuality as the context for procreation. We reduce this mystery in parody to the mechanistic and industrial processes at which we are so good, but which so well display our human limits. The ambiguity of the clonal human, as both creature and product, *Homo sapiens* hijacked by *Homo faber*, moves us decisively down the road toward what the posthumanists call the "singularity" - the envisioned state where the distinction between human being and manufactured being disappears into a seamless dress, weaving together our humankind and what we have fashioned it to be. It anticipates the union of what in Spielberg's movie "AI: Artificial Intelligence" are called "mecha" and "orga."

In his famous jeremiad, "Why the Future doesn't Need Us," Bill Joy, co-founder and former chief technology officer of Sun Microsystems, claims that genetics, robotics, and nanotechnology are the three great threats to the human race in the twenty-first century.[1] By some mixture of accident and intent they are likely to destroy or supplant the human species through some biological or mechanical meltdown - or through the triumph of machine intelligence. One does not need to buy the whole thesis to acclaim his comprehensive framing of the issues. The heart of this analysis, from inside the world of technology, exposes the dire threat to human dignity posed merely by human creativity and technological prowess.

For convenience, I refer to this analysis as the "Bioethics 2" agenda - building on from the conventional questions of medical ethics and public policy and focusing on the use of technology to control, design, and ultimately "improve" human beings at the fundamental level of genetics. To those who have been most concerned with traditional questions such as abortion and euthanasia, or the more sharply-focused discussions of resource allocation, organ donation, and the definition of death, these "Bioethics 2" questions may seem of lesser significance since in themselves they may not threaten to destroy human life. However, a fresh paradigm is emerging, which recognises that - as one might simply phrase a most profound question - taking a human life made in God's image may not in fact be as serious as making a human being in our own. Incremental technological interventions in the process of human procreation that began with artificial insemination and, subsequently, *in vitro* fertilisation and its variants, will encompass increasingly sophisticated capacities. "Genetic engineering," the blanket term used to cover interventions that make changes at the genetic level, has already begun to have limited clinical applications. While genetic alterations that benefit the individual by curing disease are to be welcomed, changes that produce "enhancements" will also be possible. Most important of all, inheritable changes, changes in the germline that will affect the genetic constitution of every subsequent human being in that family, will offer the ultimate challenge to humankind: whether or not we should seek to "improve" our human nature and take control over the kind of beings that we are.

It was of this possibility that C.S. Lewis, the English literary scholar and cultural critic, wrote with extraordinary prescience, in his essay "The Abolition of Man," in which he foresaw "man's final triumph over nature" in our triumph over our own nature. Yet, he asks, "who will have won?" What may appear to be "man's triumph over nature" is in fact "nature's triumph over man." By taking control of our own nature, we are subjecting ourselves to ourselves and turning ourselves into some*thing* that we can control and dominate. The general who rides in the triumphal procession is one and the same with the slave he pulls behind him. There is no doubt that the question of the integrity of human nature is emerging as the greatest issue to have been faced by humankind. For biotechnology will deliver into our hands a capacity to alter what it means to be human.

Even beyond genetics, we already see advances in nanotechnology and cybernetics that offer a vision of human nature with radical enhancements. In this realm of scientific advancement, we delve into the question that I categorise as "Bioethics 3." The terms "post-humanist" and "trans-humanist" have been coined by advocates of a reconstructed human nature in which the "cyborg" ("cybernetic organism"), a human-robotic amalgam, takes the place of human beings. While much of the research in these areas of science and technology may have benign purposes and offer the prospect of vast benefit to humankind, there has been little discussion of its potential significance in the reshaping of human nature. Questions raised by these technologies concerning human dignity precisely parallel those related to developments in human genetics, and there is reason to believe that some of the applications of the "Bioethics 3" technologies will raise ethical and policy issues for the human race well before genetics develops the capacities to re-shape our nature.

By curious coincidence, Dolly's birth was announced the same year the formalisation of the first two international responses intended to set policy parameters for the development of biotechnology: the Council of Europe's *European Convention on Human Rights and Biomedicine* and the UNESCO *Universal Declaration on the Human Genome and Human Rights*.[2, 3] The two events represent the attempts of the more prescient members of the international community to anticipate the policy challenges of the biotech revolution. Yet the slow rate of their progress (both these projects began in the 1980's) shows the limits to their prescience, the vastness of the challenge, and the obvious problematics of multilateral decision-making.

It has become customary for historians to see the "nineteenth century" as beginning in 1789 and ending in 1914. No doubt, in due time, they will deem the twenty-first century and the third millennium *Anno Domini* to have begun in 1997.

2. Four Contexts for Cloning

The simplest way to illustrate the significance of these developments is to set cloning itself in a fourfold context:

First, it represents the latest intervention in the process of mammalian reproduction. In the middle of the twentieth century, religious authorities and some secular opinion were deeply troubled by the application of artificial insemination (AI) to human beings, long a technique of animal husbandry. Some focused on the technique itself (in which sperm is implanted "artificially" in the womb); others focused on the use of "donor sperm" (sperm from a third party). While that debate is now behind us, the AI controversy raised some key issues that remain at stake in current debate. Moreover, AI gained wide social acceptance for the idea of medical intervention to enable conception, even if it involved a gamete from another party. AI led directly to its more sophisticated sibling, *in vitro* fertilisation (IVF), in which either gamete could now come from a third party and the fertilisation process could be conducted under controlled conditions in the lab. In England, the world's first "test-tube" baby, Louise Brown, was born as long ago as 1978. These developments raised fresh ethical issues and greatly increased the options and dilemmas involved in reproductive issues. Surprisingly, while these events generated publicity, there was little controversy outside of the Roman Catholic Church.

Against the backdrop of AI and IVF, cloning is plainly revolutionary. Indeed, one of the many ironies of current debate is the degree to which ART ("assisted reproduction technologies") involving AI and IVF and their variants radiate by contrast an aura of "normality" since they seek to replicate the natural process of conception - even though in "unnatural" and controlled circumstances. In cloning we have a completely different technique: a mechanism for *asexual* reproduction. Yet, the public mind has been prepared for this technique, but its moral sensibilities dulled to the implications. AI and IVF, aside from their respective merits and demerits, have built a bridge between procreation as natural process, the consequence of sexual union and "reproduction" as an act of human will and technique in which science and medicine combine to enable a range of "choice" in childbearing. With such techniques as the selection of gametes (for which there is a market, in the US at least: sperm on the internet, eggs through ads in college newspapers), pre-implantation genetic diagnosis (PGD) to ensure genetically attractive offspring, surrogacy, and other manipulative components, "reproduction" in the late twentieth century had already incorporated modes of manufacture to produce customised offspring. *Homo faber* is already at work on the design of himself.

The second context for cloning lies in the debate about abortion, especially in the United States where after a generation it remains one of the dominant elements of political culture and frames the debate about biotechnology. Cloning advocates fall into two groups: (1) those who favour the birth of live-born cloned children; and (2) those who may or may not favour cloned children, but who seek cloned human embryos for purposes of experimentation. Since cloning for research invariably would destroy the cloned embryo and cloning that is intended to lead to live-born children would cause many deaths along the way (Dolly was the one relatively healthy survivor out of 277), the American pro-life movement has lined up behind a ban on all human cloning. But the abortion context goes wider since the liberal abortion movement has undoubtedly paved the way for a *laisser-faire* approach to experimentation on the early embryo, itself one of the keys to *in vitro* and other reproductive technologies.

The third context is more fundamental. Much of the earlier debate about abortion focused on what seemed to many on both sides to be the central question: when does life begin? That debate has moved on. Now, what is at heart is no longer about when life begins, but what life is and whose right it is to make decisions about life and death. We might say that the debate has moved back, from embryology to anthropology - what it means to be human. An emerging struggle to redefine our assumptions about the nature of human being resonates behind the particulars of the cloning debate.

Here we note the fourth context, since in tandem with the culture's engagement in re-thinking human nature, cloning offers the first-fruit of a technology that is destined to raise the most profound of all questions for humankind. For cloning is just the beginning. The scope of biotech's possibilities has yet to be imagined. What we do know is that its capacities will press the human conscience and challenge public policy to the limit.

3. The Shape of the Biotech Debate

It is common for the many questions raised by medicine and biotechnology to be listed as if they were disconnected from one another. Yet, as we have noted, debates about abortion and euthanasia prepared the way for our discussion of issues in biotechnology both directly, by undermining our assumptions about the sanctity of life, and also indirectly, by giving shape to gathering cultural unease and confusion about what it means to be human - and, therefore, how humans should be treated.

The first major policy document in the field, like the first "test-tube baby," was produced in the United Kingdom under the guidance of moral philosopher Mary Warnock. In an infamous passage that has set the lamentable pattern for policy development in much of the world, the Warnock Report states that "instead of trying to answer" the questions "when life or personhood begins" they have "gone straight to the question of *how it is right to treat the human embryo.*"[4] It is an elementary question, but unless we know what something is, how we can know how to treat it? This admission of ignorance in relation to the nature of the human embryo is symbolic of a more general willing ignorance of the nature of human being itself that has pervaded public discussion on issues in medicine and biotechnology. If we do not know what it means to be human, how shall we know how to treat humans as we begin to develop powers over our own selves - not simply to take life but to shape it according to our wishes?

4. Legislative Debate in the United States

Since the announcement in February of 1997, Americans have been engaged in vigorous debate about human cloning. While there are individual advocates of live-birth cloning ("reproductive cloning" or the creation of a genetically identical born human being), there is a wide-ranging political consensus in Congress that it should be made illegal. Therefore, debate has focused almost entirely on the question of cloning for biomedical experimentation and research (for which the term "therapeutic cloning" has been coined by its advocates). Rapid advances in our understanding of embryonic stem-cells and their potential for "regenerative medicine" have complicated the debate and made large-scale cloning of human embryos attractive to those who see it as a panacea for chronic and degenerative disease.

A. The Federal Policy Debate

Major debate has been focused at the federal level, though there has also been state legislation. President Bush strongly urged Congress to pass a comprehensive ban on human cloning in his January 28, 2003 State of the Union Address stating that "because no human life should be started or ended as an object of an experiment, I ask you to set a high standard for humanity and pass a law against all human cloning."[5]

Several cloning bills have been introduced into the House and the Senate, though none has been passed into law. In the 107th Congress (2000-02), the House considered two rival bills, one to ban all cloning, the other to ban live-birth cloning and requiring the destruction of embryos cloned for experimental purposes. The comprehensive ban was passed by a large majority in the House, but it stalled in the Senate.[6] The same pattern is evident in the 108th Congress, although with (narrow) Republican control of the Senate, there is a somewhat enhanced chance of success in the Senate.

The House vote was very substantial (241-155) and represented support from many Democrats and most Republicans for the bill sponsored by Representatives David Weldon (Republican) and Bart Stupak (Democrat), which would impose fines of up to $1 million and imprisonment for any person who attempts to clone humans for any purpose.[7] More significant, there was support from some outspoken supporters of "abortion rights." This is highly unusual in the US context where the abortion debate is strongly polarised and has been a dominant and divisive force in politics and the wider culture. Those testifying in favour of the bill included leading pro-choice feminist Judy Norsigian, editor of the famous women's health text *Our Bodies, Ourselves*;[8] and Stuart Newman, until recently a leading figure on the environmental (and pro-choice) group, the Council for Responsible Genetics (CRG).[9]

In the Senate, a comprehensive cloning ban is proposed in S. 245, sponsored by Senator Sam Brownback (Republican) and Senator

Mary Landrieu (Democrat).[10] The opposing bill, S. 303, is worth reviewing since it indicates something of the manner in which those who support cloning for experimentation are making their case.[11] Its sponsors include Senator Orrin Hatch, a conservative Republican who is against abortion, as well as pro-choice Democrats. The bill crafts a novel and circumlocutory definition for cloning as "implanting or attempting to implant the product of nuclear transplantation into a uterus or the functional equivalent of a uterus," so it can be interpreted as legislation opposing all "cloning" while permitting cloning for research purposes since the term has been redefined. Furthermore, the bill offers an abstract, new term for the clonal embryo, referring to it as an "unfertilised blastocyst," that is, "an intact cellular structure that is the product of nuclear transplantation." However, this term is a curious neologism since blastocysts do not get fertilised; oocytes are fertilised and thereupon become blastocysts. The purpose of this coinage is plainly to seek to redefine the terms of the debate and deflect attention from the fact that the product of cloning is a clonal embryo, in order to protect the use of such embryos for deleterious experimentation.

Alarmingly, Senator Hatch is on record as construing the clonal embryo as different in nature from an embryo that results from fertilisation, suggesting that if a clonal pregnancy came to term, the resulting baby would not be the same as a normal human baby.[12]

> No doubt somewhere, some - such as the Raelians - are trying [to] make a name for themselves and are busy trying to apply the techniques that gave us Dolly the Sheep to human beings. Frankly, I am not sure that human being would even be the correct term for such an individual heretofore unknown in nature. I am a conservative and an unabashed pro-life, religious conservative at that. Or should I say, to be politically correct, I am a faith-based conservative. In any event, I would be extremely hesitant to rewrite the Book of Genesis as the story of Adam or Eve.

This notion of a sub-normal class of humans clearly contravenes the fundamental, founding declaration of the Unites States that "all men are created equal." Religious and secular critics have recognised such potential discrimination in a "new eugenics" or determinism where human cloning could implicate exclusion and exploitation similar to the historic injustices of racism and slavery. Though many agree "reproductive cloning" should be banned, once cloning of human embryos for research purposes is allowed, enforcing such a ban would be nearly impossible since the two types of cloning differ not by technique, but by intent.

B. Debate in the States

Many state legislatures have engaged in cloning legislation, some passing comprehensive bans, others, especially those with large biotech industry sectors, passing or considering permissive legislation. Venture capitalists have been hesitant to invest in the highly controversial and expensive endeavours of "therapeutic cloning," so there has been growing pressure for public funding for this work. In fact, a September 2003 study in the *Proceedings of the National Academy of Sciences* estimated that it would cost $200,000 for one patient's "therapeutic cloning" treatment. Thus, states promoting their biotech presence are under economic pressure to legalise therapeutic cloning and generate funding through taxpayers. California, the nation's largest biotech state, was the first formally to legalise "therapeutic cloning" in 2002. The state's proposed "Biomedical Research and Development Act of 2004" would explicitly encourage cloning for experimental purposes by awarding grants or loans to public or private institutions conducting stem cell research from a $3 billion tax-supported fund.[13] By contrast, at least six states have proposed legislation explicitly prohibiting the use of public funds and facilities for participating in cloning or attempted cloning.[14]

In January 2004, New Jersey signed into law a bill that authorises cloning for research and may permit the gestation of an implanted embryo clone in a womb through nine months' gestation. The shocking New Jersey bill does more than "permit human stem cell research" and allow experiments to clone human embryos. Rather, the bill effectively permits cloned human embryos to be implanted in the womb and used for experiment during pregnancy; all that it forbidden is for them to be born and survive.[15]

On the other end of the spectrum, at least five states have enacted complete bans on cloning: Arkansas, Iowa, Michigan, North Dakota, and South Dakota.[16] In addition, Rhode Island, California, Louisiana, and Virginia ban "reproductive cloning."[17] Some states ban the purchase or sale of embryos and other human tissues involved in reproductive or therapeutic cloning or ban the shipping, transferring, or receiving an embryo produced by human cloning.[18] Depending on the state, penalties for a cloning violation can be as much as $10 million for an entity or an individual.[19] Some states have enacted felony criminal statutes or provisions for professional license revocation.[20] Moreover, the constitutionality of cloning legislation has been challenged in courts where the term "experimentation" is vague or ambiguous.[21] Additionally, "reproductive liberty" challenges could be raised against statutes banning "reproductive cloning" but encouraging or permitting the cloning of embryos for research.[22]

5. The United States Position in the United Nations General Assembly (UNGA) Sixth Committee

In response to a Franco-German proposal, endorsed by the UN General Assembly in 2001, to seek an international convention to prohibit "reproductive cloning," the United States position reflected its administration's domestic policy - that is, to seek a convention that would prohibit all human cloning, and to oppose any convention that focused merely on the cloning of live-born babies.[23]

Therefore, the US worked with other states in the Sixth (legal) Committee to build support for a wider convention, noting (a) that all cloning is in fact "reproductive," and (b) that to secure a comprehensive prohibition on the birth of live-born cloned babies the use of somatic-cell nuclear transfer in human beings should be prohibited (or else the technology and the supply of clonal embryos would materially aid and ensure the future birth of clonal children).[24] Subsequently, the US co-sponsored an alternative proposal with more than 30 other nations including Spain and the Philippines.[25]

During meetings of the Sixth Committee (2003), several key developments emerged. First, the German federal government, under severe domestic pressure, withdrew from its leading role in the debate, leaving Belgium to propose a position that reflected precisely the twin-track approach it had been advocating - proposing that there be a ban on so-called "reproductive cloning" and either a ban on or regulation of so-called "therapeutic cloning."[26] Second, a resolution proposed by Costa Rica seeking a comprehensive ban began to gain momentum among a wide range of delegations and was ultimately co-sponsored by 64 states against less than half that number for the Belgian position.[27] It is my understanding that the Costa Rican delegation confidently expected the support of a further 40 nations for their resolution. However, in November 2003, the Sixth Committee voted by the narrowest of margins (80-79, and 15 abstentions) for a procedural motion to suspend the discussion until 2005, which was proposed by Iran on behalf of the Organisation of the Islamic Conference (OIC).[28] In turn, this decision was revised by the General Assembly, where after concentrated diplomatic activity it was agreed to address the matter in the 59th session of the General Assembly in the fall of 2004.

6. Special Challenges to Democracy and Debate

Behind the many particular questions of this debate lurks the great question for public policy: can our democratic institutions frame a proportionate response to the exponential demands that biotechnology will make on the fabric of human society? As government pursues its traditional priorities of national defense and adroit economic management, a stealth agenda of greater import has begun to break the surface: the question of the future of human nature. Since democratic government has evolved in a period of settled conviction about such fundamental questions, it is little surprise that our institutions find this new and fateful challenge to be traumatic.

Cloning itself was difficult enough to grasp, but the stem-cell debate that has been superimposed on cloning has made it even harder. How are ordinary folk to grasp what is at stake? What are they to make of the "experts" who rule the airwaves and the op-ed pages? And, having grasped what is going on, how are they to influence events? The more insistently we ask these questions, the bigger loom the challenges to democracy.

While at one level the basic expertise required for democratic participation is not high - in a culture that requires biology in both college and high school - the "expert" factor is of enormous political consequence.

We must confront huge questions that the democratic process was not designed to handle arising from fast-paced changes in our understanding of science. These questions challenge all of us who want to stay on top of developments that focus on controlling and changing the nature of human being itself. There is no question that the manner in which this debate is finally resolved will set the pattern for the unfolding policy debates of the so-called "Biotech Century."

Three inter-related issues on cloning policy can be identified:

A. Terminology and Ignorance.

There is no short-cut to expecting responsible citizens to develop a basic familiarity with developments in biology. One local newspaper carried a recent report on a high school senior who had just graduated despite the challenges of life in a wheelchair. "Stem-cell research" might cure her problem, she told the reporter, but "the government won't allow it." Well, those who follow the debate will note at least three basic mistakes in that statement: (1) work on adult stem-cells is hugely promising and is being funded energetically by the federal government; (2) basic work on embryonic stem-cell lines is also being funded; and (3) there is no federal prohibition on privately-funded work that involves destroying embryos to obtain embryonic stem-cells. Moreover, the federal legislation that would prohibit cloning would not affect destructive embryo research itself, merely cloning to mass-produce the embryos.

The struggle for terminology has proved a key feature in the debate. In the President's Council on Bioethics July 2002 report, *Human Cloning and Human Dignity: An Ethical Inquiry*, the Council voiced concern about "the temptation to solve the moral questions by artful redefinition or by denying to some morally crucial element a name that makes clear that there is a moral question to be faced."[29] The report adopted the terminology "cloning-to-produce children" and "cloning-for-biomedical-research" and defined "cloned human embryo" as "the immediate (and developing) product of the initial act of cloning, accomplished by successful SCNT [somatic cell nuclear transfer], whether used subsequently in attempts to produce children or in biomedical research."

The distinction between "reproductive cloning" and "therapeutic cloning," promoted by supporters of cloning for experimental purposes is dishonest and confusing, leading many to believe that these are two fundamentally different scientific processes rather than simply alternate uses to which clonal embryos can be put. The National Academy of Sciences (NAS) report provides three different meanings for "reproductive cloning," each contradicting each other and the report's own glossary definition.[30] Whether cloning for experimental or reproductive purposes, "somatic cell nuclear transfer" and "nuclear transplantation" are the terms for the common cloning procedure.[31] "Nuclear transplantation to produce stem cells" is inaccurate terminology since nuclear transplantation into an egg does not produce stems cells, but an embryo, which can be destroyed for its stems cells.[32] All cloning is plainly reproductive; the language of "therapy" suggests a rhetorical argument *per se,* which attempts to overcome public opposition and could lead to a slippery slope in permissive cloning regulation.

Nevertheless, "therapeutic cloning," coined as a soft term to make cloning respectable, has failed to foreclose the American conscience. In response, cloning proponents have shifted to the use of euphemisms in the language of S. 303, defining "human cloning" as the act of implantation of the clonal embryo and inventing the term "unfertilised blastocyst" for the clonal embryo; or, as in the case of what is now state law in New Jersey, defining cloning even as birth. Unless citizens and journalists can understand and communicate in common language that will ensure honesty and accuracy in the description of these ideas, as the biotech agenda becomes even more sophisticated, democracy will find itself increasingly dislocated and biotech will run out of control.

B. Experts Rule.

Feeding on public ignorance and fear, the biotech industry has taken on a defensive and disingenuous posture: its "experts" - whether scientists or bioethicists or plain PR people (BIO, the industry group, recently renamed its chief lobbyist "VP of Bioethics") - have fallen back on the use of mantras such as the meaningless slogan "cures now." They claim to be in favour of cures for disease; they imply that their opponents are not. So hearings are peppered with wheelchairs and the endorsement of disease advocacy organisations is cultivated as a trump card. For example, S. 303 has received strong support from the American scientific community, who has used celebrities such as quadriplegic Christopher Reeve to make their case in dramatic terms. However, this emotive appeal has not thwarted a rational ethical debate. Reeve actually stated at a Senate hearing that the duty of government was "to do the greatest good for the greatest number."

The biotechnology industry lobbies through its national trade group Biotechnology Industry Organisation (BIO) and their affiliate the "Campaign for Medical Research" (CAMR) while the public remains ignorant of the science and scared of disease.[33] The promise of human cloning advocated by BIO and CAMR has been undermined by low efficacy rates and high rates of genetic deformity in animals. Moreover, the National Institute of Health (NIH) has stated that at least 16 of the 78 of the Bush approved stem cell colonies have already died or failed to expand into undifferentiated cell cultures, rendering them virtually useless; several other colonies are showing genetic abnormalities.[34] Stem cells from adult cells and umbilical cord blood are currently used in the treatment of Parkinson's disease, sickle-cell anaemia, multiple sclerosis, spinal cord injury and other conditions. The NIH noted that the majority of the viable colonies are in foreign laboratories that have restrictions in supplying stem cells to US researchers.[35] Still, the biotech industry and its scientist-entrepreneurs tell us not to worry but to trust them to seek cures for our woes. Meanwhile, their political sponsors deny that there should be any limitation on their freedom. We should leave it all to the experts; they know best and we can trust them.

C. Biotechnology and Multilateral Discussion.

The bio issues are being considered in many multilateral fora, from the Council of Europe and other regional groupings to UNESCO and other agencies of the United Nations. It has been recognised that they are fundamental questions that affect all of humankind; and that it is hard to see how individual nations can control them. All well and good, but we know how hard it is to exercise democratic accountability in these international bodies. Indeed, it can be hard to discover what it is they are doing. This has become a pressing issue with the US decision to rejoin UNESCO, the United Nations Educational, Social, and Cultural Organisation. UNESCO's work includes the International Bioethics Committee, which is working on the long-term goal of a "universal instrument" in bioethics - that is, a convention that covers the biotech waterfront. The Declaration on the Human Genome to which we have referred is intended to be the first chapter. We may wish to observe that the declaration is couched in such worthy generalities that it would be hard to disagree with any of its proposals. Whether that is vice or virtue is for history to judge.

7. A New Politics and Cultural Influence

As the cloning debate has already shown, the questions raised by biotechnology are not the same kind of issues as have traditionally divided American politics. In contrast, the European debate has tended more readily to bring together the two points of conscience of Western culture, on the "right" and the "left." However, a new politics is emerging in which the fate of human dignity in the face of advances in genetics and cybernetics will be determined in policy discussions that cut across traditional political divides. Legislation seeking to impose a federal ban has congressional co-sponsors not just on both sides of the aisle, but on both sides of the abortion debate.

Using cloning to manufacture human embryos for experimental purposes is opposed by many sectors of society. Opposition has come from the environmental movement and from leading pro-choice feminists, who want at least a lengthy moratorium on this technology, and in many cases an outright ban.[36] Even pro-choice mainline religious institutions, such as the United Methodist Church, align with the Southern Baptist Convention and the U.S. Catholic Church in opposition to all human cloning.[37]

The very fact that opposition to cloning has come from some on the "extremes" of the pro-life and pro-choice positions on abortion dramatically illustrates merger on cloning issues in the United States, but the novelty of these questions goes deeper than a marriage of convenience. In a February, 2004 statement, feminist Judy Norsigian expressed her concern that, "because embryo cloning is the gateway to genetic modifications that go far beyond medical treatment into the realm of designer babies, we need a much broader discussion of this contentious issue." The convergence of pro-life political conservatives and pro-choice "progressives" in the United States could prove a novel and potent force. Conservatives and "progressives" share a respect for human nature and a distrust of manipulative interventions that will enable certain men and women to re-shape others. This pits them both against those on "left" and "right" who are either libertarian in their focus or who lack interest in these policy questions. It has set political "progressives" against traditional allies who take a *laisser-faire* approach to these technologies. On the conservative side, it pits those who treasure the sanctity of life against both libertarians and others who tend uncritically to favour corporate interests. It is hard to predict how this newfound alliance between those divided by their general political philosophy and their view on issues like abortion will develop. What is clear is that the bioethics questions do not fit neatly into our traditional politics.

There lies great hope in consensus building, since the two most vigorous forces in the west, the conscience - broadly speaking - of the right and the left, are beginning to find common cause; against the commodification of human nature in a reduction of human beings to products, the result of preferences in the market-place of human choice. For example, Charles Krauthammer, member of the President's Council on Bioethics, supports embryonic stem-cell research, but draws a line on intentionally creating embryos for research purposes:

> Many advocates of research cloning see nothing but "thingness." That view justifies the most ruthless exploitation of the embryo. Deliberately crafting embryos for eventual and certain destruction means the launching of an entire industry of embryo manufacture. It means the routinisation, commercialisation, the commodification of human life.[38]

In an April, 2002 White House speech, President Bush cautioned against a utilitarian calculation of human value in a time of revolutionary scientific progress:

> Advances in biomedical technology must never come at the expense of human conscience. As we seek what is possible, we must always ask what is right, and we must not forget that even the most noble ends do not justify any means . . . Life is a creation, not a commodity.[39]

As a preventive measure, in January 2004, Congress sanctioned a plan in a congressional spending bill that prohibits the U.S. Patent Office (PTO) from issuing patents on human organisms. The director of the PTO, James Rogan, voiced support for this approach on behalf of the administration as "fully consistent with our policy . . . on the non-patentability of human life forms at any state of development."[40] The appropriations rider will need to be passed every year to remain in force, and it does no more than give legislation force to current US PTO practice. It does however safeguard that practice from challenge in the courts.

8. Two Fundamental Approaches to Human Cloning Policy

Let me summarise some aspects of the arguments advanced in this debate.[41] While some biotech advocates resist any regulation, two basic approaches to cloning policy have been proposed. One approach favours a partial ban with the goal of preventing the birth of cloned babies, but permits the cloning of human embryos for research. The second approach is comprehensive, prohibiting the use of cloning (SCNT) technology in the human species.

At first sight, the partial-ban approach seems prudent. It gains support from two very different groups - those who favour a comprehensive cloning ban but who wish to attain it in two stages; and those who wish to ban the birth of cloned babies but who support the cloning of human embryos for research. But the problems with this approach are clear.

First, if the cloning of embryos for research is permitted, the only way such a policy can be applied is to require that they also be destroyed. Since biotech advocates need very large numbers of embryos for the "therapeutic cloning" model to work, very large numbers of embryos much therefore be destroyed. Policymakers with varied views on the nature of early embryonic human life have found the principle of creating and destroying huge numbers of embryonic human beings unacceptable. It is for this reason that at least two major jurisdictions - Canada and France - which permit destructive embryo research using *in vitro* embryos, have been favourably considering a comprehensive ban on human cloning, while Australia recently enacted federal legislation along exactly the same lines.

Second, if the cloning of embryos for research is permitted, it will result in the perfection of that same technology that can be used to clone embryos for implantation and live birth. This is a special problem in the context of multilateral jurisdictions, since some states may not ratify such a convention. These rogue states will then emerge as islands in which the technology devised for embryo research can be applied for the purpose of producing live-born cloned babies.

Third, if the cloning of embryos for research is permitted, according to the so-called "therapeutic cloning" model many millions of cloned human embryos will be produced in laboratories. It is inevitable that some of these embryos will be implanted and result in the birth of cloned babies. Motives will be many, all the way from criminal intent and financial reward to pro-life "rescue." One way or another, the mass production of research cloned embryos will result in the birth of cloned children.

Fourth, while some advocates seem to genuinely intend a partial ban to be the first step toward a comprehensive ban, there is widespread agreement that if a partial ban becomes policy, then the impetus for comprehensive action will be lost. The "first step" will, ironically, have the reverse effect: ensuring that it is the *only* step. For this reason, advocates of a "two-step" approach must understand that this strategy may thwart the ultimate goal of a complete ban; if they are not disingenuous, they are politically naïve. There is in fact no better way to protect the mass-production of human embryos for experimental purposes than by putting in place a "cloning ban" that gives publics around the world the impression that this egregious biotech practice has been forbidden. And at the same time, this encouragement of the mass production of embryos will make more not less likely the birth of cloned human beings.

9. In Conclusion

There has long been a simmering debate about the respective merits of the two most influential books about the future that were written in English during the 20th century. They are both alike in one respect: they are dystopias, visions of a world that has gone profoundly wrong, warnings from the future to the present.

George Orwell's *1984* sets out a vision of political oppression and control that foresees the triumph of technology in the service of totalitarianism. Big Brother is watching you. By contrast, Aldous Huxley's *Brave New World* offers a more subtle nightmare, in which technology and choice have combined to bring about a world in which, as one commentator has written, pain and suffering have been almost entirely alleviated, at the cost of everything that makes life worth living. In the closing years of the 20th century it became obvious who was right. As totalitarianisms crumbled around the world, the biotech revolution set out on its exponential path. We have reason to be grateful that Orwell seems to have been wrong. We have reason to be fearful that Huxley may have been right.

I end with this observation to make plain the stakes for which we are playing, and the historic context of this debate. This is not business as usual. Though there are technical aspects to the discussion, this is not at heart an "expert" question. The challenge of human cloning confronts the global community with a defining opportunity to declare the primacy of human dignity in the development of biotechnology. Its extraordinary benefits are mingled with threats of equal proportion. The closing years of the 20th century demonstrated that Orwell's totalitarian vision did not represent the human future. It lies in our hands at the start of the 21st to ensure that Huxley is also proved wrong.[42]

Notes

1. *Wired*, April 2000.
2. Convention for the protection of Human Rights and dignity of the human being with regard to the application of biology and medicine: *Convention on Human Rights and Biomedicine* (ETS No. 164). Open for signature by the member States of the Council of Europe, the non-member States which have participated in its elaboration and by the European Community, in Oviedo, on April 4, 1997. Entry into force: December 1, 1999. The Convention is the first legally-binding international text designed to preserve human dignity, rights and freedoms, through a series of principles and prohibitions against the misuse of biological and medical advances. Available at http://conventions.coe.int.
3. *The Universal Declaration on the Human Genome and Human Rights* was adopted by the General Conference of UNESCO at its 29th ses-

sion on November, 11 1997. Official text available at http://portal.unesco. org. As a universal instrument in the field of biology, the Declaration attempts to balance the interests in safeguarding human rights and fundamental freedoms and ensuring freedom of research.

4. Department of Health and Social Security 1984. Dame Mary Warnock was the Chairman of the Committee on Human Fertilisation and Embryology. Commencing - in 1982, the inquiry examined social, ethical and legal implications of human assisted reproduction, including artificial and in vitro fertilization, egg and embryo donation, surrogate mothers, the storage of human semen, eggs, and embryos, and scientific and ethical issues in fertility. Available at www.bopcris.ac.uk/img1984/ ref2900_1_1.html.
5. He has also acted to deny funding to human cloning for research purposes, included in his August 9, 2001 prohibition on using federal funds to support research on human embryonic stem cell lines except for a certain limited number in existence as of that date. This effectively prohibited the creation and research of new stem cell lines using federal funding.
6. See, H.R.234, H.R.534. For several reasons, including Senate procedure, it is more difficult to pass legislation in the Senate than the House.
7. HR.234, To amend title 18, United States Code, to prohibit human cloning. Sponsor: Rep Weldon, Dave [FL-15] (introduced 1/8/2003), Cosponsors: 102. Action: 3/6/2003 Referred to House subcommittee; H.R.534, To amend title 18, United States Code, to prohibit human cloning. Sponsor: Rep Weldon, Dave [FL-15] (introduced 2/5/2003), Cosponsors: 140. Action: 3/3/2003 Read the second time. Placed on Senate Legislative Calendar under General Orders. Calendar No. 23; H.RES.105, Providing for consideration of the bill (H.R. 534) to amend title 18, United States Code, to prohibit human cloning. Sponsor: Rep Myrick, Sue [NC-9] (introduced 2/26/2003). Action: 2/27/2003 Passed/agreed to in House; Available at http://thomas.loc.gov.
8. Boston Women's Health Book Collective, *Our Bodies, Ourselves For The New Century* (New York: Simon and Schuster, 1998). This book was the revision of *Our Bodies, Ourselves* (1970). Judy Norsigian is co-author and Executive Director of the Boston Women's Health Book Collective.

9. Stuart Newman was a founding member of the Council for Responsible Genetics in Cambridge, MA (www.gene-watch.org).
10. S. 245: A bill to amend the Public Health Service Act to prohibit human cloning. Sponsor: Sen. Brownback, Sam [KS] (introduced 1/29/2003), Cosponsors: 28. Action: 1/29/2003 Referred to Senate committee. Available at http://thomas.loc.gov.
11. S. 303: A bill to prohibit human cloning and protect stem cell research. Sponsor: Sen. Hatch, Orrin G. [UT] (introduced 2/5/2003), Cosponsors: 10. Action: 2/5/2003 Referred to Senate committee. Available at http://thomas.loc.gov.
12. *Judiciary Statement*: "*Cloning*," Statement of Sen. Orrin Hatch before the Senate Judiciary Committee (February 5, 2002).
13. CA SB 778.
14. IN HB 1538; IN HB 1984; IN SB 138; IN SB 151; LA HB 472; MS SB 2747; MO HB 481; MO SB 191; MO HB 163; NY AB 4533; OK HB 1130. These and other references to state legislation are cited from Andrews, 2004 (Appendix E).
15. Governor James E. McGreevey signed the "Stem Cell Research" bill A. 2840/S. 1909. The bill establishes that "research involving the derivation and use of human embryonic stem cells, human embryonic germ cells and human adult stem cells, including somatic cell nuclear transplantation, shall… be permitted." The "cloning of a human being" is prohibited, but is defined as "the replication of a human individual by cultivating a cell with genetic material through the egg, embryo, foetal and newborn stages into a new human individual." This takes the re-definition of the term cloning to a new level: the Hatch bill in the US Senate defines it as implantation; this New Jersey bill effectively defines it as birth.
16. Ark. Code § 20-16-1001 et seq. (2003) (formerly AR SB 185); Iowa Code § 707B.1-.4 (2003) (formerly IA SB 2046, became IA SB 2118); Mich. Comp. Laws §§ 333.26401-06, 333.16274, 16275, 20197, 750.430a (2003); ND Cent. Code §§ 12.1-39-01 to 02 (2003) (formerly ND HB 1424); SD SB 184 (2004).
17. Cal. Bus. & Prof. §§ 16004, 16105 (2003) and Calif. Health & Safety §§ 24185-24187 (2003) (formerly CA SB 1230; RI Gen. Laws § 23-16.4-1 to .4-4 (2003); Va. Code Ann. § 32.1-162.21 to .22 (2003).
18. See, Cal. Health & Safety Code §§ 24185(b); La. Rev. Stat. § 1299.36.2(B); Ark. Code § 20-16-1002 (A)(3); Iowa Code § 707B.4(1)(c); ND Cent. Code §§ 12.1-39-01 to 02; Va. Code Ann. § 32.1-162.22(a); Lori Andrews, n.d.

19. La. Rev. Stat. § 1299.36.3 (penalty up to $5 million for an individuals); Mich. Comp. Laws § 750.430(a)(3); RI Gen. Laws § 23-16.4-3 (fines up to $1 million for an entity and $250,000 for an individuals); Ark. Code § 20-16-1002 (b) to (D).
20. See, La. Rev. Stat. § 1299.36.4; Iowa § 707B.4 (6).
21. Lifchez v. Hartigan, 735 F. Supp. 1361 (N.D. Ill. 1990); Margaret S. v. Edwards, 794 F.2d 994 (5th Cir. 1986); Jane L. v. Bangerter, 61 F.3d 1493, 1501 (10th Cir.), rev'd on other grounds sub nom., Leavitt v. Jane L., 518 U.S. 137 (1996); Forbes v. Napolitano, 2000 U.S. App. LEXIS 38596 (9th Cir. 2000).
22. Andrews, 1998.
23. In resolution 56/93 of December 12, 2001, the General Assembly decided to establish an Ad Hoc Committee, open to all States Members of the United Nations or members of specialised agencies or of the International Atomic Energy Agency, to consider the elaboration of an international convention against the reproductive cloning of human beings and a mandate for the negotiations. Available at http://www.un. org/law/cloning/index.
24. See, International convention against the reproductive cloning of human beings Report of the Ad Hoc Committee, Official Records of the General Assembly, Fifty-seventh Session, Supplement No. 51 (A/57/ 51); Revised information document prepared by the Secretariat (A/AC.263/2002/INF/1/Rev.1).
25. See, Report of the Working Group (A/C.6/57/L.4); Corrigendum (A/C.6/57/L.3/Rev.1/Corr.1); Corrigendum (A/C.6/57/L.8/Corr.1) Draft decision (A/C.6/57/L.24) Report of the Sixth Committee (A/57/ 569).
26. Belarus, Belgium, Brazil, China, Czech Republic, Denmark, Finland, Iceland, Japan, Liechtenstein, South Africa, Sweden, Switzerland and United Kingdom of Great Britain and Northern Ireland: draft resolution (A/C.6/58/L.8).
27. Proposal by Costa Rica for a draft international convention on the prohibition of all forms of human cloning (A/58/73); Antigua and Barbuda, Benin, Costa Rica, Côte d'Ivoire, Dominica, Dominican Republic, El Salvador, Eritrea, Ethiopia, Fiji, Gambia, Georgia, Grenada, Haiti, Honduras, Italy, Kazakhstan, Kenya, Kyrgyzstan, Lesotho, Madagascar, Marshall Islands, Micronesia, Nauru, Nicaragua, Nigeria, Palau, Panama, Paraguay, Philippines, Portugal, Saint Kitts and Nevis, Saint Vincent and the Grenadines, San Marino, Sierra Leone, Spain, Suriname, Tajikistan, Timor-Leste, Uganda, United Republic of Tan-

zania, United States of America, Uzbekistan, Vanuatu and Zambia: draft resolution (A/C.6/58/L.2).

28. On November 6, 2003, the delegation of Iran, on behalf of the member States of the OIC, moved, under rule 116 of the Rules of Procedure of the General Assembly, to adjourn the debate on the agenda item until the 60th session of the General Assembly (i.e. September 2005). No action was taken on the proposals before the Committee. It should be noted that some 15 OIC states had co-sponsored the Costa Rican resolution, against one (Turkey) on the Belgian list of sponsors.
29. President's Council, 2002, xiv.
30. NAS, 2002, at 6-6, 2-8, 1-1.
31. NAS, 2002, at 1-1.
32. NAS, 2002, at 2-5.
33. BIO claims around one-third of its members outside the United States; it holds conferences and other events in Europe and elsewhere. See, www.bio.org.
34. Gillis/ Weiss, 2004.
35. Gillis/ Weiss, 2004.
36. For example, Brent Blackwelder, President of Friends of the Earth (US); Judy Norsigian, editor of *Our Bodies, Ourselves*.
37. See, Proposed Draft Resolution on Stem Cell Research for General Conference 2004: available at http://www.umc-gbcs.org; *Letter to Congressional Leaders on Weldon Amendment to Ban Patenting of Human Embryos*, Cardinal Keeler, Chairman, USCCB Committee for Pro-Life Activities, November 18, 2003; *Testimony Before Senate Commerce Subcommittee on Science, Technology and Space, Subject: Cloning and Women's Health*, by Richard Doerflinger, March 27, 2003 *Letter to Congress on the Weldon/Stupak Human Cloning Prohibition (H.R. 534)*, Cardinal Anthony Bevilacqua, USCCB Committee for Pro-Life Activities, February 25, 2003; *Letter to Senate on the Harkin/ Feinstein Cloning Bill (S. 2439), Gail Quinn, USCCB Secretariat for Pro-Life Activities, May 20, 2002*: available at http://www.nccbuscc. org/ prolife/issues/bioethic/ cloning/index.htm. By contrast, the traditionally conservative Jewish leaders in the Rabbinical Assembly Committee on Jewish Law and Standards passed a resolution in April 2003 supporting stem cell research for therapeutic purposes.
38. Krauthammer, 2002.
39. "President Bush Calls on Senate to Back Human Cloning Ban: Remarks by the President on Human Cloning Legislation." Office of the

Press Secretary: April 10, 2002 1:18 PM EDT. See at www.whitehouse.gov/ news/releases/2002/04/20020410-4.html.

40. Wilke, 2004, 42.
41. This position is in harmony with that taken by the US in the United Nations, and is sought by President Bush in domestic policy. However, I set this out as my own position, and do not claim to speak for the administration.
42. I should like to thank my research assistant, Dawn Willow of Chicago-Kent College of Law, for her assistance in the preparation of this manuscript.

References

Andrews, Lori B. "State Regulation of Embryo Stem Cell Research." In *National Bioethics Advisory Commission*, Appendix II, A1 - A13.

Andrews, Lori B. "Is There a Right to Clone? Constitutional Challenges to bans on Human Cloning." *Harvard Journal of Law and Technology* 11(3), 1998, 643.

Andrews, Lori B. Appendix E. In *Monitoring Stem Cell Research, A Report by the President's Council on Bioethics*. Washington, DC, 2004.

Department of Health and Social Security (U.K.), *Report of the Committee of Enquiry into Human Fertilisation and Embryology*, vol. L, xiv, Cmnd.9314 (HMSO), 1984.

Gillis, Justin/ Rick Weiss. "NIH: Few Stem Cell Colonies Likely Available for Research, Of Approved Lines, Many are Failing." *Washington Post*, March 3, 2004, at A03.

Krauthammer, Charles, M.D. *Human Cloning and Human Dignity: An Ethical Inquiry* (July 22, 2002). Available at www.bioethics.gov/reports/ cloningreport/appendix.html.

NAS. *Scientific and Medical Aspects of Human Reproductive Cloning*. National Academy Press, 2002.

President's Council on Bioethics. *Human Cloning and Human Dignity: An Ethical Inquiry*, July 2002, xiv.

Wilke, Dana. "Stealth Stipulation Shadows Stem Cell Research." *The Scientist*, March 1, 2004, vol. 18, issue 4, 2004, at 42.

Primordial Ownership versus Dispossession of the Body. A Contribution to the Problem of Cloning from the Perspective of Classical European Philosophy of Law

Thomas Sören Hoffmann

Abstract: Human cloning has become a matter not only of ethical preoccupations, but also of discussions within the philosophy of law. This article contains a critical examination of the three most discussed arguments in favour of cloning humans in a more or less legal context: the right to reproduction, the right to therapy, and the right to research. It can be shown that there are strong reasons for prohibiting an extension of these (subjective) rights to any kind of use of human bodiliness for the purposes of third parties or even great majorities. The paper considers the principle of a primordial ownership of the body by no one other than the „embodied subject" itself, an argument developed especially in the Kantian tradition, and shows the risks of any anonymisation and external administration of the human body (as the objectiveness of subjectivity) in societies predominated by a technological way of viewing not only the world, but also the human individual.

Key Words: "Duty" to donate eggs, Exploitation of women, Freedom of science, Human bodiliness as principle of law, Instrumentalisation of the body, Kinds of human cloning, Politics of language, Reproductive rights, Right to research, human bodiliness, Right to therapy.

More than any other bioethical topic, the problem of human cloning presents an undoubtedly epochal challenge for contemporary ethical reflection.[1] Assuming there are cases where this might suffice, in this case it is certainly not enough for bioethics simply to delimit, theoretically analyse and, in the framework of the broadest possible consensus, offer practical solutions to a problem that has arisen as a consequence of new options for human conduct. The ability to clone humans touches much more basically upon issues not only of individual human self-understanding, but also upon questions concerning the constitution of modern societies, it touches upon fundamental issues concerning ethics and law - it addresses dimensions which greatly exceed the need for a mere dispute among professional ethics experts. This was foreseeable inasmuch as the "cloned human" had an almost magical allure long before his first technological realisation, when it was nothing but a figment of imagination - it may even be no exaggeration to say that it was a singular phantasm of the twentieth century's collective consciousness. The cloned human celebrated his first appearance, as memory serves us, in 1932 with Aldous Huxley, followed later by other novels and also films, but later on also by admittedly concrete eugenic programs, as propagated in the sixties, e.g., by the geneticist and Nobel laureate Joshua Lederberg.[2] The "cloned human" is, as can be said in a first approximation, a cipher for the growing awareness that human beings entered the era of their own technological reproducibility in the twentieth century. It is an exemplary abbreviation for a lifeworld in which the parameters of humanity may require redefinition: the classical "ethics of being able to be oneself," to borrow a phrase from Jürgen Habermas, gives way to a rival concept according to which the previously familiar ethical-legal rules "for social interaction ... could be converted to norm-free functionalistic terms."[3]

Of course the topicality of the cloning problem, an emotionally volatile subject throughout the world, should not deceive us into believing that, in regard to this issue, the classical European "ethics of being able to be oneself" has not simply been steamrollered by technological innovations. On the contrary, the fact that these innovations re-examine the question of the parameters of humanity also contributes to raising awareness of precisely the most profound premises of human coexistence. Correspondingly, this paper intends to show to what extent certain central positions of classical philosophical ethics may directly apply to these new issues and also possibly contribute to a decision on the legitimacy of claims to the implementation of human cloning. We will furthermore consider whether corresponding positions belonging not only, but primarily to the circle of Kantian thought might not also be of independent importance to the situation of interculturality for which the cloning problem has been an issue from the start. An argument for this is that the primary reference point of this problematic - the human body - is the as yet "mute" and not yet culturally interpreted primordial sign of human existence, an existence which is in itself endowed with a capability of developing itself in freedom.

We will proceed by critically examining the three most important arguments put forward in the literature in favour of the option of cloning humans, and then draw upon the philosophy of law's basic principle of the primordial ownership of the body, which originates conceptually in the tradition of natural and (modern) rational law. The fact that this basic principle or axiom - an axiom which could be summarised as: "The origin of all your rights is your bodily existence in space and time" - is qualified (if not neutralised) today by the phenomenon of anonymisation and external administration of bodiliness itself, serves as a basic description of the main problem before which the issue of human cloning stands. As means of explanation at this point, allow me to mention that the concept of a "primordial ownership of a person" ("Urbesitz der Person") in the immediate "*having* of a body" originates with the philosopher of law Hans Reiner, who sees in it the essential content of classical natural law rights, such as exclusive self-possession of one's own body, entitlement to freedom from bodily harm, and the right to life summarised.[4] This idea is much older and can be easily derived from the rational law theories of Kant and, to an even greater degree, Fichte. Especially in Fichte's *Grundlage des Naturrechts*, the body is portrayed as an elementary principle of law: just as all relations of law are relations between *bodily* instantiated subjects, so law as a whole is a body of rules concerning objective recognition of bodily spheres which are principally "holy" to one another and only thus become aware of their own determination to freedom. Also: just as the body is the elementary and irreplaceable experience of a suitable coincidence of world and self and therefore of real freedom, there is also no empirical self-consciousness except as a reflex of a recognised bodily plane of action, which in turn is only limited by planes of action of other bodily subjects.[5] Kant's principle that "each human [has] an inherent right to be at some place on earth,"[6] is in the same way captured by Fichte in the idea: all rights of a human being are a function of his unconditional having himself at the "place" of his body, whereas law itself is the system or cosmos of these "places" as just as many "perspectives" of reason set into a finite way of being. The question towards which the problem of cloning leads is to what extent there may be any other possessor of human bodiliness apart from the subject which reflexively finds itself at the place of its body, while preserving the respect of bodily instantiatedness of subjectivity which is fundamental to rules of law as a whole. More precisely, the question is whether, as a result of the new biotechnological options, the phenomenon of an "expropriated body," to which access by a third party is possible or allowed, does not already exist.

1. Cloning of Humans

"Cloning" is often discussed in multiple meanings, without adequately differentiating between the various meanings of the term. Ambiguities result in some cases because different methods of asexual reproduction are called "cloning" - namely, both artificial multiple replication via embryo splitting (a procedure first applied to humans in 1993) as well as reproduction of an already matured individual by transferring his genetic programme, contained in the cell nucleus, to an unfertilised and denucleated egg cell. Other ambiguities arise when one and the same cloning procedure is extrinsically associated with its purpose and, according to the possible purposes, "types of cloning" are distinguished which cannot be differentiated based upon the cloning procedure as such. It is clear that at this point the politics of language have already set in, with the primary intention of preventing reference to good or reputedly good (primarily therapeutic) purposes, e. g. an intrinsic judgement on the cloning process as being a morally possibly dubious act of production.[7] At any rate, considerable semantic problems arise when "*reproductive*" cloning is to be legitimised, particularly for possible *therapeutic* use (whereby the idea is primarily to "use" someone's "time-shifted twin" as a source of organs and tissue or to "treat" infertility)[8] or when, as part of a programme of "therapeutic cloning," not only the reproduction of "matching" tissues or organs, but also research on the causes of disease, and even indeed the organism mechanisms, is to be furthered (the cloning process should therefore not lead directly to the creation of a cure, but only indirectly suggest types of therapy). When in the American state of New Jersey "therapeutic cloning" includes the option of implanting the clone in a woman's uterus and only the carrying of the child to full term is prohibited as "reproductive" cloning, it becomes apparent how arbitrary this nomenclature can be in individual cases.[9] For our part, we will restrict ourselves in the following to that cloning technology which first achieved a clear breakthrough with a famous Scottish sheep and then successfully with humans in the experiments conducted by Woo Suk Hwang and Shin Yong Moon,[10] i.e. cloning by somatic cell nuclear transfer. Instead of differentiating between application "types" of this technology, which in truth overlap multifariously,[11] we will examine the three most important moral or legal rights with which the corresponding application is actually to be justified: the right to reproduction, the right to therapy, and the right to free and unfettered research.

2. The Right to Reproduction

Undoubtedly, the epochal uncertainty mentioned earlier, which arose out of fear of the "cloned human," almost exclusively refers to the creation of artificial doppelgangers and multiple copies, i.e., to the technologically reproduced individual. At this stage - and without being already justified convincingly - resistance to and sympathy towards an effective ban are at their greatest. Nevertheless, the technology of cloning by somatic cell nuclear transfer is welcomed - if not advocated - by a minority consisting not only of ideological eccentrics; interest was great especially with respect to "reproduction technology." One of the main arguments offered here in favour of cloning humans is an argument originating in the "freedom of reproduction"[12] - an argument that in its stronger version draws on positive "reproductive rights" and in a weaker version on "freedom of reproduction," which is intended to imply at least a non-obstruction of access to all available reproductive technologies. "Reproductive rights" or "freedoms," in the meaning presupposed here, are certainly a novelty in law when viewed historically. This should come as no surprise, considering they could not have taken on a real substance prior to the development of techniques of artificial insemination; at the same time, a connection to the subjective rights to marry and start a family, as they are set down in several documents of basic rights (ECHR Art. 12; EU Basic Rights Charter Art. 9), is ruled out. For the latter rights are, at their core, rights to protect against state intrusion on corresponding personal freedoms, intrusion which can occur e. g. through marriage bans for certain groups of people, but also by a public eugenics programme.[13] An actual "right" to reproduction per se, i.e., an entitlement guaranteeing the procreation of descendants in any possible way, is not contained in these protective rights.

But what could such a "right," taken in the strictest sense, imply in the first place? Firstly, a "claim right" to reproduction would have real meaning only in the sense of a *duty* of the state to grant each individual the opportunity to have a physically related descendant. In this respect a "right to reproduction" would be a state-mediated right to the coming-into-existence of another individual: a right, in brief, to ultimately bring about the existence of an individual by all available means. Thus formulated, it becomes clear where the conflict between a veritable right to reproduction and the legal principle of the primordial ownership of the body would have to lie: it occurs on two levels, namely on one level in that an individual, which belongs essentially to himself, lays claim to an object that is basically his bodily substrate; it lies - especially with regard to the creation of such an individual by a cloning procedure - on the level that this individual comes into being as an artefact, as a "res artificialis" instead of as an original place of freedom.[14] If the law, as is the case when construed in the classical-modern sense, represents a body of rules of recognition which *presupposes* the physical existence of individuals as members of the legal community, both the force of existence and the artificial redefinition of the individual must be fully omitted.[15] There can be no legally relevant postulate of existence of individuals in reference to single or future individuals, for this would have to include an asymmetrical entitlement of individuals to others who would, so to speak, be obligated to exist. It is abundantly clear that potential parents' wish to have children, which is certainly not trivial as concerns the law, cannot mean a legally valent "postulate of existence" regarding descendants. In the interpretation of law as rules of recognition, parents are not the *reason* for the existence of their children, even if they are in the physical sense the "cause" of their procreation. As rules of recognition, the law sees no one as the "reason for the existence" of another, except for the case in which each law abider is as such "reason" for the existence of the cooperative of law abiders and therefore also for my existence as a law abider. While a "right to reproduction" would be a legal entitlement allowing one to view a certain physical existence as a function of one's own will, according to the principle of the primordial ownership of one's own body, the physical existence of every human fundamentally breaks the will of any other human being directed *towards his existence* - which is simply what is meant by the idea of rules of law considered as rules of recognition. According to the perspective of classical rational law, "reproduction" is considered in this respect only as "freeing" and not as "coercive reproduction." A child, says Kant, is a "citizen of the world," not the "legacy" or "property" of its parents:[16] it has, though, by virtue of being physically present, though, already entered into spiritual relations guaranteed by law, the core of which is to be recognised objectively.

The weaker form of reproduction claims refers not to a "right" but to a "freedom" to reproduce. This means, especially within the framework of the cloning problem, not only that a decision concerning descendants is left up to the individual. What is meant here is rather a "free choice of means of reproduction," i.e., where possible, also a choice of reproduction via asexual means. Freedom in the choice of genetic endowments of the child is also intended, however, beginning with the choice of sex and extending to other psychophysical characteristics and to the choice of genetic optima, which could be achieved by optimising genetic material.[17] The corresponding position is considered "liberal" inasmuch as it advocates decision-making leeway for the individual, whereas the state has no positive duties in this context, only that it should not interfere with or limit the individual's choice of preference. Yet, we must also consider whether there are not good reasons for corresponding restrictions, and, indeed, precisely those which follow more or less directly from the concept of the rule of law as a body of rules of recognition. In order to answer this question, we distinguish the two previously mentioned aspects: (a) the choice of means of reproduction and (b) the choice of the unborn child's "qualities" by the "reproducer."

(a) A real choice regarding the type of reproduction has existed, as already indicated, since the first opportunities of artificial insemination, be it *in utero* or *in vitro*. Therefore, this choice is not an entirely new aspect that has entered the scene with "reproductive" cloning. What *is* new is the ability to create via cloning a being which genetically is neither autonomous nor singular, but heteronomous and merely secondary: this is a being which - as with any technological production - is dependent upon a given *prototype* and can in principle be "serialised," i.e., reproduced in as many copies as desired. One of the most frequently cited arguments against reproductive cloning is that it violates the "right to uniqueness," to a "distinctive identity," or even to "contingency." The objection has been raised that on the one hand it cannot be "clearly determined" "whether humans have a fundamental interest in possessing a unique genome."[18] This question can certainly be negated as long as humans do not construct their relation to themselves via those technologically mediated descriptions in whose context alone the concept of the genome possesses its well defined meaning. Rather, the question should be: "whether for humans it represents a restriction of their relation to self to find themselves confronted with one or several doppelgangers experienced as such or even with an 'original' version of themselves." It certainly ought to be pointed out that there will never be an ideal doppelganger, as (already according to Leibniz's principle of the indiscernibility of identicals) there can never be two beings existing in different places and at different times which are in a strict sense "identical." In this sense it is in fact often emphasized that personal identity does not result "deterministically" from one's genes, but that it always substantially results from a concrete life history, from individual personal interaction, etc. However, it should be mentioned that identical twins have to deal with the problem of "doppelgangerness" and that they can clearly overcome this problem without great difficulty. Nonetheless, it is precisely the twin example which is helpful: for this case concerns two individuals of the *same origin*, that is to say a prototype contingently existing twice, *not* a prototype-*dependent* being which, in principle, is capable of being reproduced many times over and which must understand itself from a constitutionally heteronomous physical substratum. It is the problem of the "clone" to have to comprehend itself from a "borrowed," not from its own, primordial bodiliness. Emancipation from this borrowed state is impossible, and it is therefore able to see itself residing only under the constant reservation of heteronomousity.[19] Here, we are not so much concerned with empirical-psychological questions, questions about what it means for the development of one's personality to be aware that one is the replica, e.g., of a genius or a more dubious personality: questions which not even the experiment may answer due to the basic principle disallowing all types of experiments with human beings (and in this case the experiments would even include experiments with a person's entire life!) More crucial here is the general aspect that in the case of reproductive cloning via the prototype orientation, a dependency structure is established which hypothesises the existence of the clone to a much higher degree than is the case for naturally procreated individuals. Indeed, this is truly an expression of the complete dispossession of bodiliness which - even where a government without any stimulation merely allows the practice - leads to the body being seen as an object of societal administration, instead of serving as the primary point of reference for legal recognition. There is a parallel to all theories of human dignity which construe the latter as a title conferred by society and not as a demarcation of societal claims; however, the relation of recognition can only exist based upon this demarcation. The law must allow concrete bodiliness - taken here in the sense of "primordial ownership of one's body" - to serve as a limit, protecting the individual from abuse with reference to the right of disposal; concrete bodiliness serves as a sensual sign marking point of origin of law for the individual, an origin which the individual himself is as abider and symbol of the law.[20] It is quite clear, as required by law as an elementary body of rules of recognition, that a (hypothetically) encountered clone as a being that has a "human face" would also be interpreted based on its Fichtean "holiness": that its body would also be granted the same "conditions" of being viewed not as a thing, but as a symbolised subject. However, the question here is not hypothetical, it is categorical: is the claim to a "freedom of choice with reference to the means of reproduction" reason enough for the government to tolerate physical structures of dependency that would be far more intense than those occurring in a natural generational relationship? This could be rebutted by the argument that there can never be a reason for the law to accept acts which indeed claim a (negative) "freedom," but which in essence put limitations on freedom for both an *individual* human being and structurally for the community as a whole with reference to third parties - *individually* in that they claim the power of determination as it applies to the physical substratum of a third party; *structurally* by creating a new type of dependency relations which circumvent the rule of law as a body of rules of relations of recognition between bodily primordial places of freedom. Not every claim to be (negatively) free to act actually invalidates the right of qualified freedom or the task of the law to ensure the maximum degree of freedom for all. Rather, the reverse is true: the preservation of the state of law - which Kant calls the "suprema lex" - marks the final limits for arbitrary freedom of acts.[21]

(b) The second question, the choice of "qualities" of the unborn child, is often discussed publicly with reference to the term "designer baby." Inasmuch as this question merely demonstrates the actual new dependency relationships between procreator and procreated *ad oculos*, it is already answered by the basic principle previously discussed. This issue does, however, contain one additional aspect, i.e., it raises the problem of the standards guiding the choice of prototype. As in (a) this means, in a moral respect, that (prospective) persons only come into being under the demands of a determining prototype; additionally, it is of importance here that this prototype exhibits characteristics of perfection which should be able to be "freely" determined. It is not merely of historical interest to point out that Kant broke with the tradition of natural law by excluding any appeal to aspects of perfection from the vicinity of legal justification.[22] Kant saw very clearly that an actual founding of the legal relation between humans based upon *practical* reason could only be constructed resting on the idea of freedom and of relations of freedom between them; this foundation cannot be contaminated with expectations of perfection of either a theoretical or descriptive nature. The belief that the use of certain characteristics to define what a human is should be banned, which is at the core of the idea of human dignity, follows precisely from the standpoint that Kant had worked out; any attempt to use "objective" criteria derived from concepts of perfection to get at the essence of a free being and then use these criteria to determine when something should be recognised as such a being, violates the principle of dignity. In other words: being "free to choose" a child's qualities should always be judged as a situation in which the child's existence and the dispossession of his body are serving another person's purposes, and we should uphold the principle of self-purposefulness and the rights of the future member of the legal community against the "freedom to choose" qualities of the unborn. With the demand, in addition to the issue of reproduction, to be also able to choose *in concreto* the characteristics of one's child, the violation of the principle of "primordial ownership of the body" will only be intensified. Thus, it should be rejected *a fortiori* based upon the arguments presented with reference to (a).

3. The Right to Therapy

In the wake of Hwang and Moon's experiments, a debate on cloning was initiated in the United States in which arguments for rights to cloning therapies were rendered with almost unprecedented vehemence; these debates were reminiscent of religious preachings promising salvation. We found a proliferation of emotional appeals made to the nation in the editorials of America's gazettes - the country where so many religious appeals are made; in Germany, there were articles in which "therapeutic cloning" was presented as a moral, even as a "religious commandment."[23] Especially the exaggerated hopes suggesting that mass healings of the sick and infirm were imminent give us good reason to make all the more of an attempt to establish conceptual clarity here.[24] We ask at this point: what could a (passive) "right to therapeutic cloning" or even an (active) "duty" to undertake such measures actually mean, and what is actually at stake here?

If "rights to therapy" were to be established at a level deeper than in the basic area of private law (a contract for treatment by a physician), then these could be taken to include the primordial right to be free from all bodily harm. "Therapy rights" lay claim to more than just a negative entitlement to an injunction requiring third parties to refrain from acts impairing one's bodily integrity; rather, they intend a *performative right* to the most complete restoration possible of an already endangered integrity. Thus, in general they can only be portrayed as a kind of protection contract much like a "citizenship contract." These protection contract would, however, not simply be against infringements by third parties, but would purport to protect against all naturally occurring bodily failures.[25] Within the framework of the welfare state, we do in fact consider the "citizenship contract" or the community of law abiders to be a type of "security" which can be used when an individual is unable to remedy emergency situations that directly concern his existence as a citizen or member of the legal community. Examples include the "right to work" (which Fichte supported), other cultural achievements that would be impossible for anyone to accomplish alone, and the duty to perform assistance to individuals. Yet, seen from the idea of law, the welfare state is not an immediate *end in itself*: rather, it serves to maintain the rule of law (Kant's "suprema lex") and the legal community, especially with reference to maintaining of its own psycho-physical requirements. It is important to note this because it includes a point which limits the possible legal postulates of the individual. If the welfare state is not to be considered as having immediate value or embodying an imperative but, rather is to be understood in terms of its role in preserving legal reality (which, of course, can only be thought with reference to the self-preservation of the members of the legal community), this means that the individual's positive, legally fixed protection and provision entitlements here are never in the same sense "infinite," as are, for example, the right to freedom from bodily harm or the right to the freedom of conscience, etc. They all are relativised by higher legal principles, particularly by each individual's right to primordial ownership of the body in symmetrical relation to all other individuals. This means also that the welfare service must be "doled out fairly," i. e. in a way reflecting the entire system. This in turn means that the social systems of a state can only be expanded proportionally to the other state systems relevant to existence; there must be a hierarchy of priorities for the type and scope of services provided.

Applied to "rights to therapy," this can only be taken to mean that an individual can have such a right only within the framework of a legal system as a whole and *in concreto* within the framework of the public measures of provision: this right is not immediate and unrestricted. The scope of the corresponding rights is on the one hand a priori (namely by virtue of principles of law higher than the concrete right of protection), and on the other hand limited by the actual, legally defined system governing the distribution of emergency goods. This limitation of scope eliminates any unrestricted choice of therapeutic methods - for all choice is limited by legal principles as well as by concrete rules of fairness as regards the distribution of resources. As much as we expect a public health system to be based on publicly responsible and reasonable medical practices, and not on arbitrary remedies, we should be able to assume with the same justification that the same system reflects the rule of law as a rule of coexistence. In this way the fundamental rule guaranteeing the individual freedom from bodily harm necessarily has a selective effect on potential treatments. Deserving special attention here is the fact that a public health system is not, on the contrary, to be turned into a system for the redistribution of bodiliness dispossessed by this process - tendencies which did not first become a topical problem with consumptionist research on "surplus" IVF embryos. Rather, these problems were already found in transplantation medicine and have been played out by "moral philosophers" in scenarios culminating in the option of complete dispossession, namely the killing of healthy persons to the benefit of a majority of the sick in a "lottery of survival."[26] A similar tendency with reference to the condition of "social compulsatoriness of the body" and,[27] related to this, the societal administration of “anthropoid biomass” can only arise when the individual “therapy rights” derived from the principles of the legal community, which only a general (active) “therapy duty” can fulfil, are *not violated* with reference to the principle of a person’s “primordial ownership of the body,” thereby suggesting an immediate, infinite right to emergency help. The government’s therapy policies then produce a needs-based analysis aimed to achieve the “largest number of successful therapies for the largest number of patients.” In so doing, the “non-treatable” patients come to be seen as a disposable mass - and are treated as such. It is in this context that we can make sense of the suggestive phrase “respect for (spare) embryos by using them in research,” a phrase by which in fact the utilisation of the individual “biomass” is described and transfigured as a form of “enhancement.” It is also this vantage point which explains why advocates of this right to therapy via cloning not only do not see the large number of artificially acquired egg cells for what it is, namely, a mode of dispossession of the female body in the interests of anonymous machineries of therapy, and even consider the process to be “honourable.”[28] In order to reject “therapeutic” cloning, one need not even necessarily touch on the question of the “status” of the embryo or the issue of preserving its dignity in acts involving its destruction. It suffices fully to take into consideration the fact that here a kind of physical presence - a kind of existence which every human being has passed through - is put at society’s disposal, thereby abolishing the principle of “primordial ownership of the body.” As the example describing access to the female body shows, this rapidly leads to a dynamic which, in developing beyond the consumption of embryos for purposes of therapy, dispossesses bodiliness in general or calls into question “primordial ownership” of the individual’s body. Society becomes a “collective organism” in an almost literal sense - it may even become an arena in which a new battle for the “bio-resources” of the human body is to be waged.

4. The Right to Research

Just a few weeks after the presentation of Hwang and Moon’s cloning results, a row developed in the United States immediately following President Bush’s decision not to reappoint two members of the Council on Bioethics on completion of their two-year period of office and to replace them with other candidates. This led to an open letter of protest to the President organised by Arthur Caplan, which was signed by 173 people from the fields of ethics and medicine. Also, a fierce statement by one of the two members not reappointed, cell biologist Elizabeth Blackburn, was published in the *New England Journal of Medicine* (and in a rather unusual step was also pre-published). This very personal article makes, as stated in the title, the accusation of a "political distortion of science," an accusation which was of course dismissed in clear terms both by the chairman of the President's Council, the biologist and moral philosopher Leon R. Kass, and by officials in Washington.[29] It is not so much the details of this flurry of papers which interest us here, but rather the argument which appeared here in this especially sensational form, namely that "free research" does not tolerate any political intervention. European scientists and scientific organisations have taken a similar stand on the current American position on bioethical issues.[30]

The concept of a right to "free" science is in fact one of the most important achievements of modern times, and, indeed, it is this achievement which, despite the substantial guarantees provided by basic rights, remains an endangered asset - for various reasons. Yet, it must also be seen here what a right to "free research" can actually mean - and what it cannot. Like all basic rights, a regulation such as Art. 5 Par. 3 of the German Basic Law defines the relation of recognition to an object not merely as being set; recognition must take place according to the object's nature based on the idea of liberty: in the case we are dealing with, the recognition of the (in itself) free act of knowledge, as given systematic expression in the framework of science. From the corresponding basic right it follows firstly that the state must refrain from interfering in matters of free knowledge and desists from replacing or brushing it aside, using something resembling an ensemble of "state truths." Basically, this means that the state does not prevent free the development of ideas, their dissemination and teaching, and that it does not censor the same even in the interest of the state; at best, the government reserves certain secondary rights to raise objections, for instance those which result from the duties of academic teachers with civil servant status which have a bearing on the constitution.

Yet two things certainly do *not* follow from this fundamental liberation of thought in public recognition of its own freedom: it does not follow from a "right" to free research that the state is bound to promote any and all research programmes or plans, including active advocacy in which it might create the external conditions required for the realisation of such research. *Nor* does the maxim follow that the state is not allowed to formulate its own judgement regarding possible *applications* of scientific research and also certain *research practices*. The state should, as part of its cultural responsibility, create the general conditions for the free cultivation of science. This is done by giving *knowledge* a place to thrive in state-supported universities, an area that is independent of the short-sighted interests dictated by trends, markets or political parties; in so doing, the state supports one of the few self-serving and thereby immediately meaningful occupations of humankind, at least as seen by long European tradition. This demands abstaining from "commissioned research." Likewise - when viewed from the standpoint of the freedom of research - it does not include a duty on behalf of the government to award grants to individual research programmes and areas. Thus, even liberal advocates of human cloning concede that politics should be free not to subsidise cloning programs from public funds, which is in keeping with the American President's decision of August 9, 2001.[31] But this is not enough: as mentioned, it does not follow from the right to freedom of scientific enquiry that the government has an obligation to promote or approve any possible application of knowledge already gained. On the contrary, the largest and most meaningful ethical scientific debate of the twentieth century, i.e., that centred on the atomic bomb, made one thing very clear: how hazardous it is to have a policy of public funding of research according to mere political goals, especially when policy decisions concerning scientific and technical options are made without any ethical reflection on the only way of their later use. To put this rather bluntly, the government has no "duty to accept" each and every project offered by science and technology. On the contrary, we have here much more a duty to carry on a public debate in order to foster the development of public judgements and a common will of the people, and this debate has to centre on legal issues, not merely on the always one-sided benefits for one or the other party in society.

The freedom of science may well be restricted to some extent by these considerations. But if we take into account the fact that modern conditions really change the character of research done especially in the natural sciences, then these restrictions are intensified: natural science does not simply produce "theoretical knowledge," but rather "acts of knowledge" - an aspect that has been emphasised, e. g., by those following the methodical constructivism introduced by Paul Lorenzen.[32] Scientifically constitutive acts of knowledge do not interest the state as long as their "objects" are truly objects, that is to say lifeless matter as such. Ethically relevant limits have been reached in animal experiments, and in cases concerning acts of knowledge relating to humans (e. g., in experiments without consent, in the experimental dispossession of the body, in research on the defenceless for the benefit of others), these limits have already been crossed. At this point, state intervention becomes a practical duty: it is the highest basic right of the individual, the right to one's own body, which must override the right to free research. It is obvious what the decisive standpoint in a state assessment of options for scientific application has to be: it can only be the standpoint of compatibility of these applications or their implications with the fundamental legal principles of the rule of law as a body of rules of recognition itself. The task of a "national ethics council," for example, cannot be reduced to participating in the research process, nor can it consist in merely making complex scientific problems more easily understandable to the general public, for instance, by promoting acceptance of a particular decision. What is eliminated as well is, in a stricter sense, a moral judgement of the practices in question; the state cannot directly take sides within a plurality of morals because it is in itself not a moral entity. However, it is the state's duty to prevent research practices and applications that are incompatible with the ideals and the rule of law overall. It is probably true to say that this element first entered the general consciousness with the scientific-ethical debate in the twentieth century on human experiments - experiments which were dealt with much more casually at the beginning of the same century than would now be the case due to our more acute general awareness of these issues. The accompanying debate continues to centre on the (Kantian) ban on the instrumentalisation of persons. This ban, which for Kant includes any type of violation of bodily integrity for the self-pilferage of organs due to one's ethical relationship to oneself,[33] has a legally relevant side: an existing entity exhibiting the possession of freedom, i.e. a (human) physical constitution conceived of as being involved in the process of self-organisation towards liberated existence, must be directly endowed with innate rights of freedom and can be seen, therefore, as a member of the legal community. Because the law is bound constantly and without exception directly to the *bodily* entities of personality - an idea that is directly related to the legal axiom of the primordial ownership of the body -, any encroachment on the body affects the *core* of the rule of law - the elementary empirical condition of possibility of law.

Applied to the example of cloning, this is not just an issue when cloned embryos are killed, for instance, in the interests of basic research, for the cultivation of tissue or the testing of medicines. Rather, it is an issue each time an external administration of foreign bodiliness takes place, i.e., each time a third party has access to what is the primordial property of each individual and which makes up the condition for the possibility of its singularity. It is precisely here that the law must object - and also point out that the subjective rights of reproduction, of access to therapy or of free research cannot exist when the elementary basis the subject alone possesses in the legal world has been jerked out from under the subject's feet. There are signs today, as the example of "right" to donation of young women's egg cells clearly demonstrates, that the beginning of a societal dispossession of the body, which has been initiated with the embryos or is still to be initiated, will not be the last step. Access to the bone marrow of children yet to be created, who are apparently to be born not as persons but as biomass with the "duty" to "help" sick siblings, as a recent case in Norway demonstrated, is an example of a case that is already taken for granted by large segments of the population.[34] Cases in which "living donations" of organs is made possible as well as situations in which drug experiments are conducted on children for the benefit of others are to going to grow in number. It is this situation in which the law - like no other authority - must guarantee that those limits which have already been established with the *individual* ownership of one's own body as the primordial fact of law remain respected. The right to freedom of research is, as are all other concrete freedoms, restricted in such cases: and that means restricted ultimately by the right of law unto itself, by the right of freedom and, thus, by the right of every human being to be at the core a world-being for itself and, simultaneously, a worldly self.

Notes

1. Honnefelder, 2004; Lauritzen, 2001; Oduncu, 2002.
2. Lederberg, 1963, 264f; Stevens, 2000, 14-6; Kass, 2001, 141f.
3. Habermas, 2001, 32.
4. Reiner, 1964, 33f.
5. Hoffmann, 2003, 488-509.
6. Kant, *Vorarbeiten zur Rechtslehre*, AA XXIII, 279.
7. Sexton, 2000.
8. Robertson, 1997; Brock, n.d., 98f.
9. To the corresponding "Stem Cell Research bill" A. 2840/S. 1909 from New Jersey cf. also the contribution by Nigel M. de S. Cameron in this volume.
10. Cf. Woo Suk Hwang et al., 2004, *Evidence of a Pluripotent Human Embryonic Stem Cell Line Derived from a Cloned Blastocyst*, www.sciencexpress.org / 12 February 2004. As early as November of 2001, successful embryo clonings in the laboratories of the private firm Advanced Cell Technology were reported; however, the embryos did not develop beyond the six to eight cell stage at that time.

11. In an ABC interview on February 13, 2004, Woo Suk Hwang stated that the employed "technique cannot be separated from reproductive people cloning and therapy cloning," and for just this reason initiatives to ban "reproductive" cloning were now necessary, http://www.abc.net. au/am/content/2004/s1044228.htm.
12. Letteron, 1997; Liu, 1991; Robertson, 1994, 80-1; Annas, 1997; Warnock, 2002.
13. An example of the latter is the Republic of Cyprus, which requires a genetic certificate for permission to marry to prevent thalassemia cases.
14. Kant, *Metaphysik der Sitten. Rechtslehre*, Anhang erläuternder Bemerkungen 3, AA VI, 360.
15. This also means that the arguments based on the unpredictably high risks of "reproductive cloning," well-known from animal production via cloning, can be left out of consideration here; even if these risks could be technically controlled, this would change nothing as regards a deontologically justified cloning ban.
16. Kant, *Metaphysik der Sitten. Rechtslehre* § 28, AA VI, 281.
17. Brock, n.d., 97.
18. Gordjin, 1999, 18.
19. Habermas, 2001, 107f.
20. Simon, 1989, 294f.
21. Kant, *Briefentwurf an Jung-Stilling*, March 1789, AA XI, 10.
22. Hufeland review, AA VIII, 127-130.
23. Cf. the leading article by the biotechnologist Cibelli, 2004; for Germany cf. only Oeming, 2004.
24. Cf. the campaign "Now is the Time to Help Save Lives!", *Coalition for the Advancement of Medical Research*, March 4, 2004, in which it was mentioned that "therapeutic cloning" has a healing potential for diseases from which "over 100 million Americans" suffer, www.camradvocacy. org.
25. Concerning the theory of the citizenship contract as including a protection contract cf. Fichte, *Grundlage des Naturrechts* § 17, ed. M. Zahn, Hamburg 1979, especially 191-198.
26. Harris, 1978, 81-87.
27. Cf. Schneider, 2003, 271: "Der Klonforschung ist eine Tendenz zur Sozialpflichtigkeit des weiblichen Körpers immanent." - The idea of social compulsatoriness of the body is also leading to all licenses of research on patients benefiting others, e.g. on patients or children not

capable of giving their consent, and also to pressuring society with respect to organ donation duty.

28. In Hwang and Moon's experiments, 242 egg cells were used to produce a single line of stem cells, which had been received from 16 young women following hormone treatment. 213 embryos were produced via cloning, of which 30 reached the state of blastocysts and upon which the inner cell mass could be isolated 20 times.
29. Cf. Blackburn, 2004b. In very similar wording and content Blackburn had also stated her grievances to the press, cf. Blackburn, 2004b. In the same newspaper the head of the President's Council on Bioethics commented on the occurence, particularly in connection with the replacement of Blackburn by the neurologist Ben Carson (Kass, 2004).
30. Cf. Stanford, 2004, based on interviews with European scientists, including the president of "Euroscience," Carl Johan Sundberg.
31. Cf. Brock, n.d., 97: "If there is such right [sc. the right to freedom of scientific inquiry], it would presumably be violated by a legal prohibition of research on human cloning, although the government could still permissibly decide not to spend public funds to support such research."
32. Janich, 1996.
33. Kant, *Metaphysik der Sitten. Tugendlehre*, § 6, AA VI, 423.
34. Reuter, 2004.

References

Annas, G. J./J. A. Robertson. "Human Cloning. Should the United States Legislate Against it?" *ABA Journal* 83, 1997, 80-81.

Blackburn, Elizabeth. "A 'Full Range' of Bioethical Views Just Got Narrower." *Washington Post*, March 7 (2004a), Page B02.

Blackburn, Elizabeth. "Bioethics and the Political Distortion of Biomedical Science." *New England Journal of Medicine* 350; 14, April 1, 2004; prepublished on March 15, http://content.nejm.org/ cgi/content/ abstract/ NEJMp048072.

Brock, Dan W. *Cloning Human Beings: An Assessment of the Ethical Issues Pro and Con*, Lauritzen op. cit.

Carson, Ben/Leon Kass. “We Don’t Play Politics With Science.” *Washington Post*, March 3, 2004, Page A27.

Cibelli, Jose. “Wake up America.” *Wall Street Journal*, March 1, 2004.

Fichte, J. G. *Grundlage des Naturrechts* § 17, M. Zahn (ed.). Hamburg, 1979.

Gordijn, Bert. “Das Klonen von Menschen. Eine alte Debatte - aber immer noch in den Kinderschuhen.” *Ethik in der Medizin* 11, 1999, 12-34.

Habermas, Jürgen. *Die Zukunft der menschlichen Natur. Auf dem Weg zu einer liberalen Eugenik?* Frankfurt/Main, 2001.

Harris, J. “The Survival Lottery.” *Philosophy* 50, 1975, 81-87.

Hoffmann, Th. S. *Philosophische Physiologie. Eine Systematik des Begriffs der Natur im Spiegel der Geschichte der Philosophie.* Stuttgart-Bad Cannstatt, 2003.

Honnefelder, L./D. Lanzerath (eds.). *Cloning in Biomedical Research and Reproduction. Scientific Aspects - Ethical, Legal and Social Limits.* Bonn, 2004.

Hwang, Woo Suk et al. *Evidence of a Pluripotent Human Embryonic Stem Cell Line Derived from a Cloned Blastocyst*, www. sciencexpress.org, 12, February 2004.

Janich, P. *Konstruktivismus und Naturerkenntnis. Auf dem Wege zum Kulturalismus.* Frankfurt/Main, 1996.

Kant, I. *Briefentwurf an Jung-Stilling*, March 1789, Gesammelte Schriften (Akademie-Ausgabe = AA), Berlin 1902ss., 1789, XI, 10.

Kant, I. *Hufeland review*, AA VIII, 125-130.

Kant, I. *Metaphysik der Sitten. Rechtslehre*, AA VI, 203-372.

Kant. I. *Vorarbeiten zur Rechtslehre*, AA XXIII, 253-370.

Kass, Leon R. *Life, Liberty and the Defense of Dignity. The Challenge for Bioethics*. San Francisco, 2002.

Lauritzen, P. (ed.). *Cloning and the Future of Human Embryo Research*. Oxford/New York, 2001.

Lederberg, J. "Biological Future of Man." In *Man and His Future: A Ciba Foundation Volume*, G. Wolstonholme (ed.). Boston: Little, Brown, 1963.

Letteron, R. *Le droit de la procréation*. Paris, 1997.

Liu, A. *Artificial Reproduction as Reproductive Rights*. Dartmouth, 1991.

Oduncu, F. (ed.). *Stammzellforschung und therapeutisches Klonen*. Göttingen, 2002.

Oeming, M. "Ist Genforschung ein religiöses Gebot? Du sollst Leben retten! Ein Plädoyer für das therapeutische Klonen aus biblisch-theologischer Sicht." *Welt am Sonntag,* February 15, 2004.

Reiner, H. *Grundlagen, Grundsätze und Einzelnormen des Naturrechts*. Freiburg/München, 1964.

Reuters. "Boy may help overturn strict Norway bioethics law." from March 18, 2004.

Robertson, J. A. *Children of Choice: Freedom an the New Reproductive Technologies*. Princeton, New Jersey, 1994.

Robertson, J. A. *A Ban on Cloning and Cloning Research is Unjustified.* Testimony presented to the National Bioethics Advisory Commission (March 1997).

Schneider, I. "'Reproduktives' und 'therapeutisches' Klonen." In *Bioethik. Eine Einführung*, M. Düwell/Kl. Steigleder (eds.). Frankfurt/Main, 2003, 267-275.

Sexton, S. *If Cloning is the Answer, what was the Question?* Dorset, 2000.

Simon, Josef. *Philosophie des Zeichens*. Berlin/New York, 1989.

Stanford, N. “Euros concerned for US science. Scientists worried that politics is damaging science in the US and the world.” March 9, 2004 (http://www.biomedcentral.com/news/20040309/02/)

Stevens, M. L. Tina. *Bioethics in America. Origins and Cultural Politics*, Baltimore and London: The Johns Hopkins University Press, 2000.

Warnock, Mary. *Making Babies. Is there a Right to Have Children?* Oxford, 2002.

Human Cloning as a Challenge to Traditional Health Care Cultures

Ilhan Ilkilic

Abstract: There is almost no bioethical discussion today in which the meaning and importance of culture is not mentioned. However, the normative and practical implications for moral judgements and a successful life in a pluralistic society are seldom discussed and certainly not conceptualised. This article first discusses a number of essential changes in our understanding of health and sickness as well as in medical thinking and practice in the wake of new developments in medical technology. It then considers what these changes mean for a multicultural society and addresses the challenge that new developments in bio-medicine such as stem cell research and reproductive and therapeutic cloning pose for both the individual and for society. Finally, some practical suggestions are formulated for dealing with these bioethical problems in a pluralistic society and consideration is given to the possible role and significance of health literacy.

Keywords: Concept of health, Multicultural society, Health literacy, Health communication.

1. Identity and Plurality in Medical Traditions

The history of medicine bears witness to the fact that medical knowledge, health care rules and practices, and therapeutic strategies neither emerged nor were applied and developed in isolation within individual cultures. Medical historians tell of a relatively early continuous exchange of ideas and knowledge between cultures in the field of medicine. An example is the astonishing similarity of approaches to humoral physiology in ancient Egyptian, Indian and Greek medicine, as well as their adoption and further development in the Islamic Middle Ages and their influence on Western Christian medicine up to the nineteenth century.[1]

The well-known Greek historian Herodotus (fifth century B.C.) and, some centuries later, the Greek historian Diodorus (first century B.C.) made significant contributions to the transfer of medical knowledge. In their works they give enthusiastic accounts of Egyptian health rules and methods for preventing and treating illness.[2] Hippocrates of Cos (460 - 377 B.C.) and Galen of Pergamon (129 - 199 A.D.) are two of the most important Greek physicians who, through their works or works attributed to them, influenced Islamic and Western Christian medicine over several centuries. In the *Bait al-Hikma* ("School of Wisdom") founded by the Abbasid caliph al-Ma'mun (813 - 833), a large number of Greek, Syrian and Persian works on philosophy, medicine and other disciplines were collected and translated into Arabic, among them the works of Hippocrates and Galen. In the wake of this intensive period of translation, all scientific disciplines began to flourish and produced many influential doctor-philosophers such as Ibn Sina (980-1037, Latinised Avicenna) and Ibn Rushd (1126-1198, Latinised Averroes).

The transfer of medical knowledge between the cultures was by no means limited to scholarly literature which could only be understood or accessed by doctors and philosophers. Relatively early, medical writings for laypersons were translated and received with great interest. One of the most famous of these is without a doubt the Arabic work "Taqwīm as-Sihha" (Health Chart) by Ibn Butlān (died 1066). In the European Middle Ages, this text, under the title "Tacuinum santitatis," was one of the most widely read and valued works on health and health care. The text enjoyed a renaissance in the sixteenth century in a translation into German by Michael Herr, a German doctor and author. This work discussed not only food and nutrition in relation to health care, but also factors such as housing, sleeping, seasons, the weather, sport, music, and sexual activity.[3]
Since in many periods the inhabitants of a city were heterogeneous with respect to their tribe, nation or religion, the relationships between doctors and patients were simultaneously encounters between adherents of different religions and nations. In particular in the early Abbasid period (750 - 1250), the doctors in Baghdad were for the most part Christians and Jews.[4] The Abbasid caliph Harun ar-Rashid (786 - 809) said the following about his Christian physician Djibril ibn Bakhtishū, who belonged to the famous family of physicians Bakhtishū from Djundishapur:[5] "The well-being of my body and my constitution depends on him, and the well-being of the Muslims depends on me, therefore the well-being of the Muslims depends on his well-being and maintenance."[6]

Throughout the history of medicine, this plurality of cultures can be found in many different periods and cultural traditions, a fact which can be analysed and evaluated on the basis of various different questions and methods. How, for example, does the translation of one of Galen's tractates into Arabic in the ninth century differ from the translation of an English article on stem cell research into Arabic in 2004? How does the situation in which a Muslim in eleventh century Baghdad is attended by a Christian doctor differ from that in which a Muslim patient in present-day

Berlin is treated by a Christian doctor? What characterises these differences and what do they mean for the areas in which medical conflicts arise and for the bioethical discussion in a pluralistic society?[7]

2. Basic Concepts of Health and Care Independent of Culture

From the classical works of medicine, we learn that health was a highly valued good in many cultural traditions and that an appropriate lifestyle was needed in order to maintain it. Being healthy was a sign of physical and psychological harmony; it required a life of moderation and temperance that found expression in the dietetic schema of the six non-natural conditions of life: the patterns of light and air, food and drink, movement and rest, sleeping and waking, physical secretions, and the emotions (*sex res non naturales*).[8] Two goals were central to the art of medicine: the maintenance of health and the healing of illness. Although the one was no less important than the other, the maintenance of health and the associated measures attained greater significance. At least until the nineteenth century, medicine remained primarily a discipline of health care and only secondarily a discipline of healing the sick.[9]

Alongside considerations in the area of natural philosophy, this attitude was certainly also shaped by the limited options for the therapy of many illnesses and the costs of medical consultation and treatment. Since it was difficult to regain health once it was lost, the layperson had to take responsibility for this important good and lead a life conducive to health in order not to fall sick. As early as in the ninth century, Ishāq b. ar-Ruhāwī recommends in his Arabic work “Adab at-Tabīb” (*the ethos of the doctor*) that the layperson take the necessary steps before becoming ill. Ar-Ruhāwī, an expert in Greek medicine and philosophy, makes use of vivid, easy-to-understand examples in a chapter written for medical laypersons.[10] Similarly, the Western Han Dynasty health care expert Huang Di Nei Jing advises laypersons: “The sages do not wait until the sickness is there to cure the sickness, they cure it before it takes place .. if one only waits until the sickness is there and then uses medicine to cure it, that is no different from waiting until one is thirsty and then starting to dig a well.”[11]

3. Treating Disease or Caring for Health?

In many cultural traditions, humoral physiology and pathology, combined with approaches drawn from natural philosophy, served as the basis of teaching about health and illness. According to this theory, the nourishment taken in and turned to liquid was transformed by means of the metabolic process into four cardinal humours: blood, phlegm, choler, and melancholy.[12] A harmonious mix of these fluids gives a person health. A disruption, on the other hand, is taken to be the cause of illness. The character of the illness corresponds to the quality of the humours and the proportion in which they stand to one other. In the course of time, a person develops a particular disposition on the basis of his or her particular humoral make-up and the environment: sanguine, phlegmatic, choleric, or melancholic.

Since health depends on an optimal and balanced mix of these bodily fluids, it is only possible to maintain and regain health through appropriate nutrition. Plants, animal products and meat have their own qualities such as warmth, cold, or moisture, and these should be taken into account when deciding what to eat. Attention should also be given to the temperaments of the healthy and the sick (sanguine, phlegmatic, choleric, or melancholic), the climatic and geographic conditions in which they live, the seasons, and the specific features of the illness.

On the basis of this medical theory, dietetics was not only important for maintaining health, but also for therapy. Illnesses were treated according to the qualities of foodstuffs with the goal of achieving a balance and harmony of the fluids. In this way dietetics became a part of medical treatment alongside medication and surgery. In fact, it even took first place among therapeutic methods. Sā'id ibn al-Hasan (d. 1072), for example, speaks of a certain hierarchy of therapies which can also be found in many other authorities: "If he [the physician] can heal by means of foods, he must avoid medication, and if he can heal by means of medication, he must avoid the scalpel unless it is absolutely necessary."[13]

4. New Challenges

The discovery of antibiotics, advances in anaesthetics alongside major successes in surgery, and new means of visualising the body through x-rays, ultrasound, CT and MRT have led to decisive progress in the diagnosis and therapy of disease. These developments in modern medicine in the last 100 years were accompanied by a change in medical thinking and practice. The classical patient-doctor relationship and the relationship of the patient or medical layperson to his or her own health and illness have also changed and taken on new facets.

Not the natural philosophy approach, but the natural science approach has remained the paradigm of modern medicine, which itself is based on the principle of causality.[14] The illness is perceived by the doctor not as the variation of a harmony or balance understood in terms of natural philosophy, but rather in the context of scientifically objectifiable facts, as measurable deviations from a species-specific biological functioning. In descriptions of illnesses according to this paradigm, biological, biochemical and physical components are presented in their causal relationships in a more or less deterministic fashion, and the clinical picture has an empirical-statistical normative character. The patient's complaints are registered by the doctor and explained in terms of causes possible within the framework of current medical concepts. In order to implement a quantifiable therapy, the patient's complaints have to objectified. The medical decision for or against a specific method of examination or therapy tends to be a judgement made on the basis of costs and benefits. The conceptual approach on which such judgements are based, their premises and the specific conclusions drawn from them are subject to paradigm changes in medical practice. The activities of the doctor are now concentrated on pathology, so that modern medicine takes on more the character of a science of illnesses rather than one of health. Medical practice is characterised more by therapy than be prevention.

The rapid developments in medicine which have led to a decisive increase in doctors' technical abilities also have a number of consequences for medical laypersons and patients as those directly affected by medical practice. Many illnesses which were once terminal or which involved great suffering can now be cured or more effectively treated. Health remains an important good, but it is easier to regain than it once was. At the same time, measures for maintaining health and a lifestyle conducive to health in the age of high-technology medicine have today lost their earlier importance or are completely neglected. The great successes in modern medicine give rise to high expectations and convey to healthy medical laypeople the impression that they no longer need pay much attention to their health and to a healthy lifestyle. The repair mentality in modern medicine is not only supported by new paradigm changes but also by "users." The cost explosion in health care in the last twenty years and limited financial resources call for new conceptual approaches on the part of physicians and the medical profession as well as on the part of patients. Given that the annual costs of therapy for nutrition-related illnesses in Germany currently run to 70 thousand million Euro, the importance of health education, prevention and individual responsibility for the health system is all the more obvious.

5. Who is the Moral Agent for Health Care in a Globalised World?

The classical principles of medicine such as not to harm (*nil nocere*) and to do good (*bonum facere*) still remain central principles of medical thinking and practice today despite changes in structure. From the perspective of patients and medical laypersons, health remains a high good and illness a phenomenon involving suffering, even if the borders between the two states are becoming increasingly blurred. New bioethical problems with a certain complexity tend to arise in connection with the application of biomedical technologies. These completely new options for technical intervention influence people's ways of dealing with their own bodies and their understanding of health and sickness. They make it possible to act on the body in new ways, not all of which can be understood and evaluated simply in terms of traditional notions of corporeality. Stem cell research using embryos and reproductive and therapeutic cloning raise new questions about the beginning of human existence which go beyond traditional concepts of human nature. Traditional notions of kinship and their significance for social existence are called into question by reproductive medicine. In the light of pre-natal diagnosis, a pregnant woman may decide whether or not her unborn child should live. Strategies of prolonging life and definitions of brain death challenge the classical definitions of death in different cultures. It is not easy, on the basis of traditional concepts of the human being and religious convictions, to answer the questions concerning human identity and the borders between individuals raised by organ transplantation and the decoding of the human genome. The discussion of these problems obviously transcends the framework of the natural sciences and is directly connected with normative concepts such as human dignity, person, the understanding of health and illness, the integrity and identity of the human being, and the freedom of research. Clearly, these notions and their normative implications cannot be discussed in a cultural vacuum.

In this book, different arguments concerning reproductive and therapeutic cloning as well as stem cell research in different cultures are presented and discussed. The positions taken in these papers are presented and discussed in the context of arguments which are influenced both by religion (Judaism, Buddhism and Islam) and by the bioethical attitudes which dominate in specific countries (China, Korea and Japan). The religiously and geographically determined arguments in these papers are anything but homogenous. We encounter an even more complicated and problematic situation, however, when we find Buddhist, Hindu, Jewish, Christian and Islamic attitudes to bioethical issues in *one* society. In the process of globalisation, in which we are caught up, this situation is by no means an academic fantasy, but rather daily reality. This reality, characterised by the existence of diverse value systems in a single society, gives rise to new, complicated questions about bioethical issues on different levels. On the philosophical level, the universal validity of moral arguments about these bioethical questions can be discussed.[15] The questions of whether moral norms are culturally invariant and whether the notion of the good is the same in all cultures are among the fundamental problems of ethics and are not the topic of this article. Rather, I want to deal here with the social and individual dimensions of bioethical questions in a pluralistic society.

With respect to bioethical conflicts in a pluralistic society, Hans-Martin Sass asks: "Is there a bioethics which is equally binding on everyone and in which everyone can place his or her trust?"[16] He is convinced that in the bioethical discussion there are models of explanation and the construction of identity which provide an alternative to the Manichean battle between the cultures. Instead of the "clash of cultures" model, Sass argues for communication and co-operation between the cultures which would offer a methodologically and conceptually more flexible and better model for understanding the current cultural and ethical conflicts and confrontations.[17] How can successful communication and co-operation between the cultures on bioethical topics be brought about? How can the bioethical conflicts in their many dimensions be made accessible to laypersons? How can the layperson be supported in forming an opinion about his or her cultural identity? How can civil participation in socio-political decision-making processes on bioethical issues be realised? The following suggestions, formulated within the framework of a series of theses, do not provide answers to these questions, but rather consider the necessary first steps in a pluralistic society.

The shaping of lay opinion and public participation in socio-political decision-making processes require an effective information policy.

Obviously, the relevant scientific knowledge is needed to form an opinion and to take part in the discussion about the bioethical problems of stem cell research and reproductive and therapeutic cloning. Only then can a decision be made and the relevant competence developed. The most recent studies show, however, that the information deficit of laypeople in Germany on topics relating to human genetics and new biotechnological procedures is significant.[18] Social class and linguistic and cultural barriers limit access to information even more. For this reason, appropriate measures must be taken to improve information such as the compilation of brochures and information material, well-researched discussions of these areas of conflict on radio and television, and the creation of informative and well-founded websites *etc.*

An effective information policy in a pluralistic society should take cultural aspects into account.
It is not enough to translate information materials developed for the native population into the languages of minorities. Even if they are didactically and linguistically well-prepared, they also need to take cultural and religious aspects into account. Care must be taken to present positions objectively because these materials are intended to provide the necessary information for reaching a decision and not to prejudice the reader's judgement.

A bridge needs to be built between laypeople and experts which is sensitive to cultural issues.
Information materials can only close the knowledge gaps of a layperson to a certain degree; they cannot replace interactive dialogue and individual consultation. Educational seminars in easily accessible locations can build bridges locally between laypeople and experts and provide answers to specific questions. A telephone hotline or an internet gateway can also support an interactive dialogue.[19] Here the linguistic abilities of the layperson and the scientific and intercultural competence of the expert must be taken account of.

Intercultural and interreligious discussion forums on bioethical topics can promote interreligious dialogue and intercultural understanding.
Intercultural and interreligious discussion forums on bioethical topics can achieve an organised social exchange of opinions. They will also contribute to mutual understanding between people of different beliefs and opinions and promote understanding and tolerance. If people are convinced of the importance of a social discussion and the need for participation, they should be more willing to make a contribution. The formation of an opinion and participation in such discussions presupposes a certain amount of information about the function and goals of these biomedical techniques,

which cannot be acquired without cooperation and commitment. Everyone should come with the necessary openness to dialogue, including respect for the opinions of others and an interest in other perspectives. Ultimately, the goal is not to dictate one's own opinion to others, but rather to learn from each other and to think critically about and to clarify one's own position.

In England, Switzerland, Denmark and Germany, a number of public conferences on bioethical topics have been organised with the aim of promoting public participation in the debate about bioethics and biomedicine. Randomly selected German citizens were invited to conferences on genetic diagnosis in Dresden in 2001 and stem cell research in Berlin in 2004. The participants in these conferences informed themselves about the bioethical issues, asked questions of experts in different areas, and finally formulated a position paper. These conferences, which were in general very positively received, have so far taken no account of culturally specific aspects. They can nevertheless serve as precedents for further intercultural and interreligious discussion forums and public conferences.[20]

6. The Role of Health Literacy

The concept of "health literacy" was defined by *Healthy People 2010* as "the degree to which individuals have the capacity to obtain, process, and understand basic health information and services needed to make appropriate health decisions." This notion was used for the first time in the English-speaking scientific debate in 1974.[21] The idea of health literacy with its normative implications has not been adequately discussed in the bioethical discussions hitherto.[22] What conditions must be met in order for an individual to develop his or her own opinion and how can these be realised? What role is played by his or her own understanding of health and illness, of being directly or indirectly affected by an illness, and by cultural influences? How can individual decision-making and reflective competence with respect to bioethical issues be promoted in a person's own cultural context? These issues need to be further explored and researched. They are the topic of a subproject on health literacy, its transcultural tradition and future.[23]

A discussion of these questions and of the specific connection between health literacy and bioethical problems requires additional argumentation. In this discussion the term health literacy should be used in a broader sense than its definition by "Healthy People 2010." Health literacy (*Gesundheitsmuendigkeit*) is a competence based on taking responsibility for one's own health. It enables an individual to weigh up his or her behaviour in health issues in certain circumstances with respect to scientific knowledge and personal preferences, and thereby leads to the development of a personal approach to dealing health and illness. Since it is based on self-responsibility for health issues, health literacy cannot be reduced to a motivation which only allows an individual to assent to an option put forward by medical practitioners according to the current state of scientific research and medical experience. Rather, it involves rather processing the information provided by experts and weighing it up according to individual value systems and cultural influences in order to come to a responsible decision. It therefore requires a layperson, whether healthy or sick, to address the question: "What do health and sickness mean to me?" Health literacy concentrates not only on individual health, but also causes the individual to consider the health-related consequences of his or her actions for society both on the medical level (e.g., correct behaviour in the case of infectious disease, AIDS, hepatitis, SARS) and on the economic level (the financial implications of one's actions for the welfare state).

Health literacy has a bearing both on the present and on the future. A healthy lifestyle and corresponding hygienic and dietary measures have to do with the current state of health and with its future maintenance. With regard to the relevance of health literacy for the future, the individual should think very critically about medical-technical intervention and their consequences, which will probably only directly affect him/her in the later stages of life. These include life-prolonging measures in the areas of intensive care and palliative medicine as well as intravenous feeding, the removal of organs, brain death criteria and so on. Living wills and powers of attorney for medical care make it possible for people to be treated according to their own value systems in stages of their lives in which they are likely to be unable to take decisions. These so-called living wills, however, require laypersons to concern themselves with the state of an illness and with bioethical problems before these arise.[24]

The relationship of health literacy to the future also implies the necessity of forming an opinion on new biomedical research such as stem cell research and reproductive and therapeutic cloning, and their bioethical consequences. Medical knowledge and information about the related moral questions are essential for taking a position on stem cell research on embryos. This requires a personal process of thinking and evaluation. This process is further connected with demanding questions of conscience in which cultural paradigms, individual value systems and concepts of humanity and life play an important role.

7. New Health Care Cultures and Health Communication

The notion of a "multicultural society" refers to a situation in which different cultures exist alongside each other in one society. A "pluralistic society" goes one step further and implies not only the existence, but also the presence of different cultures in social life and in decision-making processes on critical questions. The development and integration of cultural value systems as elementary features of a pluralistic society give rise to different questions about bioethical issues on academic, social and individual levels. An appropriate social discussion in a pluralistic society of bioethical issues, such as stem cell research and reproductive and therapeutic cloning, presupposes communication and co-operation between cultures, which in turn must first be realised and established by means of appropriate strategies. The concept of "health literacy," with its normative implications, provides a basis for understanding the different positions and an individually responsible orientation among them.

Notes

1. Schipperges, 1959; Steger, 2004.
2. Herodot, 1995, 276; Berghoff, 1947, 6; Schipperges, 1985, 74.
3. Zotter, 1988.
4. Ullmann, 1970; Meyerhof, 1984; Ebied, 1971.
5. Browne, 1921, 18-42.
6. Bürgel, 1967/1, 11.
7. Ilkilic, 2002.
8. Bergdolt, 1999; Hartmann, 2003.
9. Ilkilic, 2004a, 2004b.
10. Ar-Ruhāw□, 1985.
11. Ni, 1999, 31.
12. Ullmann, 1970, 97.
13. Sā'id ibn al-Hasan, 1968, 103.
14. Anschütz, 1987, 64.
15. Engelhardt, 1982, 64-78; Macklin, 1996, 1-22; Angell, 1988, 1081-83; Kreß, 2003; Schicktanz, 2003, 263-82.
16. Sass, 2003a, 1.
17. Ibid., 1-3.

18. Ilkilic, 2002b.
19. Schröder, 2003; Jähn, 2004.
20. www.buergerkonferenz.de. See also at www.bioethik-diskurs.de/Buergerkonferenz.
21. Simonds, 1974/2; Schröder, 2003, 2004. See also http://www.nlm.nih.gov/pubs/cbm/hliteracy.html.
22. Sass, 2004.
23. Sass, 2004/14, 13-22.
24. Sass, 2003b.

References

Angell, Marcia. “Ethical Imperialism, Ethics in International Collaborative Clinical Research.” *The New England Journal of Medicine*, Vol. 329, Nr. 16, 1988, 1081-1083.

Anschütz, Felix. *Ärztliches Handeln, Grundlagen Möglichkeiten, Grenzen, Widersprüche*. Darmstadt, 1987.

Ar-Ruhāw□,Ishāq b. ‘Al□.*Adab at-Tabīb*. Faksimile. Stuttgart, 1985.

Bergdolt, Klaus. *Leib und Seele. Eine Kulturgeschichte des Lebens*. München, 1999.

Berghoff, Emanuel. *Entwicklungsgeschichte des Krankheitsbegriffes*, 2nd Edition. Wien, 1947.

Bürgel, Johann Ch. “Die wissenschaftliche Medizin im Kräftefeld der islamischen Kultur.” *Bustan. Österreichische Zeitschrift für Kultur, Politik und Wirtschaft der islamischen Länder*, 1967, 9-19.

Browne, Edward G. *Arabian Medicine*. London, 1921.

Ebied, Rifaat Y. (ed.). *Bibliography of Medieval Arabic and Jewish Medicine and Allied Sciences*. London, 1971.

Engelhardt, H. Tristram JR. “Bioethics in Pluralist Societies.” *Perspectives in Biology and Medicine*, 26/1, 1982, 64-78.

Hartmann, Fritz. *Patienten als Gehilfen ihrer Ärzte*, Medizinethische Materialien Heft 145. Bochum, 2003.

Herodot. *Historien*, Griech.-Deutsch, Josef Feix (ed.). vol. I. Zürich, 1995.

Ilkilic, Ilhan. *Der muslimische Patient. Medizinethische Aspekte des muslimischen Krankheitsverständnisses in einer wertpluralen Gesellschaft*. Münster – London, 2002a.

Ilkilic, Ilhan, Marcus Düwell and Sigrid Graumann. *Information und Aufklärung über Chancen und Risiken der Humangenetik und neuer gen- und biotechnischer Verfahren*, report of project, funded by the BZgA (Federal Center for Health Education). Tübingen, 2002b.

Ilkilic, Ilhan. *Gesundheitsverständnis und Gesundheitsmündigkeit in der islamischen Tradition*, Medizinethische Materialien, Heft 152. Bochum, 2004a.

Ilkilic, Ilhan. *Gesundheitsmündigkeit / Health Literacy, Arbeits, klassische Texte*, 2004b. See www.Health-Literacy.org, under bibliography.

Jähn, Karl and Eckhard Nagel. *e-Health*. Berlin, 2004.

Kreß, Hartmut. *Medizinische Ethik, Kulturelle Grundlagen und ethische Wertkonflikte heutiger Medizin*. Stuttgart, 2003.

Macklin, Ruth. "Ethical Relativism in a multicultural Society." *Kennedy Institute of Ethics Journal*, 8/1, 1996, 1-22.

Meyerhof, Max. *Studies in Medieval Arabic Medicine. Theory and Practice*, P. Johnstone (ed.). London, 1984.

Ni, Peimin. "Confucian Virtues and Personal Health." In *Confucian Bioethics*, R. Fan (ed.). London, 1999, 27-44.

Sā'id ibn al-Hasan. *Kitāb at-Tašwīq at-Tibbī, Übers. und Bearb. des Kitāb at-Tašwīq at-Tibbī des Sā'id ibn al-Hasan, ein medizinisches Adabwerk aus dem 11. Jahrhundert*, S. E. Taschkandi (Übers.). Bonn, 1968.

Sass, Hans-Martin. *Menschliche Ethik im Streit der Kulturen*, Medizinethische Materialien, Heft 132, 2nd ed. Bochum, 2003a.

Sass, Hans-Martin and Kielstein, Rita. *Pateintenverfügung und Betreuungs*, 2nd ed. Münster, 2003b.

Sass, Hans-Martin. *Ambiguities in Biopolitics of Stem Cell Research for Therapy*, Medizinethische Materialien, Heft 151. Bochum, 2004.

Sass, Hans-Martin. "Asian and Western Bioethics - Converging, Conflicting, Competing?" *Eubios Journal Asian Intern Bioethics*, 2004/14, 13-22.

Schicktanz, Silke. "Die kulturelle Vielfalt in der Bioethik-Debatte." In *Kulturelle Aspekte der Biomedizin, Bioethik, Religionen und Alltagsperspektiven*, S. Schicktanz, Ch. Tannert and P. Wiedemann (ed.). Frankfurt a.M. New York, 2003, 263-282.

Schipperges, Heinrich. *Die Assimilation der arabischen Medizin durch das lateinische Mittelalter*. Wiesbaden, 1959.

Schipperges, Heinrich. *Homo Patiens. Zur Geschichte des kranken Menschen*. München, 1985.

Schröder, Peter. *Vom Sprechzimmer ins Internetcafé: Medizinische Informationen und ärztliche Beratung im 21. Jahrhundert*, Medizinethische Materialien, Heft 137. Bochum, 2003.

Schröder, Peter. *Health Literacy, Actual Literature*, 2004. See www. Health-Literacy.org.

Simonds, S. K. "Health Education as Social Policy." In *Health Education Monography*, 1974/2, 1-25.

Steger, Florian and Jankrift, Kay Peter. *Gesundheit - Krankheit. Kulturtransfer medizinischen Wissens von der Spätantike bis in die Frühe Neuzeit*. Köln, 2004.

Ullmann, Manfred. *Die Medizin im Islam*. Leiden, 1970.

Zotter, Hans. Das Buch vom Gesunden Leben. Die Gesundheitstabellen des Ibn Butlan in der illustrierten deutschen Übertragung des Michael Herr, nach der bei Hans Schott erschienen Ausgabe Strassburg 1533. Graz, 1988.

Let Probands and Patients Decide About Moral Risk Stem Cell Research and Medical Treatment

Hans-Martin Sass

Abstract: This article analyses conceptual and moral deficiencies in generalisations and failure to differentiate between fertilisation resulting from two haploid sets of nuclei and somatic cell nuclear transfer or reprogramming of somatic cells. The result of one is traditionally called embryo, the product of somatic cell transfer or reprogramming should be called differently: embryoid or pseudo-embryo. Given the scientific and moral uncertainties and risk associated with embryoids and reprogrammed somatic cells, it would be immoral to use these products in human reproduction. The article describes different real or possible moral, cultural, and religious evaluations of stem cells and embryoids and, in light of these controversies, calls for respecting human dignity in terms of respecting individual moral choice in donating or selling one's own somatic cells for clinical research or receiving therapeutic products derived from stem cell research or containing human stem cells. Probands and patients should be allowed to make moral choices based on their individual conscience and politicians should prescribe themselves a moratorium on legislating and regulating clinical stem cell research and use.

Keywords: Embryo, Embryoid, Cell reprogramming, Human dignity, Moral choice, Moral risk, Political moratorium, Stem cell.

1. Conceptual and Moral Challenges in Evaluating New Technology

New technology always has carried two different kinds of risk, technological risk and moral risk. The elimination or reduction of technological risk falls primarily into the domain of the expert and might result in better construction, safer procedures, and norms for quality control and safety; resulting risk and uncertainties need to be communicated to customers and consumers, making them a partner in risk reduction and in the handling of uncertainty. Moral and cultural risk is related to moral and cultural need, utility, harm or possible use for unwarranted goals. In culturally closed societies moral harm and utility of new technology is determined by authorities holding power over interpreting Godly commands, natural laws or human rights. In open societies, rich in different values and allowing for different visions of good and moral life defined by individuals and moral communities, the evaluation of moral harm is more complicated. Sometimes politicians are tempted to paternalistically define harm based on so-called “leitkultur” of the traditionally leading tradition, their assumed authority to interpret that leitkultur, or simply on the free or presumed priorities of their special constituency; less often do politicians and regulators accept that citizens make their own educated choices and restrict themselves to providing risk information and education. Of course, certain procedures and products will need to be outlawed as they clearly violate civil rights or very likely will severely harm citizens; but the determination of a clear violation or prospective violation of civil and human rights and of severe harm is not always easy to make.[1]

As far as reprogramming technologies of human cells and of human cloning are concerned, scientific evidence and additional uncertainties will not allow to use either one of these technologies in producing embryonic constructs. Embryonic constructs are not embryos in the traditional sense as they are not derived from the merging of two nuclei of haploid genetic property. No medical oversight or regulatory body would approve experimenting with embryonic constructs for reproductive purposes; no quality standards can yet be written; even topics and requirements for such quality features can not be formulated today. However, the actual situation of scientific ignorance in cell programming and nuclear transfer should not exclude ethical and religious discourse on using these technologies in the future for reproductive purposes; such a discourse would be useful, even warranted for self-understanding and self-evaluation of individuals, communities, cultures and for eventually preparing for future national and international legislation and regulation. It has been argued that some people, particular in traditional Asian culture favoring male offsprings, would somatic cell nuclear transfer techniques to produce babies, if originally developed for therapeutic purposes. However, such a suggestion underestimates cultural family quality standards of potential users of re-programming technology, expecting a “dream child” or at least “any normal child” and not a product resulting from an embryonic construct of unknown and questionable genetic mix-up and disorder. Idiots might in the future exploit existing stem cell technologies as they do with other technologies for crazy activities; criminals might use these and other technologies, as Cain slaughtered his brother Abel and sadists use electricity for torturing their victims. But rarely have cultures and legal systems deterred and fought idiots and criminals making instruments and methods of potential multiple purpose use illegal.

The potential use of cell reprogramming and somatic nuclear cell transfer for therapeutic purposes and medical research represents a different set of technical and moral risk. Saving of life, the curing of diseases or at least the alleviation or reduction of pain and suffering has been one of the prime and undisputed moral goods in all cultures and in demand by individuals, communities and societies; experts in these fields have been gratefully honored and praised.[2] Medical research and medical treatment finds religious and humanist support everywhere and is asked for and demanded by citizens as being vulnerable and mortal beings. It is out of question that medical research and treatment need to be "safe" and need to involve "informed consent or contract" of probands or patients, as probands or patients might decline participation in some or all research or refuse certain forms treatment based on their individual understanding of moral or medical risk.

Moral risk perceived in replacing the nucleus of a human egg cell with the nucleus of a mature somatic cell for the production and farming of stem cells and of cell specific tissue includes the construction of a totipotent cell construct in vitro, potentially capable of developing into a fetus and potentially even into a newborn when transplanted into a human womb. As these quasi totipotent cell constructs potentially contain some or all properties and potentials of genuine totipotent cells and blastocysts in utero, their moral standing rightly is in question and of concern and needs to be evaluated. The results of such an evaluation, as the contributions in this volume document, falls along the lines of different approaches towards the moral interpretation of early human life, contraceptive devices and abortion. But rarely do these assessment go the extra mile to differentiate between the natural properties of embryonic constructs in vitro and embryos in vivo or those produced in vitro with the intention to being implanted. Individuals and moral and religious communities have to find methods and consensus on how to evaluate potential moral harm in the development of cloning techniques and stem cell. Governments and parliaments are tempted to regulate or legislate moral intuition and to steward research and treatment based on a perceived need to protect civil rights; but how should one define the civil rights of embryonic constructs in the in vitro dish environment constructed in order to produce cells, tissue, or other remedy for therapeutic purposes of citizens. Needed is a prudent moral balance of the undisputed virtue of caring for the sick and suffering with a potential threat to embryonic constructs as potentially potential human persons. History and the contributions in this volume tell that answers will not be uniform, rather different based on different moral intuitions and traditions; therefore it would be preferable that individuals rather than governments should be given the authority to decide these d other complex moral issues of developing and applying new technology.

2. Immoral Use of Moral Generalization

Not helpful in the moral assessment of any new form of technology are technoenthusiastic or technophobic generalizations often found in poorly informed or manipulated public debate, in the media and among biopoliticians. Those generalizations carry inherent conceptual and moral risks and deficiencies. Some of those moral debates and political fights over the moral, medical, cultural and legal evaluation of new technology could be avoided if conceptual and ethical clarity would allow for making distinctions, differentiations, and in applying moral principles and visions more closely (a) to specific moral agents among individuals and communities and (b) to concrete situations in which these moral agents would have to make decisions. Technoenthusiasm in regard to cell reprogamming and cloning techniques does not recognize already known risky and uncertain shortcomings of cloned cells derived not from the recombination of two haploid sets of nuclei. Technophobism argues that cell reprogramming is inherently evil because it violates "human dignity"; technoenthusiam expects wonderdrugs and cures for everything from yet unproven techniques.

It is not the technology or tool, it is a specific use of technology which makes acts moral or immoral. Cain killed his brother Abel with an axe. Axes can be used for different purposes, to eradicate wilderness, to kill animals for food or in defense, to kill aggressors in protecting oneself and one's family and community, also to kill one's brother or friend. Instruments of different kind are associated with and constituting human culture; they can and have been used for civilized and uncivilized, human and inhuman goals. We all would have been hurt, if similar standards would have been applied in approving Thomas Edison's invention of electric circuitry and transportation of electricity over long distance into private households. The use of electricity in public and private buildings is regulated by safety instructions and rules. But electricity also is one of the most vicious and most efficacious means to torture people. We have not and will not make electricity illegal; but we have laws, regulation, and education in place to criminalize electro-torture together with all other forms of torture.

Medical interventions have to follow defined quality standards, physicians need to be licensed, trained and submit to continuous education programs and professional and governmental regulations. New interventions need to be certified by professional organizations, often painfully long processes. Physicians would violate medical and ethical standards if they would use cell-programming techniques in embryonic construct research for use in reproductive medicine. Similarly naive and simplistic as the "slippery slope" model is the argument that stem cell research has not yet yielded any results and therefore could easily be banned. Those arguments are ludicrous and morally deficient. When has research ever been undertaken, if results were visible and easy to reach? We need more and differentiated debate and we need citizens and patients suffering from yet untreatable or poorly treatable diseases involved; we have to respect moral intuitions of different kind and origin as an expression of human dignity and responsibility to make one's own choices and stand up for those choices. Values have consequences; it would be unethical to have others bear the consequences of one's own value priorities and choices; politicians have no authentic authority to paternalize citizens by making citizens bear the consequences of choices made by politicians in matters of moral intuition and vision.

3. Making Differences and Differenciations

Ethics is about making differences in planning and judging moral actions. "Respect for life," a prime moral intuition supported by many cultures and religions, is a maxim, not an inflexible fundamental rule, an "ideal," as Birnbacher would say, a vision,[3] which needs to be taylored to the situation. There is a difference between breaking off a flower and throwing it away and breaking off a flower and giving it to a person one loves, and there are only a few among us who would not flowers to loved ones in respect for life. Killing flying insect just for fun, is not an expression of respecting life, but killing a mosquito on my skin who just starts or has finished to bite protects me from nuisance and pain, sometimes even from the transmission of infection. Killing large numbers of mosquitos by insecticides in order to prevent the transmission of malaria or west nile virus infection has become a modern standard in hygiene and public health. Some worldviews hold that killing animals for protein intake is disrespectful of life, and that enough non-animal protein is available for morally sensitive and cultivated people. Sources in Rabbinic tradition hold that killing a mosquito on Shabbat is a larger sin than aborting a fetus. Killing a mosquito is not immoral, but depending on the situation is ambiguous and arguments pro and contra might be controversial. "Killing" an embryonic construct cannot even be compared to the abortion of an embryo, and probably should not even be compared morally to the killing of a mosquito.

Ethics is about making distinctions and basing moral intuition and interpretation on differentiated analysis and evaluation of facts and scenarios. New scenarios require particularly careful assessment of scientific facts and medical evidence without jumping to early conclusions. Recent longitudinal studies of humans born using in-vitro fertilization techniques confirm that genetic as well as environmental factors seem to influence genomic development and the formation of active polymorphisms in individual DNA codes; thus in-vitro fertilization has definite and certain shortcomings and contains health risks not associated with the natural way of "making babies"; more "natural ways" of assisted reproduction such in-vivo insemination (GIFT and other methods) include less risk as the fallopian tube environment is more "natural" than a petri dish. Animal studies have confirmed that DNA "is both inherited and environmentally responsive. Behavior is orchestrated by an interplay between inherited and environmental influences acting on the same substrate, the genome."[4] Daoists and Western natural law philosophies rightly associate means of modifying, manipulating or even fighting and suppressing natural forces as having possible factual as well as moral risk. On the other hand, Western and Eastern worldviews also see the cruelty, immorality and inhumanity in untamed natural forces, - a yet unsolved and probably unsolvable ideational battle and dialectical challenge in most cultures.[5]

Opponents of embryonic stem cell research for therapy arguing that genetic code defines individuality and character, use an outdated model of individuality as recent scientific evidence provides insight into an "interplay of hereditary and environmental influences on genomic activity and individual behavior" already during very early stages of development and gestation.[6] They do not take into account the different environments of test tubes, petri dishes, fallial tubes and the uterus, nor do they recognize the constructional shortcomings of cloning techniques.

Denker has summarized a number of insufficiently known facts about normal embryogenesis and early pattern formation. We know much less about the formational properties of embryonic stem cells insolated in different types of nutrition medium in vitro: "Colonies of embryonic stem cells forming in vitro must be expected to lack, as a rule, the simple but ordered asymmetries of the embryonic disc of normal embryos that are derived from the asymmetries of the egg and the zygote system";[7] but limited knowledge today cannot exclude that an early stage of further embryonic formation "can start spontaneously as a rare event even in standard cultures in human embryonic stem cells,"[8] comparable to yet not well know developmental pattern resulting in twins or Siamese twins.

Enbryonic constructs, also called "embroyid bodies" and "coined for the mouse system and refers to the fact that when mouse terato-carcinoma / embryonal carcinoma cells (EC) or ESC's (embryonic stem cells) are kept in mouse ascites or in vitro"; they can develop into further developmental stages, but "are lacking other extra-embryonic cells which play important roles in embryo implantation and yolk sac formation."[9] Given these and other results from animal studies, one may safely argue that embryonic constructs or embroyid bodies or isolates embryonic stem cells derived from human blastocysts or blastocyst constructs in vitro are different from embryos in utero. They were constructed in vitro for further manipulation into tissue or other remedies; they never will be aborted, i.e. by interventional means separated from a womb, from a mother. They never had been in received in a womb nor did they ever have a mother or a father in the traditional biological and moral sense. They were constructed by transfering a somatic cell nucleus into the denucleid cytoplasm of a human oocyte in an artificial medium outside the womb for medical purposes and for healing, not for implantation and reproduction. Embryonic constructs therefore are not embryos, neither in the biological nor in the moral sense. They are pseudo-embryos at most. As they have no mother or father in the traditional sense, biologically by merging two haploid oocytes, morally by having genetic properties from two persons. How could one ever argue that these pseudo-embryos need to enjoy the protection and respect of genuine embryos whether created in vivo or, in case of infertility, in vitro. These man-made constructs should not be allowed to be implanted into a womb and criminal law in most civilized countries already makes such acts criminal deeds. If existing laws to not outlaw reproductive cloning, then those laws should be modified or supplemented. Embryonic constructs produced by technologies and procedures existing today are just transitory products for the farming and further developing embryonic stem cells; the only value and utility of those stem cells is for therapeutic purposes.

Moral theology and philosophical arguments using natural law theory are cutting conceptual and analytical corners as they do not recognize different biological and associated moral properties of embryonic constructs as pseudo embryos when compared to embryos derived with or without medical assistance, in vivo or in vitro, from two haploid sets of human DNA. As reprogramming human cells has never been a scientific possibility and option until recently, cultural, philosophical and religious traditions do not provide ready-made concepts for cultural and ethical evaluation of cell reprogramming and cell cloning; Buddha, Lao Tse, Jesus, Mohammed or Plato had no reason nor a chance to think about it.

Scientists so far agree that available methods of cloning embryos are not perfect and do produce "embryonic constructs" rather than embryos. Those embryonic constructs definitely do not qualify for reproduction, as evidenced by animal studies. But as already mentioned, it is a worthwhile intellectual and ethical exercise today to discuss medical, ethical, cultural and legal issues of prospective future cloning techniques for human reproduction. Those techniques will have to be approved by professional oversight bodies such as Chambers of Physicians, using existing and yet to be refined quality control and quality assurance standards. The features of such future quality standards are not yet known and it might take another generation or two until those quality requirements even can be formulated and guaranteed.

Also, humankind or some moral communities and national or international legislation might not accept cloning for human reproduction even though no medical harm would be involved. Existing technologies are only good enough for harvesting pseudo-embryonic and other stem cells for further scientific modification targeted at the development of certain cell lines or tissue for therapeutic purposes. Of ethical importance in in-vitro cloning is the construction method of pseudo-embryos, the developmental environment and the purpose, they are grown for. It is unscientific and irresponsible to just apply Newtonian laws of physics to specific issues in modern physics; it is similarly unprofessional and logical and moral malpractice to apply fundamental moral laws to each and every situation without specification, differentiation, and evaluation, to simply use the same arguments in evaluating the moral recognition of pseudo embryonic constructs in vitro and of embryos in vivo.

4. Allowing Moral Choices and Avoiding Moral Asymmetry

Ethics is about making choices; there are easy choices and there some very hard choices to be made by individuals or communities of different kind. Disputes on moral decision making are already common in religiously and culturally homogeneous moral communities and legal systems; qualitatively and quantitatively they increase in dimension in pluralistic societies rich in diverse individual and communal cultures and religious and humanist systems of reference and belief. The already mentioned multitude of voices from different cultural and moral traditions and a recent report of the German National Ethics Council on cultural and moral diversity in the global cloning debate demonstrate quite clearly different cultural and religious attitudes and arguments in regard to reprogramming of human cell lines,[10] cloning and the moral assessment of various forms of unborn human life, reincarnated human life or afterlife. The minutes of the Ethikrat also give a view into the variety of different positions within the same cultural or religious tradition, similar to differences in Christian traditions.

Respect values and visions of individuals and moral communities and at the same time agree on a consensus to accept dissensus in central matters of conviction, morals, and worldview. In Europe, at last since the Age of Reason, models of conflict solution have been introduced and in many areas successfully used by applying principles of "tolerance" and "self-determination" together with "respect for humans by respecting individual choice and conscience." Spinoza in 1670 argued that law and order would not break down if individual systems of belief would be allowed; rather, law and order and the shear existence of a peaceable and harmonious society would be endangered if individual choice and conscience would not be respected in religious and moral matters within an integrated framework of security and responsibility. We rightly would feel severely violated in our individual civil and human rights if someone would request us to believe that Jesus, indeed, walked on water or rose from the grave, that real trans-substantiation of bread and wine into the blood and flesh of Jesus Christ occurs when Holy Supper is spend under correct liturgical conditions. We today would find it not only uncivilized, but rude and offensive, if someone would force Hindi to eat beef, or Jews and Muslims to eat pork, or vegetarians to eat any animal products, competent adult Jehovah's Witnesses to forcefully receive other people's blood.

But there are not only modern post-enlightenment arguments for respecting human dignity in honoring different values and visions of fellow humans of different tradition, religion, or culture, but also religious insights supporting value plurality and arguing for freedom of believe and exercise of related actions. It was Moses Mendelsohn, enlightened and conservative Jewish rabbi, who grounded tolerance and individual moral autonomy and cultural and moral diversity in God's will and creation: "Brethren, if you want true peacefulness in God (Gottseligkeit), let us not lie about consensus when plurality seems to have been the plan and the goal of providence. No one among us reasons and feels precisely the same way the fellow-human does. Why do we hide from each other in masquerades (Mummerei) in the most important issues of our lives, as God not without reason has given each of us his/her own image and face."[11] Ilkilic refers to Paret when similarly arguing that Muslims have seen the diversity of positions and fatwas in Islam as a special and particular blessing of Islam by God.[12]

Should a moral or could a legal request for abortion by a devoted Jewish female in her fourth month of pregnancy be honored in a Frankfurt hospital in Germany, even if physicians and nurses would be found, sharing her religious belief as expressed many times in Talmudic teaching that morally to be respected human life only begins after birth (1. Sabbath 107b; 9. Sabbath XIV, 4; 1. Nidda 44b; 1. Sanhedrin 72b and VIII, 9; also Exodus 21:22f)? Different countries have different laws on abortion, based on legislative majority decisions or supreme court rulings. In Germany, abortion is illegal because according to a supreme court decision all unborn human life shares the protection of the Constitution, but killings of early human life by hormonal or mechanical antinidatives is not even discussed as possibly a criminal act, and abortion of embryos up to three months will not be prosecuted if certain consultative and other means recommended by the same court are fulfilled. It is a challenge to the moral and cultural environment and legal system of open societies to apply norms to facts and procedures without hurting individual judgment and conscience and without differentiation and situational sensitivity and responsibility.

In regard to evaluating facts and goals of stem cell research for therapeutic purposes, Avram Steinberg summarized that Judaism had different legal approaches to moral issues and does not accept absolute statements and judgments. To destroy a blastocyste is bad as it destroys life, he holds. But when destroying a blastocyst in order to safe life, then one has to ponder whether or not the saving of life of a suffering patient is a virtue higher than the vice of destructing a blastocyst. But the destruction of blastocysts for the production of cosmetics or the development of biological weapons carries other moral cost-benefit risk.[13]

Mary Mahowald recently argued that citizens have a basic right to "self-preservation," i.e. "preservation of one's own bodily integrity of life, provides a persuasive, but not necessarily adequate justification for cloning and embryo stem cell retrieval a negative moral right."[14] She proposes a model of moral symmetry comparing embryonic constructs, at a stage after embryonic stem cells have been harvested from the blastocyst, to "surplus embryos" in infertility treatment or to ventilated brain dead organ donors after successful explanation and suggests they be "allowed to die rather than actively terminated."

Ethics is about avoiding asymmetric moral judgment. Diversity in moral evaluation of scientific facts and procedures also includes different moral approaches towards blastocysts depending on novelty of the situation and the role of biopolitics in bioethics. Human blastocysts and morulae are destroyed at a rate of about 50% naturally; modern hormonal antinidatives, still called contraceptives, and intrauterine devices kill blastocysts, morulae and pre-embryos. While the natural abortion rate of about 50% or higher can be understood as natural, the destruction of blastocysts by hormonal or technical devices is intentionally done by human intervention and is associated with enjoying unprotected sex. Opponents of embryonic stem cell research for therapeutic purposes have yet to explain why they do not object as strongly against women's choice in sexual activity and why they have not yet called for research programs to "safe" early human life, i.e. blastocysts, from "premature death." Asymmetry in ethics is unethical and ruins moral judgment and moral behavior. Patients deprived of hope for better remedies feel that moral asymmetry is applied to treat blastocysts used for potentially developing better medicine if compared to the fate of blastocysts intentionally aborted by hormonal or mechanical devices or "naturally." Asymmetry in moral argumentation occurs easily when theory and real-life are distinctly separated from each other, when intellectual debates lack virtues of care, love, solidarity, and respect for fellow humans. Real-life cases of patients in need of better medicines might break the cycles of academic disputes over theories and general principles, some of them controversial between cultures and within cultural traditions and religious teachings. A paralyzed mother in a wheelchair in need of neurological treatment, a father who has lost substantial parts of his left ventricular heart muscle, these are not theories, even not just "cases"; these are not embryonic constructs in vitro, these are real-life fellow humans, citizens caring for other fellow citizens and in need of better medicines and better medical care.[15]

5. Respecting the Dignity of Choices Made by Probands and Patients

The actual debate on ambiguities surrounding human cloning and cell reprogramming occurs along the lines of controversy about the pro and contra of abortion.[16] In European countries, sharing a "common moral heritage" as is often quoted by politicians and regulators, abortion legislation and practice is quite diverse. Pregnant citizens requiring an abortion for whatever reasons and based on their religious or moral convictions may, if they are rich enough, travel to a neighboring country and get the service which is refused in her home country. Similar differences exist for preimplantation services and other medical interventions.

It is a situation similar to the half-way solution of value-and-belief conflicts at the European Muenster peace treaty of 1648 after 30 years of cruel battlings and killings all over central Europe: cujus regio ejus religio', when the feudal sovereign was given freedom of value-and-vision, but not the underling.[17] However, in more liberal states the underling then and now was and is allowed to emigrate. It took a few centuries until such governmental tutelage disappeared on issues of transsubstantiation, not because the rulers gave in, but because no one cared any more so much about whether or not the Holy Trinity was one person in three or three persons in one; thus, Spinoza's vision, Mendelsohn's belief, and the liberal moral and legal traditions in Judaism and Islam, understanding differences of opinion as God's special blessing found a late real-life justification. Today we have this kind of freedom of choice and following one's conscience for the rich and well connected, not for the poor and the less flexible and less well connected. Living in American or European countries where therapeutic research on embryonic stem cells and embryonic constructs is severely restricted, if not illegal, is not only unfortunate, but unjust to many whose conscience makes them choose differently than biopolitics and so called "leitkultur" (traditionally dominant culture) and their contemporary protectors and defenders have determined.

Given cultural diversities and moral convictions and visions in open, democratic and pluralistic societies, the clash of cultures in controversial fields of morality, value, vision, and belief is not so much the difference between traditional Christian, humanist, Jewish, Muslim, Hindi or Confucian cultures but the clash of conflict solution culture within each of these traditions, among them, and within the political and regulatory culture of dealing with issues of conflicting visions and values in open societies.[18]

Richard McCormick SJ and Josef Fuchs SJ have pointed out, abortion is not only a challenge to traditional Christian beliefs rooted in a particular moral faith, also a challenge to secular humanism. For Fuchs, Christian morality "is 'human' not distinctly Christian,"[19] while McCormick hints even stronger at a humanistic test for Christian ethics and actions: "One need not to be a Christian to be concerned with the poor, with health, with the food problems, and justice and rights. But if one is a Christian and is not so concerned, something is wrong with that Christianity. It has been ceased to be Christian because it has ceased to what his founder was - human."[20] If someone is not concerned with improving health and wellbeing of suffering fellow humans and feels a higher obligation towards embryonic constructs in vitro than towards fellow humans in hospitals, than her or his humanity and belief system needs to accept similar scrutiny and questions.

6. A Moratorium for Biopolitics in Cloning Research

Discrepancies within legal systems and ethical deficiencies in moral traditions and communities are nothing new, nor is the fact that different communities have preferred terminological or orientational tools based on their specific cultural tradition Nevertheless, most moral communities of world religions and other systems of belief seem to be preconditioned to support the visions and mission of universal ethics. As far as bioethics and clinical ethics are closely related to physical suffering and pain, to survival and health as the bases for other goods, values, and virtues, codes of professional conducts and lay person's expectations in all cultures throughout history contain a core of principles in professional and lay ethics and in virtues, which is quite stable and not depending on cultural differences or the fashion fads of cultural preferences or political priorities. Taoist reasoning would provide strong arguments for supporting and allowing harmony and individual and communal life while fighting or avoiding disorder, harm, or various forms of imbalance. Buddhist thinking centers on avoiding suffering and giving and supporting life with a minimum of suffering, if possible. Jewish, Christian, and Muslim traditions of care, love, and protection from harm also request the respect of human dignity and human and civil rights.[21]

But there are many controversial issues dealt with differently in different traditions and controversial among citizens of pluralistic societies. Issues of abortion, euthanasia, basics and limits of care for others, many issues in social justice are controversial and will be controversial. As conflicts in evaluating these and other issues deeply related to very personal visions and interpretations of culture, dignity, and tradition, uniform solutions should not be enforced as they will only suppress the dignity of the individual conscience, vision and system of belief. When consensus in theories cannot be achieved, then consensus need to be achieved on how to proceed in the presence of dissent. Consensus on procedural matters becomes even more important as the people of different cultural traditions have to work together globally and as more and more societies become multicultural loosing traditional patterns of ideational coherence.[22] An old European principle, the principle of subsidiarity, has been reintroduced into modern social theory and ethics and could very well be used in bioethics and other areas of applied ethics.[23]

As moral and political controversies and ethical conflicts in part are rooted in the dignity and the right of individual choice, uniform solutions would suppress individual dignity and vision and one should leave complex decisions on good and evil to smaller moral communities, i.e. family and neighborhood, only thereafter with decreasing authority to act as moral agents to societies, politicians, and judges. Subsidiarity,[24] first developed as a principle for direct and individual care of the socially needy by individuals and moral communities and targeted against totalitarian welfare programs, would be a most helpful principle in bioethics whenever and as long as philosophers, ethicists, theologians, politicians, church and state bureaucrats, physicians, groups in society, and educated citizens disagree on principled actions. They all could agree that respecting the dignity of individual moral choice and responsibility is an unalienable right of each and every fellow human.

European cultural traditions, honoring human dignity in the respect for the individual's conscience, are not only based in the Age of Reason concept of self-determination and self-responsibility, but also in religious teaching calling for following one's conscience, as a recent Roman-Catholic encyclical confirms: "Like the natural law itself and all practical knowledge, the judgment of conscience also has an imperative character: man must act in accordance with it. If man acts against this judgment or, in a case where he lacks certainty about the rightness and goodness of a determined act, he stands condemned by his own conscience, the proximate norm of personal morality."[25] Except for the sexist language, the wide majority of cultures and ethical traditions could agree to this Vatican statement. Religious, communal, political, or cultural pressure in the name of universal ethics, except in the most evident and rare cases, would be counterproductive to recognize and to support human dignity and rights.

Muslims living in traditionally Muslim countries or anywhere else in the world would know or would have been told by their spiritual advisor that abortion after the 40th day post conceptionem would not be allowed or would need very strong reasoning and arguments which in singular cases would permit to violate the protection for formed or ensouled unborn human life. But if relatively undifferentiated primitive cells, having the capacity to give rise to more differentiated and specifically formed cells and tissue are used in research targeting at developing new therapies for suffering patients, then there would be no religious or moral problem.[26]

Not everything beyond universal and basic common morality needs to be arranged uniformly, in particular if personal visions and values differ and respect and compassion for the dignity of other people's beliefs, principles, virtues and visions would be hurt rather than supported by uniform solutions and rules. The Buddhist concept of personal dharma and the Confucian understanding of different obligations in different interpersonal roles are quite close to the principle of subsidiarity. Common moral priorities can be formulated in the presence of cultural diversity, indeed,[27] but more complex moral issues should be left to personal responsibility in respect for the dignity of the individual conscience. Döring discussing Chinese values in the debate on stem cell research argues that "cultural understanding starts with accepting the individual as a member of humankind's culture"; he points out how modern Chinese thinkers modify traditional terminology from family orientation towards emphasis on recognition and respect for the individual.[28]

Some have argued that respecting citizen's conscientious individual choice in culturally and morally controversial issues would lead moral reasoning and moral deeds "downwards" by finding a consensus on the lowest common denominator.[29] The contrary is true: Paternalizing individual citizen's moral choice is an insult to the dignity of the individual conscience and a disrespect for visions and values of individuals and small moral communities within a society rich in different visions and values. We have come a long way to accept individual refusal of medical treatment involving the acceptance of blood infusions based on strange religious beliefs; we are in the process to accept refusal of lifeprolonging medical intervention of competent adults after appropriate information reject lifesaving or lifeprolonging treatment; we accept that women vote their conscience in deciding pro or contra abortion after mandatory consultation and advice targeted pro life; we allow the unchallenged use of hormonal and mechanical devices for the destruction of blastocysts and other stages of early human life. But biopolitics is not ready yet to let citizens decide pro or contra providing their own oocytes for cloning research.

7. A Preliminary Suggestion

Given a multitude of medical and moral risk and uncertainty in using somatic cell nuclear transfer for human reproduction, professional self-regulation and national legislation, supported by international declarations, should be banned. But bioethicists should continue to debate moral and cultural parameters and issues of human dignity and civil rights in order to prepare themselves and the public in case safe and efficacious cloning for reproduction one day might be feasible. In case cloning for reproduction never will be a viable choice for responsible parenting, such a discussion as an end in itself will contribute to a re-appreciation of our understanding of human dignity and human rights.

Given reasonable hope that cell reprogramming and human somatic cell nuclear transfer sooner or later might result in a cure or treatment for a multitude of diseases and disorders, thereby providing medical treatment for many suffering or dying fellow human beings, morally required and politically prudent priorities in action would include the following: Bioethicists need to break down the walls of the ivory towers of academic disputes and closed-door debates with biopoliticians; they have to inform citizens and patients in plain language about technical and moral advantages and disadvantages, risks and uncertainties citizens and patients and listen to their concern, hopes, fears and visions.

Biopoliticians need to prescribe for themselves a moratorium on regulation and legislation as long as no major medical or moral harm or abuse or undue pressure on probands and patients has occurred; they should trust the common sense and moral sense of citizens and patients as prime moral agents for their health, wellbeing and risk-taking; they are responsible to their constituency, citizens, not to embryonic constructs in petri dishes.

Probands, patients and their doctors should be allowed to proceed in communication-in-trust and cooperation-in-trust in developing new medicines and treatments for saving life and reducing suffering. As no one should be forced to accept blood donated by others, even in lifesaving situations, or to consume animal proteins or medicines containing pig tissue or manufactured using pig tissue against their will, similarly nobody should be required to donate her oocytes for research purposes or to accept medical treatment based on cell reprogramming or the transfer of somatic cell nuclei, if their moral intuition or religious instruction guides them to refuse those procedures or products.

Central to the proposed set of preliminary suggestions is the respect biopoliticians owe values and visions of citizens in their involvement in medical research and in accepting or refusing certain forms of treatment. Biopoliticians should not play doctor in medical research and in medical treatment as long as routine and standard features in research and treatment such as quality control, informed consent or contract, and other reasonable and well established risk reduction features are observed. They also have neither authenticity nor are they given authority to doctor values and visions of citizens.

Notes

1. Sass, 1989.
2. Sass, 2004a.
3. Birnbacher, 2004.
4. Robinson, 2004, 397.
5. Sass, 2004a; Roetz, 2004.
6. Robinson, 2004, 398; Jaenisch, 2003.
7. Denker, 2004, 16.
8. Ibid., 18.
9. Ibid., 4.
10. Nationaler Ethikrat, 2003.
11. Mendelsohn, 1819, 201.
12. Ilkilic, 2004.
13. Nationaler Ethikrat, 2004: 31.
14. Mahowald, 2004, 65.
15. Sass, 2004b.
16. Walters, 2004; Sass, 1991.
17. Sass, 2003.
18. Ilkilic, 2004; Sass, 2003.
19. Fuchs, 1980, 19.
20. McCormick, 1980, 172.
21. Sass, 2003, 2004a.
22. Sass 2004a.
23. Vatican, 1931; Sass, 2003.
24. Vatican, 1931.
25. Vatican, 1993: art.60.
26. Ilkilic, 2004, 71f.
27. Sass, 2003a.
28. Döring, 2004.
29. Roetz, 2004.

References

Birnbacher, D. "Das Dilemma des ethischen Pluralismus." In *Weltanschauliche Offenheit in der Bioethik*, E. Baumann, A. Brink, A. May, P. Schröder, C. Schutzeichel (eds). Berlin: Duncker und Humblot, 2004, 51-64.

Denker, HM. "Early human development: new data raise important embryological and ethical questions relevant for stem cell research." *Naturwissenschaften* 91:1-21 (DOI 10.1007/s00114-003-0490-8), 2004.

Döring, Ole. "Was bedeutet ethische Verständigung zwischen den Kulturen? Ein philosophischer Problemzugang am Beispiel der Auseinandersetzung mit der Forschung an menschlichen Embryonen." In *Weltanschauliche Offenheit in der Bioethik*, E. Baumann, A. Brink, A. May, P. Schröder, C. Schutzeichel (eds.). Berlin: Duncker und Humblot, 2004, 179-212.

Fuchs, SJ, Josef. "Is there a Christian morality?" In *The Distinctiveness of Christian Ethics*, ChE Curran, RM McCormick (ed.). New York: Ramsey, 1980, 3-19.

Hwang, Woo Suk; Ryu, Young June; Park, Jong Hyuk; et al. "Evidence of a pluripotent human embryonic stem cell line derived from a cloned blastocyst." *Science* Febr 12, 2004.

Ilkilic, Ilhan. *Der muslimische Patient. Medizinethische Aspekte des muslimischen Krankheitsverständnisses in einer wertpluralen Gesellschaft*. Münster: Lit, 2002.

Ilkilic, Ilhan. "Der moralische Status des Embryos im Islam und die wertplurale Gesellschaft." In *Weltanschauliche Offenheit in der Bioethik*, E. Baumann, A. Brink, A. May, P. Schröder, C. Schutzeichel (eds.). Berlin: Duncker und Humblot, 2004, 163-178.

Jaenisch R, Bird A. "Epigenetic regulation of gene expression: How the genome integrates intrinsic and enviromental signals." *Nature Genetics* 35(Suppl), 2003, 245-254.

Mahowald, MB. "Self-Preservation: An argument for therapeutic cloning and a strategy for fostering respect for moral integrity." In *American Journal of Bioethics* 4(2), 2004, 56-66.

McCormick, SJ, Richard. "Does religious faith add to ethical perception?" In *The Distinctiveness of Christian Ethics*, ChE Curran, RM McCormick (ed.). New York: Ramsey, 1980, 156-17.

Mendelsohn, Moses. *Jerusalem oder ueber religioese Macht und Judentum*, Ofen: Burian, 1819.

Nationaler Ethikrat. *Der Umgang mit vorgeburtlichem Leben in anderen Kulturen.* Wortprotokoll der Jahrestagung am 23.10.2003. Berlin: Nationaler Ethikrat, 2003.

Robinson, GE. "Beyond nature and nurture." *Science* 304, 2004, 307-399.

Roetz, Heiner. "Muss der kulturelle Pluralismus einen substantiellen ethischen Konsens verhindern? Zur Bioethik im Zeitalter der Globalisierung." In *Weltanschauliche Offenheit in der Bioethik*, E. Baumann, A. Brink, A. May, P. Schröder, C. Schutzeichel (eds.). Berlin: Duncker und Humblot, 2004, 213-232.

Sass, Hans-Martin. "The moral a priori and the Diversity of Cultures." *Analecta Husserliana* 20, 1986, 407-422.

Sass, Hans-Martin. "Technical and cultural aspects of risk perception." *Research in Philosophy and Technology* 9, 1989, 115-124.

Sass, Hans-Martin. *Can there ever be a Consensus in the Abortion Debate?* Bochum: ZME, 1991.

Sass, Hans-Martin. "Würde des Gewissens und europäisches Binnenethikrecht." In *Strafrecht, Biorecht, Rechtsphilosophie.* Festschrift für Hans-Ludwig Schreiber. Heidelberg: Mueller, 2003, 783-92.

Sass, Hans-Martin. "Asian and Western Bioethics - Converging, Conflicting, Competing?" *Eubios Journal of Asian and International Bioethics* 14, 2004a, 13-22.

Sass, Hans-Martin. *Ambiguities in Biopolitics of Stem Cell research for Therapy. Who cares for Gretchen Mueller?* Bochum: Zentrum fuer Medizinische Ethik, 2004b.

Schueller, SJ, Bruno. "The debate on the specific character of a Christian ethics." In *The Distinctiveness of Christian Ethics*, ChE Curran, RM McCormick (ed.). New York: Ramsey, 1980, 207-233.

Vatican (Pius IX). *Quadrogesimo Anno.* Rome: Vatican, 1931.

Vatican (John Paul II). *Veritatis Splendor*. Vatican City, 1993.

Veatch, Robert M. "Common morality and human finitude: a foundation for bioethics." In *Weltanschauliche Offenheit in der Bioethik*, E. Baumann, A. Brink, A. May, P. Schröder, C. Schutzeichel (eds.). Berlin: Duncker und Humblot, 2004, 37-50.

Walters, L. "Human embryonic stem cell research: an intercultural perspective." *Kennedy Institute of Ethics Journal* 14(1), 2003, 3-38.

Human Reproductive Cloning: A Test Case for Individual Rights?

Florian Braune, Nikola Biller-Andorno, Claudia Wiesemann

Abstract: In most societies the freedom to reproduce is limited, for varying reasons. Cloning creates a kind of test case for the denial or recognition of individual rights and the way these are balanced against other values or interests. The moral stances taken towards reproductive autonomy and the doctor-patient relationship more generally reveal a lot about the role of the individual and individual human rights in different social and cultural contexts as well as on a global scale.
Informed consent is one of the central pillars of current international bioethics and is generally regarded as a valuable, meaningful element in the cross-cultural discourse on bioethics. However, in order for this concept to unfold its potential, it has to be applied in a way that its culturally varying restrictions are not ignored but spelled out and analysed.
The almost universal rejection of human reproductive cloning could lead to the impression that there is global agreement on where to limit individual rights and why. But as a closer look reveals, the reasons for the rejection of cloning are as varied as the legitimisation and the implementation of the concept of individual autonomy and informed consent that are shaping the doctor-patient relationship both in reproductive medicine and in medicine in general. For instance, in both China and Hong Kong informed consent is considered a valid bioethical standard but not, however, by emphasising individual rights.

Key Words: Reproductive autonomy, Individual rights, Informed consent, Doctor-patient relationship, Cross-cultural discourse on bioethics.

1. The Role of Individual Rights in Human Reproduction

The right to self-determination is of great importance to Western societies and is commonly seen as the basis of democracy. Interference from the political as well as the societal side can only be justified when the basic rights of others have to be protected.[1] A prominent tradition in medical ethics thus holds that individual autonomy is the basis for medical decision making. Likewise, reproductive freedom is based on the idea of reproductive autonomy. If this premise is accepted, any ban on assisted reproduction will have to be rejected as long as no one is damaged or otherwise hurt. If the principle of autonomy is accepted, strong reasons are needed to deny anybody the potential benefits of modern reproductive medicine.

In democratic societies guidance or regulation by state authorities is limited by basic laws. In Germany, for example, state interference with individual rights may be justified only if the law in question does not violate constitutional rights.[2] However, the development of modern biotechnology, particularly of human cloning, leads to vexing questions concerning the role of human freedom and individual self-determination, questions that are ultimately linked to the cultural understanding of the human being and his/her role in society.[3] On a global scale, policy makers have to meet a particular challenge: on the one hand reproductive cloning seems to be a scenario where most people would opt for a legal ban; on the other hand they have to keep in mind the individual and academic freedom usually granted by democratic constitutions. Currently, governments throughout the world have to decide how to regulate human therapeutic and reproductive cloning. Such a task has to address the individual as well as the social dimension of ethics. The way in which individual freedom is balanced against other considerations sheds light on different cultural concepts of morality and the human self.

In different societies, individual rights, particularly with regard to human procreation, are restricted to a greater or lesser extent: Abortion, for example, has been practiced virtually without limits in China.[4] It is allowed within certain limits in the USA: abortion has been a divisive issue in America since a Supreme Court ruling (Roe v Wade) gave American women the constitutional right to abortion in 1973. Abortion is tolerated to a great extent and not penalised in Germany, although in most cases it is officially considered illegal. In the Republic of Ireland it is still illegal except where this is a real and substantial risk to the life (as distinct from the health) of the mother.

The particular problems associated with human reproduction usually provoke a wide range of solutions taking into account both individual reproductive rights and other considerations. Human reproductive cloning seems to be an interesting exception. Obviously, a large majority of states represented in the United Nations would have voted for a ban on the technology, even though the latest draft resolution failed:

> The Sixth Committee in its report on the item International convention against the reproductive cloning of human beings (document A/58/520) recommended on 6 November that the question should be included in the

> provisional agenda of the Assembly's sixtieth session in 2005. On the same date, a motion to adjourn the debate on the item until the Assembly's sixtieth session was carried by a recorded vote of 80 votes in favour to 79 against, with 15 abstentions.[5]

Most delegates from the Western world would have agreed to a ban on reproductive cloning, notwithstanding the important role that individual self-determination in medicine usually plays in these countries. And neither the People's Republic of China nor other countries like Singapore or Malaysia supported a different approach based on the often cited "Asian values," contested as they are.[6]

In the case of human reproductive cloning, the freedom to reproduce is limited by the majority of societies for varying reasons. Some refer to the potential damage to mother and child, as cloning is still thought to be a risky procedure in humans. These "technical" problems might be overcome sooner or later. Others argue for an unconditional ban even in the event that reproductive cloning were safe because it is considered to question the essence of human existence. These opponents, in particular, have to justify why the decision should not be left to the informed parents themselves, if they accept the principle of reproductive autonomy at all. Thus, cloning creates a kind of test case for the denial or recognition of individual rights and the way these are balanced against other values or interests. The moral stances taken towards reproductive autonomy and the doctor-patient relationship more generally reveal a lot about the role of the individual and individual human rights in different social and cultural contexts as well as on a global scale.

2. Informed Consent: Individual versus Community Orientation and Cultural Variability

Over the last 30 years, lively debates about one of the most prominent concepts in bioethics, informed consent, have sensitised us to the challenges of searching for a cross-culturally meaningful language in bioethics.[7] Today, informed consent forms the basis of a large number of international declarations and guidelines which claim universal validity across different cultures of the world. It seems quite remarkable that, for example, a document like the Charter of Fundamental Rights of the European Union (2000), which is primarily related neither to bioethics nor to medicine, makes explicit reference in Article 3 to the concept of informed consent. At the global level, too, there are a considerable number of bioethics documents that rely on the notion of informed consent, for example the Declaration of Helsinki from the World Medical Association (Edinburgh 2002). However, only some of these suggest that the notion of patient autonomy underlying the concept of informed consent may be a cultural variable. The CIOMS Guidelines, for example, are such an exception, as they explicitly recognise the need to develop ways of obtaining informed consent that are appropriate to the respective cultural setting, insisting, however, on obtaining individual informed consent.

> The Guidelines take the position that research involving human subjects must not violate any universally applicable ethical standards, but acknowledge that, in superficial aspects, the application of the ethical principles, e.g., in relation to individual autonomy and informed consent, needs to take account of cultural values, while respecting absolutely the ethical standards.[8]

Although the concept of informed consent is frequently invoked, it has not remained unquestioned.[9] Particularly its individualist presumptions have been criticised from a culturally based point of view. Firstly, it has been argued that decision making might not be a purely individual process. Secondly, it has been criticised that some cultures place a high value on community-shared decisions, even though they generally agree upon the value and necessity of informed consent. This family or community-oriented approach might seem a typically "Eastern" solution at first glance. The idea of particular Asian values was propagated by the concept's leading promoter, Lee Kuan Yew, Singapore's senior minister. This concept has met with criticism from a number of prominent Asian thinkers. Malaysia's former deputy prime minister, Anwar Ibrahim, claimed that "no Asian tradition can be cited to support the proposition that in Asia the individual must melt into a faceless community."[10] Moreover, a closer look reveals that community involvement in the consent process has also been called for by "Western" bioethicists.[11]

In the context of increasing efforts to harmonise standards and to search for common values, it is a remarkable fact that countries, cultures, and societies differ so widely, not only on a range of traditionally controversial bioethics topics like the treatment of the human embryo, but also on the interpretation of a fundamental principle like informed consent and the role of the individual in society.

In the mentioned documents, informed consent constitutes the general ethical and legal precondition for interventions for diagnostic, therapeutic or research purposes. Its function is generally to safeguard the autonomy of the individual patient, even when he or she is de facto not able to make such decisions. However, in practice the relevance of individual decision making is restricted by several factors resulting either from social arrangements or cultural traditions preceding any individual decision in the medical setting or from fundamental problems inherent to the informed consent model.

Every society, even the most liberal one, restricts the range of medical choices for an individual on the basis of commonly shared values or traditions. In Western as well as in Eastern society, social and societal ideals play an important role in the management of bioethical conflicts. Countries and cultures have developed different ways of balancing the interests of society and the individual. Thus, everywhere, social agreements determine the relationship of individual and community decision making. This becomes particularly evident when challenging topics like human reproductive cloning are discussed. It is apparent that virtually no country in the world - for different cultural reasons, maybe - will leave the decision to give birth to a cloned human being only to the woman involved.

Informed consent is one of the central pillars of current international bioethics and seems to be a meaningful element in a genuinely cross-cultural bioethics. However, in order for this concept to unfold its potential, it has to be applied in a way that its culturally varying restrictions are not ignored but spelled out and analysed. International documents rarely reflect an awareness of the problem of different interpretations of such concepts in specific cultural contexts. Cross-culturally well informed bioethical research can contribute to the improvement of this situation.

3. China and Hong Kong: Informed Consent and Doctor-Patient Relationship

A central implication of the respect for individual autonomy is that any medical intervention has to be preceded by the informed consent of the persons involved; this means that information on risks and weighing of risks is a key part in any doctor-patient relationship. But how informed consent is obtained depends strongly on the prevailing and socially accepted models of medical decision making. Particularly in times when these models are in flux it is important to analyse the individual and familial reactions of those personally involved.[12] In a rapidly changing society like China it is not an easy task to balance the imperatives of modernity with aspects of Chinese tradition. Family decision making is still common as the concept of the "individual" in its Western meaning has not been popular until recently. The communist legacy and its emphasis of the community rather than the individual, but also a traditional group-related social structure and an awareness of the self in relation to others rather than as autonomous and independent have to be taken into account. It is still an open question whether and how family-based decision making will adapt to the conditions of modernity, focusing on the individual as agent of medical decision making. One indicator of a change in mentality is that today the first generation of children from single-child families is approaching adulthood.[13] For them individuality is a personal experience, and an increase in decisions made by one individual alone seems to be inevitable simply due to the lack or reduced number of other family members involved.

However, in the health care system the family is regaining importance. The basic social security system founded by China´s ruling Communist Party after 1949 is currently being replaced by an individually oriented health care insurance system. But the development of a citizens' insurance scheme that is built upon individual contributions will take time and is a challenging task in a country as big as China. Until the scheme is up and running, the costs of medical assistance will have to be paid mainly by the family. With the rising costs of medical treatment, health care in case of illness is a severe financial burden to most of China's inhabitants. Hence, it is all the more obvious that the family still reserves a right to decide whether or not the costs are bearable. As Yali Cong observes, decision making in health care problems thus remains a family issue:

> Usually, the decision is made by all the family members, but there are some families where the decision is made by the leader, usually the grandfather, elder son, or other special person.[14]

What conceptual framework might be suitable to take the cultural variability of moral norms into account? One paradigm that has been widely received and commented upon by both Western and Non-Western authors rests on four commonly acknowledged ethical principles - respect for autonomy, non-maleficence, beneficence and justice.[15] Such a principle-oriented conception refrains from fixation on ethical theories; it facilitates the identification and structuring of ethical problems, and eventually the achievement of a consensus. The "four principles" approach to biomedical ethics has been continuously developed and elaborated as a universal bioethics method since the 1970s. Despite its wide acceptance and popularity, its cross-cultural plausibility has been challenged on several counts. A closer look might reveal that the differences are surprisingly small. It is often a matter of which principle is stressed. In some contexts, for instance, Asian authors stress the principle of beneficence more than respect for autonomy. But this does not mean that autonomy is dismissed altogether.[16]

Another consideration relates to the fact that decision-making models vary according to the respective societal structures. This rather important factor, however, is often not represented at a normative level. The documents in question usually stress a more or less individual informed consent, regardless which decision-making model is in fact socially accepted. Legal documents often represent political ideals rather than social necessities. For instance, two of the relevant documents regarding reproductive matters in China - "Ethical Principles and Management Proposals on Human Embryonic Stem Cell Research" (2001) and "Ethical Guidelines for Human Embryo Stem Cell Research" (2001) - "share a general esteem of human life" and "emphasise informed consent,"[17] although unlimited abortion and family decision making is a reality in China today. However, the formulation of normative documents can be considered as a first step towards addressing a formerly unregulated area.

A significant contextual factor for the implementation of informed consent in medical practice, however, is the doctor-patient relationship. And here we do in fact encounter some substantial variation. Individual decision making is closely related to the particular understanding of the self, which can focus more on the individual or on the relational dimension. Even in urban areas of China like Beijing, the situation can still be characterised as a cooperative familial style of decision making, although the old patriarchal style has mostly been left behind:

> In summary, the traditional patriarchal style has been changed to the style of family co-decision making, but it has not completely reached individual autonomy yet. That is one of the main reasons why family autonomy,

> not individual autonomy, still plays a dominant role in decision making today.[18]

The role of the physician in charge, however, is somehow paternalistic with respect to the family and to the patients themselves: "patients and family share a common psychology: they both hope that the doctor will recommend a particular treatment that they would like to follow."[19] However, a family member will help to arbitrate if there is disagreement between doctor and patient. The patient's attitude towards his/her own family and to the doctor is traditionally determined by "responsibility" rather than by "rights."[20] Traditional Chinese values like filial piety are still valid and also play an important role in family decision making.[21]

China's cultural heritage is without doubt most influential in East Asian debates and has for a long time also inspired Western thinkers.[22] Cultural identity as well as cultural exchange can be studied in the Chinese context. China and Hong Kong depict two political and cultural entities that share the same cultural heritage but differ both in their political tradition and in their stance towards Western culture.[23] The example of Hong Kong, too, is particularly appropriate to illuminate how the struggle to preserve identity on the one hand and the need to incorporate new perspectives on the other hand have to be balanced in modern societies. Hong Kong can probably show what it may mean to have a distinct cultural identity and, at the same time, to live in a globalised world. It is an example of a Chinese culture strongly influenced by Western ideas. In recent history Hong Kong was somewhat surprisingly confronted with the topic of its own identity. The return of the territory to China concluded colonial history but did not settle this question: as a Special Administrative Region - a status that is granted for the next 50 years from 1997 onwards - it still persists in a transitional phase. During colonial times Hong Kong struggled for its emancipation from the British, and, in the meantime, no one can doubt that it has made the transition into a modern Chinese society, adopting new cultural identities and accepting challenges that such a new situation implies.[24] However, Hong Kong does not equal China. Post-colonial developments and inherited liberal traditions mark a particular Hong Kong identity. The world is offered a new kind of vision of what being "Chinese" can look like. The Hong Kong experience may teach us that there is not just cultural diversity but also diversity within culture, given that nearly all (95%) of Hong Kong's population is ethnic Chinese.[25] It can also show us that Eastern and Western cultures have for some time been merging into new cross-cultural ways, also with regard to bioethical problems.

With the "Human Reproductive Technology Bill" enacted by the Legislative Council in 2000 and the publication of the supplementary "Code of Practice on Reproductive Technology and Embryo Research" in 2002, Hong Kong has taken a critical and decisive stance towards the regulation of bioethical problems. Regulations include the prohibition of cloning embryos and the restriction of reproductive technology to legally married couples. According to the Chairman of the Council on Human Reproductive Technology, Che-hung Leong, "the safe and informed practice of reproductive technology in a way which respects human life and the role of family is emphasised."[26] Interestingly the Code, however, makes explicit reference to the Declaration of Helsinki, thus endorsing the concept of individual informed consent. However, it is commonly assumed that in Hong Kong, the legal system has little impact on decision making in medical practice. There is no legislation regarding patient autonomy.[27] The role of the individual in medical decision making is still shaped by strong familial structures.[28] In practice, a particular model of reconciling individual and familial interest is in place. A recent study has shown that, compared to China, "familism ... takes a different shape because the most preferable arrangement is that the patient and the other family members make the decision together as a single unit."[29] The role of the physician is limited to that of moderator. Traditionally, the physician´s role has been more paternalistic, and medical decision making has been dominated by the doctor.[30] Hence, medical paternalism has only recently lost its persuasiveness although the family continues to be the main reference for health care decisions. In the field of bioethics, Hong Kong has thus developed an original approach shaped as much by the local circumstances reflecting its own identity as by relevant developments abroad.[31]

4. Concluding Remarks

The almost universal rejection of human reproductive cloning could lead to the impression that there is global agreement on where to limit individual rights and why. But as a closer look reveals, the reasons for the rejection of cloning are as varied as the legitimisation and the implementation of the concept of individual autonomy and informed consent that are shaping the doctor-patient relationship both in reproductive medicine and in medicine in general. For instance, in both China and Hong Kong informed consent is considered a valid bioethical standard but not, however, by emphasising individual rights. Rather, the doctor-family-patient relationship as practised in China and Hong Kong emphasises the importance of family decision making.

An international bioethics that aims to be genuinely cross-cultural would profit from a closer look at these different ways of legitimising and of concretising the general concept of informed consent. If the resulting international regulations, including those on human reproductive cloning, succeed in taking such a broader vision into account, they will stand a much better chance of being perceived as substantive contributions to common moral standards and not just as political compromises or as Western ideas imposed on the "rest of the world."

Notes

1. Hesselberger, 2003, 61-66.
2. Hesselberger, 2003, 71-72.
3. Wiesemann, 1997, 2001.
4. The Criminal Code of the People's Republic of China enacted by the National People's Congress on 1 July 1979 does not contain any provisions under which abortion, performed with the consent of the pregnant woman, constitutes an offence.
5. United Nations Press Release GA/10218, 2003.
6. Legett, 2003; Kersting, 2000, 224-225; Sen, 1997.
7. Faden and Beauchamp, 1986.
8. CIOMS, 2002, 8.
9. Levine, 1991, 207-213.
10. *The Economist*, 1998.
11. Thomasma, 2000, 47-57.
12. Fan and Tao, 2004.
13. *The Economist*, 2004.
14. Cong, 2004, 152.
15. Beauchamp and Childress, 2001.
16. Tsai, 1999, 320; Becker, 2004.
17. Döring, 2004, 43.
18. Cong, 2004, 163.
19. Ibid., 172.
20. Ibid., 168-169.
21. Ibid., 163.
22. Cheng, 2002, 352-355.

23. Welsh, 1997, 564-565.
24. Becker, 1997, 21-22.
25. Becker, 2003, 263.
26. Code of practice on human reproductive technology, 2002.
27. Ip, 1998, 449.
28. Ibid.
29. Chan, 2004, 200.
30. Ip and Yam, 1997.
31. Becker, 2003, 281.

References

Beauchamp, Tom L. and James F. Childress. *Principles of Biomedical Ethics*, 5th edit. Oxford: Oxford University Press, 2001.

Becker, Bert. "Hongkong und die Demokratie." *Asien*, 64, 1997, 7-26.

Becker, Gerhold K. "Bioethics with Chinese Characteristics: The Development of Bioethics in Hong Kong." In *The Annals of Bioethics. Regional Perspectives in Bioethics*, John F. Peppin and Mark J. Cherry (eds.). Lisse, The Netherlands: Swets & Zeitlinger Publishers, 2003, 263-284.

Becker, Gerhold K. "Chinese Ethics and Human Cloning: A View from Hong Kong." (forthcoming).

Charter of Fundamental Rights of the European Union *Official Journal of the European Communities* C364, 2000,1-22.

Chan, Ho Mun. "Informed Consent Hong Kong Style: An Instance of Moderate Familism." *Journal of Medicine and Philosophy*, 29, 2004, 195-206.

Cheng, Chung-Ying. "Recent Trends in Chinese Philosophy in China and the West." In *Contemporary Chinese philosophy*, Chung-Ying Cheng and Nicholas Bunnin (eds.). Oxford: Blackwell Publishers, 2002, 349-364.

Code of practice on human reproductive technology published (2002) Available at http://www.info.gov.hk/gia/general/200212/30/1230142.htm.

Cong, Yali. "Doctor-Family-Patient Relationship: The Chinese Paradigm of Informed Consent." *Journal of Medicine and Philosophy*, 29, 2004, 149-178.

Council for International Organizations of Medical Sciences (CIOMS) *International Ethical Guidelines for Biomedical Research Involving Human Beings*. Geneva: CIOMS, 2002.

Döring, Ole. "Chinese Researchers Promote Biomedical Regulations: What Are the Motives of the Biopolitical Dawn in China and Where Are They Heading?" *Kennedy Institute of Ethics Journal*, 14, 2004, 39-46.

Ethical Principles and Management Proposals on Human Embryonic Stem Cell Research (2001). Translation of the regulation available at http://www.ruhr-uni-bochum.de/kbe/.

Ethics Committee of the Chinese National Human Genome Center at Shanghai "Ethical Guidelines for Human Embryo Stem Cell Research." *Kennedy Institute of Ethics Journal*, 14, 2001, 47-54.

Faden, Ruth R. and Tom L. Beauchamp. *A History and Theory of Informed Consent*. Oxford: Oxford University Press, 1986.

Fan, Ruiping and Julia Tao. "Consent to Medical Treatment: The Complex Interplay of Patients, Families, and Physicians." *Journal of Medicine and Philosophy*, 29, 2004, 139-148.

"Golden boys and girls." *The Economist print edition*, Feb 12th, 2004.

Hesselberger, Dieter. *Das Grundgesetz. Kommentar,* 13th edit. Bonn: Bundeszentrale für politische Bildung, 2003.

Human Reproductive Technology Bill (2000). Available at http://www.info.gov.hk/gia/general/200212/30/1230142.htm.

Ip, Mary, Timothy Gilligan, Barbara Koenig, and Thomas A. Raffin. "Ethical Decision-making in Critical Care in Hong Kong." *Critical Care Medicine*, 26, 1998, 447-451.

Ip, Mary and Loretta Yam. "Critical Care Rationing in Hong Kong." *Current Opinion in Critical Care*, 3, 1997, 322-328.

Kersting, Wolfgang. *Politik und Recht.* Weilerswist, Germany: Velbrück Wissenschaft, 2000.

Leggett, Karby. "China Has Tightened Genetics Regulation - Rules Ban Human Cloning. Moves Could Quiet Critics of Freewheeling Research." *Asian Wall Street Journal*, Oct 13th, 2003, A1.
Levine, Robert J. "Informed Consent: Some Challenges to the Universal Validity of the Western Model." *Law, Medicine and Health Care*, 19, 1991, 207-213.

"Nuremberg Code." In *Ethics Codes in Medicine*, Ulrich Tröhler and Stella Reiter-Theil (eds.). Brookfield, USA: Ashgate, 1998, 353-354.

Oppitz, Ulrich-Dieter. *Medizinverbrechen vor Gericht. Das Urteil im Nürnberger Ärzteprozeß gegen Karl Brandt und andere sowie aus dem Prozeß gegen Generalfeldmarschall Milch.* (Erlanger Studien zur Ethik in der Medizin, Vol. 7), Andreas Frewer and Claudia Wiesemann (eds.), Erlangen and Jena: Palm und Enke, 1999.

Scharping, Thomas. "Bevölkerungsentwicklung und -politik." In *Länderbericht China*, Carsten Herrmann-Pillath and Michael Lackner (eds.). Bonn: Bundeszentrale für politische Bildung, 2000, 358-375.

Schubert, Gunter. "Das politische System Hongkongs." In *Einführung in die politischen Systeme Ostasiens*, Claudia Derichs and Thomas Heberer (eds.). Opladen, Germany: Leske und Buderich, 2003, 134-137.

Sen, Amartya. "Human Rights and Asian Values." *The New Republic*, July 14th - July 21st, 1997.

Thomasma, David C. "A model of community substituted consent for research on the vulnerable." *Medicine, Health Care and Philosophy*, 3, 2000, 47-57.

Tsai, Fu-Chang "Ancient Chinese medical ethics and the four principles of biomedical ethics." *Journal of Medical Ethics*, 25, 1999, 315-322.

United Nations.Press Release GA/10218. (2003) Available at http://www.un.org/News/Press/docs/2003/ga10218.p2.doc.htm.

"What would Confucius say now?" *The Economist print edition*, Jul 23rd, 1998.

Welsh, Frank. *A History of Hong Kong*. London: HarperCollins Publisher, 1997.

Wiesemann, Claudia. "Das Recht auf Selbstbestimmung und das Arzt-Patient-Verhältnis aus sozialgeschichtlicher Perspektive." In *Geschichte und Ethik in der Medizin* (Medizin-Ethik, Vol. 10), R. Toellner and U. Wiesing (eds.). Stuttgart and Jena: Gustav Fischer, 1997, 67-90.

Wiesemann, Claudia. "Selbstbestimmte Patienten? - Die Nutznießer der Medizin und ihre Rechte." *Das Gesundheitswesen*, 63, 2001, 591-596.

World Medical Association (WMA) (2002) Declaration of Helsinki, Available at http://www.wma.net/e/policy/b3.htm.

Epilogue
Cross-Cultural Discourse In Bioethics: It's a Small World After All

Nikola Biller-Andorno

In many parts of the world bioethics has expanded considerably over the last few decades, not only as an academic enterprise, but also as a prominent set of issues in the policy arena. Discourse in bioethics is not limited to class rooms and scientific conferences but involves politicians, regulators and many other stakeholders as well as society at large. Issues like the use of embryonic stem cells, patenting of drugs or organ selling touch upon deep moral convictions and views about our very existence as humans.[1] As the stakes are high, debates can become fierce when societies are struggling to find norms in controversial areas. Reaching consensus - at least as soon as results are supposed to be legally binding - may be difficult or downright impossible, as the example of the debate on human reproductive cloning at the United Nations has shown.

In this policy discourse, philosophical questions about the possibility, nature and limits of cross-cultural bioethics - and attempts to answer them - risk being pushed aside or instrumentalised for political purposes: For instance, the temptation to quickly draw up international or even global declarations on bioethics may prevail over a more strenuous and protracted quest for areas of cross-cultural agreement and disagreement; or raising doubts about evident universal standards in bioethics and thinking out loud about possible limits may be interpreted as an indicator for a "relativist" and thus "liberal" position that intends to undermine existing "high" ethical standards, dichotomising or marginalising positions in an undue manner.

If the process of debating bioethical issues and maybe reaching normative conclusions is a very sensitive and complex one even within a particular society, it is even more so in a cross-cultural context. In the light of the increasing importance of international collaboration in a globalised world there is widespread agreement about the necessity of harmonising standards in bioethics. But at the same time, this endeavour runs the risk of being perceived as moral imperialism. And in fact, when funding, be it for research or for developmental aid, is linked to the fulfilment of certain ethical standards, moral values are being imposed on those who cannot afford to reject the money.

The increasing discontent with this situation has contributed to a clash between "Western" and "non-Western" ideas that reverberates in theoretical discussion on cross-cultural bioethics. "Western" bioethics is perceived as secular, individualist, rationalist and universal in its claim and is contrasted with a different approach that aims at the integration of religious values, the particularities of human relationships, and regional or local perspectives.[2] It is not surprising, then, that some of the core values of "Western" bioethics are being dismissed by some "Non-Western" authors as irrelevant or non-existent in their culture. A prominent example of such a rejected notion is "autonomy," one of the cornerstones of Western bioethics. Seemingly lacking an adequate equivalent in languages like, for example, Urdu, the concept is taken to be an empirical one and to imply a strictly individual decision-making process. At least some of the disagreement on the importance and universality of ethical concepts may be due to a misunderstanding: In the case of autonomy, "Western" authors have argued that the goal of empowering patients to make "autonomous" decisions does not necessarily imply that health care providers and relatives have to be excluded from the process. It has also been underscored that autonomy is in fact a mid-level principle that does not in itself constitute an ethical theory and that it needs to be interpreted in concrete cases.[3] And even a rather strict universalist position would grant that this process of interpretation and weighing against other principles can indeed look different in different cultures, leading to distinct local moral worlds. Instead of pronouncing alternative standards or a completely different sets of values (cf. the "Asian values" debate) a more promising strategy might be to claim full-fledged participation in the discourse on the priority and understanding of ethical key concepts, whether or not they are of "Western" origin.

What else can a cross-cultural discourse aim at beyond conceptual clarifications and fostering mutual understanding? Another task would be a self-critical inquiry into the possibilities and limits of a cross-cultural bioethics. The question - if there can be a cross-cultural bioethics, and if so, what it would look like and where its limits might be - is a largely unresolved or, at least, a controversial one. There is no agreement on how to approach this question methodologically - should we, for example, try to develop a comprehensive theory of cross-cultural bioethics or should we by way of an empirical approach look for principles or maybe values that we all seem to share? Answers differ widely as well: Whereas some argue in favour of a universalist position, claiming the existence of more or less concrete common moral standards for all,[4] others reject such a notion pointing to the cultural variability of moral norms.[5] There is, of course, a variety of positions in between, among them the notion of a nucleus of a common morality, which may be surrounded by areas in which we "agree to disagree."[6] Longstanding participants in this debate like Daniel Callahan have observed a shift in focus from the universal to the particular over the past few decades;[7] looking at current developments we might be at the point where the pendulum is set to swing back to the universal. Whereas in some areas, like the treatment of the human embryo, there may continue to be disagreement, other practices, like organ trafficking, are being recognized undisputedly as abusive or exploitative. It seems worthwhile to investigate the possibility of articulating a clear, if limited, set of core values or principles with regard to bioethics, similar to - or maybe based on - the formulation of human rights.

A further contribution of a cross-cultural discourse on bioethics could consist in the formulation of procedural rules. As soon as sensitive issues are discussed and norms are to be defined, the discourse is at risk of being hijacked by those who are not interested in a patient, tolerant search for mutual understanding and compromise, but who aim solely at imposing their opinion on others, be it in an openly aggressive or a more subtle, manipulative way. In order to keep discourse from being ruined this way, participants should agree to a set of defined rules, which would include simple measures like a transparent presentation of one's own position, respectful treatment of participants who hold a different view, and restraint from polarising and manipulating the topic and the agenda according to one's own purposes. Discourse is not immune to abuse by those who despise it and use it for their own purposes and power games. With clear rules, however, it would be easier to exclude disrespectful individuals or groups from participation. A lively and meaningful discourse on cross-cultural bioethics cannot be created by force. It is a fragile institution that depends on the goodwill of participants. It is, however, at the same time, a precious expression of a genuine interest in the moral universe of others as well as the will to work out ways to live together in spite of differences in moral judgement.

However justifiable the criticism of Western moral imperialism may be, it would be unfortunate if its consequence were the development of or insistence on parallel moral worlds. Although ready-made, global solutions to bioethical issues may be rightly looked at with suspicion, concerns and alternative suggestions should be voiced from within a discourse on bioethics, not from outside. It may be a small world, after all, but its richness and complexity should not be glossed over precociously. Instead of yielding quick and simple solutions, discourse may turn out to be an end in itself.

Notes

1. Roetz, 2004.
2. e.g. Moazam and Zaman, 2003; Marshall and Koenig, 2000.
3. Beauchamp and Childress, 2001.
4. Macklin, 1999.
5. Baker, 1998; Christakis, 1992.
6. Beauchamp, 1998.
7. Daniel Callahan, 2000.

References

Baker, Robert. "A Theory of International Bioethics: Multiculturalism, Postmodernism, and the Bankruptcy of Fundamentalism." *Kennedy Institute of Ethics Journal*, 8, 1998, 201-232.

Beauchamp, Tom L. "The Mettle of Moral Fundamentalism: A Reply to Robert Baker." *Kennedy Institute of Ethics Journal*, 8, 1998, 389-402.

Beauchamp, Tom L. and James F. Childress. *Principles of Biomedical Ethics,* 5th edit. Oxford: Oxford University Press, 2001.

Callahan, Daniel. "Universalism and Particularism. Fighting to a Draw." *Hastings Center Report*, 30, 2000, 37-44.

Christakis, Nicholas. "Ethics Are Local: Engaging Cross-Cultural Variation in the Ethics for Clinical Research." *Social Science & Medicine*, 35, 1992, 1079-1091.

Macklin, Ruth. *Against Relativism. Cultural Diversity and the Search for Ethical Universals in Medicine*. New York/Oxford: Oxford University Press, 1999.

Marshall, Patricia and Barbara Koenig. "Bioethiques et Anthropologie: Situer le <<Bien>> dans la Pratique Médicale [Intersection of Bioethics and Anthropology: Locating the 'Good' in Medical Practices]." *Anthropologie et Sociétés*, 24, 2000, 35-55.

Moazam, Farhat, Riffat M. Zaman. "At the interface of cultures." *Journal of Clinical Ethics*, 14, 2003, 246-258.

Roetz, Heiner. “Der Mensch als Mitschöpfer. Bioethik und kulturelle Differenzen.” *Neue Zürcher Zeitung*. 20. Februar 2004.

Notes on Contributors

Michael Barilan was born in Jerusalem and received his medical degree from the Technion, Haifa. He serves as a senior physician in an internal medicine ward in a teaching hospital very near the Israeli-Palestinian division. Barilan is also affiliated with the department of behavioral sciences, Sackler Faculty of Medicine, Tel Aviv University. He publishes in the areas of bioethics and medical humanities, particularly in relation to social history, religion and art.

Gerhold K. Becker, currently Visiting Professor at the Graduate School of Philosophy, Assumption University, Bangkok, retired in 2004 as Chair Professor of Philosophy and Religion and Founding Director of the Centre for Applied Ethics after more than eighteen years of teaching and research at Hong Kong Baptist University. From 1996-2004, he was a member of the Hong Kong Government's Council on Human Reproductive Technology and from 2000-2004 he served as chairman of the Council's Ethics Committee. He is associate editor of Rodopi's *Value Inquiry Book Series* and editor of its special series *Studies in Applied Ethics.* Representative of his research interests in applied ethics (particularly bioethics) are the following book titles *The Moral Status of Persons: Perspectives on Bioethics* (2000), *Changing Nature's Course: The Ethical Challenge of Biotechnology* (1996) and *Ethics in Business and Society: Chinese and Western Perspectives,* (1996).

Nikola Biller-Andorno, M.D., Ph.D., is professor of medical ethics at the Charité - Joint Medical Faculty of Humboldt and Free University in Berlin.
Prior to joining the Charité she worked as an ethicist with the World Health Organization for over two years. Her special interests are the theoretical considerations relating to cross-cultural bioethics and the universalism-particularism debate and the implications for policy making. She is a member of the research group "Culture-transcending Bioethics."

Florian Braune, M.A., is a member of the research group "Culture-transcending Bioethics" and currently working on a research project devoted to the concept of informed consent and its concrete application in international bioethics at the Department of Medical Ethics and History of Medicine, University of Göttingen, Germany. As a political scientist, he is interested in the decision-making processes in public health care systems on a comparative basis. He focuses on the study of justice and conflict in bioethics from a culturally sensitive and cross-cultural perspective.

Nigel M. de S. Cameron, Ph.D., is research professor of bioethics at Chicago-Kent College of Law in the Illinois Institute of Technology and president of the Institute on Biotechnology and the Human Future (www.thehumanfuture.org). Originally from the United Kingdom, he also chairs the Centre for Bioethics and Public Policy in London as well as the Council for Biotechnology Policy in Washington, DC. Recent activities have included US congressional testimony on cloning and embyronic stem cell policy issues, representing the US as bioethics adviser at the United Nations General Assembly discussion of a cloning convention, and serving as Scholar-in-Residence at UBS Wolfsberg (Switzerland).

Chin Kyo-Hun is Professor of Philosophy at the Department of National Ethics Studies, Seoul National University, Korea. He is President both of the Korean Bioethics Association and the National Advisory Commission for Bioethics. He serves on the editorial board of the Korean Society for Hermeneutics and is President of the Korean Society for Philosophical Anthropology. Prof. Chin was awarded the degree of Dr. phil. in Philosophy at the Universiy of Vienna, Austria

Ole Döring, Ph.D., MA phil., has studied Philosophy and Sinology. Since 1996, he has been conducting research projects about culture-transcending bioethics, with a special focus on biomedicine and the health sector in China. His main interest lies in exploring the conceptual language of a culturally alert philosophical ethics and the foundations of the related practical understanding. He is currently a Research Fellow at Bochum University in the research group "Culture-transcending Bioethics," Project Fellow at the federal Institute of Asian Affairs in Hamburg, and Visiting Scholar at Hunan Normal University in Changsha, Henan Province, PR China.

Marion Eggert, PhD, is Professor of Korean Studies at Ruhr University Bochum, Germany, and a member of the research group "Culture-transcending Bioethics." Her main interests lie with Korean cultural history, especially processes of identity formation and re-configuration, including invention and use of cultural traditions in premodern and modern Korea.

Thomas Eich studied at the Universities of Bamberg, Damascus and Freiburg. He received his Ph.D. from Ruhr University Bochum, Germany.

The topic of his doctoral thesis was an analysis of the social networks of a Syrian Mystic in the late Ottoman Empire and the development of his writings. Since January 2003 he has worked on a research project about Islamic Bioethics at the Department of Oriental Studies at Ruhr University Bochum in the research group "Culture-transcending Bioethics."

Christofer Frey is a retired Professor for ethics (and systematic theology) at the Protestant Theological Faculty of Ruhr University Bochum. He is continuing his research (and partly his teaching) with an emphasis on the following topics: foundation of ethics, philosophy of religion, and bioethics. His interest in the Korean bioethical discussion is concentrated on such themes as 'universalisation,' 'human dignity' in an syncretistic world, and the influence of traditions. He is a member of the research group "Culture-transcending Bioethics."

Thomas Sören Hoffmann, Ph.D., is professor of philosophy at Bonn University and from 2003 to 2005 was a member of the research project "Utilitarian Culture versus Normative Culture. Intra-cultural Differences in Western Bioethics" within the research group "Culture-transcending Bioethics" at Ruhr University Bochum. His special interests are the philosophy of Kant and German idealism, practical philosophy, the philosophy of law, and bioethics.

Dr. Robert Horres, is Professor of Japanese Studies at Tübingen University, Germany, and a member of the research group "Culture-transcending Bioethics." He studied Japanology, Economics and Comparative Religion in Bonn and Tokyo. His doctoral dissertation is on Japanese space development policy. He has also held visiting appointments at the National Institute for Science and Technology Policy, Tokyo, and Tokyo University. Research interests include theories and policies relating to technology and society in Japan (space development policies; biotechnology ; bioethical conflicts and politics in Japan, information society, technology assessment), humanities and information science, and computer based knowledge management (CAI, Multimedia Projects and E-Learning Concepts) for Japanese Studies.

Ilhan Ilkilic M.D., Ph.D, studied medicine, philosophy and Islamic-Sciences in Istanbul, Bochum and Tübingen. He worked as a researcher on the "Health Literacy" project as a member of the research group "Culture-transcending Bioethics." He is currently coordinator of the "Public

Health Genetics" project at the Institute for the History, Philosophy and Ethics of Medicine, University of Mainz. His special interests include Islamic bioethics, transcultural bioethics, lay health literacy, and public health genetics.

Phillan Joung, Ph.D., studied German literature and philosophy at the University of Münster, Germany. She is currently a Research Assistant in the interdisciplinary project "Culture-transcending Bioethics" at Ruhr University Bochum, Germany. Her main interests are the Korean bioethical discourse and comparative cultural studies.

Masahiro Morioka is Professor of Philosophy and Ethics at the College of Integrated Arts and Sciences, Osaka Prefecture University and a visiting professor at the Graduate School of Humanities and Sociology, University of Tokyo. He obtained his Ph.D. from the University of Tokyo.

Siti Nurani Mohd Nor holds the degrees of M.Sc. and Ph.D.. She studied at the Kennedy Institute of Ethics, Georgetown University, Washington DC. She has been a participant at the meetings on the ethics of the medical profession of the Joint Commonwealth Medical Association Trust, the Malaysian Medical Association Conference on Medical Ethics and Human Rights, and the International Islamic Medical Association. She is Associate Professor at the Department of Science and Technology Studies, Faculty of Science, University of Malaya, Kuala Lumpur, Malaysia and coordinator for professional ethics and bioethics at the University of Malaya.

Dr. Hans Dieter Ölschleger is senior lecturer at the Center for Modern Japanese Studies, University of Bonn, Germany and a member of the research group "Culture-transcending Bioethics." He studied cultural anthropology, sociology and archaeology of South America and wrote his doctoral dissertation on the economy and cultural ecology of the Ainu, the indigenous people of Hokkaidô. Since a prolonged stay in Japan (1988–1995) his main research interests cover modernization in Japan, modern Japanese society, minorities in Japan, and the history and sociology of Japanese communities in the Americas.

Qiu Renzong is Professor at the Center for Bioethics, Beijing Union Medical College (PUMC), at the Health Science Center, Beijing University, and at the Central China University of Science and Technology. He is an emeritus senior research scholar at the Institute of Philosophy (IP) of the Chinese Academy of Social Sciences (CASS), Beijing. He is a lifetime member of the Kennedy Institute of Philosophy, USA, a member of the International HapMap ELSI Committee, a member of the Ethics Committee and the Expert Committee on Health Policy and Management of the Chinese Ministry of Health, and a member of HIV/AIDS Expert Advisory Commission of the Government. He is President of the Asian Bioethics Association.

Pinit Ratanakul, who received his Ph.D. at Yale University, is Professor of Philosophy and Director of the College of Religious Studies at Mahidol University, Bangkok, Thailand. He is the author of Bioethics: An Introduction to the Ethics of Medicine and Life Science, and a contributor to the Enclyclopedia of Bioethics. He has published widely on bioethics from Theravada Buddhist perspectives.

Heiner Roetz studied sinology and philosophy at J. W. Goethe University in Frankfurt/M. Since 1998 he is Professor for Chinese History and Philosophy at Ruhr-University, Bochum. Major research fields and teaching subjects are classical Chinese philosophy, history of Chinese ethics, Chinese culture and human rights, comparative philosophy, cross-cultural hermeneutics and cross-cultural bioethics. He is the speaker of the research project "Culture-transcending bioethics," funded by the German Research Foundation (DFG).

Abdulaziz Sachedina is Professor of Religious Studies at the University of Virginia, Charlottesville. He is a member of the International Association of Bioethics and of the American Academy of Religion. He serves on the Steering Committee, Center on Religion and Democracy, the Advisory Board, Center for the Study of Islam and Democracy, and the Advisory Board, Center for Bioethics, UVa. He is Director of the Organization for Islamic Learning. Abdulaziz Sachedina studied in India, Iraq, Iran, and Canada and obtained his Ph.D. from the University of Toronto. He has been conducting research and writing in the field of Islamic Law and Theology for more than two decades. In the last ten years he has concentrated on social and political ethics, including Interfaith and Intrafaith Relations and Islamic Biomedical Ethics. He is an American citizen born in Tanzania.

Hans-Martin Sass, Ph.D., Professor of Philosophy, Ruhr University, is a Dir ector of the Zentrum fuer Medizinische Ethik, Bochum, Germany, and a senior research scholar at the Kennedy Institute of Ethics, Georgetown University, Washington DC, USA. He has published widely in philosophy and bioethics and holds academic appointments at the Medical Faculty of Ruhr University and Georgetown University, People's University of China, Beijing, Peking Union Medical College, and other universities. He is a member of the research group "Culture-transcending Bioethics."

Schlieter, Jens, PhD., studied Philosophy, Comparative Religious Studies, Buddhology /Tibetology at Vienna and Bonn. Currently research staff member at the Indology Department, University of Bonn; member of the research group "Culture-transcending Bioethics." Publications on Buddhist Bioethics, Indo-Tibetan Buddhism, and Comparative Philosophy.

Christian Steineck is a post-doc researcher at Bonn University's Center for Research on Modern Japan, where he is currently working on the project 'Bioethics in Japan' within the research group "Culture-transcending Bioethics." He holds a doctoral degree in philosophy and an M.A. in Japanese Studies both from Bonn University. His publications include Grundstrukturen mystischen Denkens (2000) and two volumes with translations and analyses of Japanese Pure Land- and Zen-Buddhist texts respectively (Quellentexte des japanischen Amida-Buddhismus, 1997, and Leib und Herz bei Dôgen, 2003), and articles on philosophy of technology, bioethical discussions in Japan, and modern Japanese philosophy.

Dr. Heinz Werner Wessler is a senior lecturer at the Indological Department, Center of Asian Studies, University of Bonn, Germany. In teaching and research, he has been working on the history of classical and modernHinduism, Sanskrit and Hindi texts. His doctoral dissertation is on the concept of time and history according to the Vishnupurana. At present, his main research project is on contemporary Hindi Dalit literature. His interest is in questions relating to religion and society, politics, and culture in India.

Claudia Wiesemann, MD, is head of the Department for Medical Ethics and History of Medicine at the Medical Faculty of Goettingen University and president of the German Academy for Ethics in Medicine. Her current fields of research are medical ethics from a cultural perspective and the ethics of the parent-child relationship. She is a member of the research group "Culture-transcending Bioethics."

Beyond Boundaries of Biomedicine

Pragmatic Perspectives on Health and Disease

Wim J. van der Steen, Vincent K.Y. Ho, and Ferry J. Karmelk

Amsterdam/New York, NY 2003. XI, 292 pp.
(At the Interface/Probing the Boundaries 4)

ISBN: 90-420-0816-4 Paper € 60,-/US$ 72.-

Cultural forces shape much of medicine including psychiatry, and medicine shapes much of our culture. Medicine provides us with beneficial treatments of disease, but it also causes harm, increasingly so in the form of overmedication enhanced by the pharmaceutical industry. The book explores boundaries of medicine and psychiatry in a cultural setting by building bridges between unconnected literatures. Boundaries have to be redrawn since effects of the environment, biological, social and political, on health and disease are undervalued. Potential beneficial effects of diet therapies are a recurrent theme throughout the text, with particular emphasis on omega-3 fatty acids. Deficiencies of these acids in common diets may contribute to many chronic diseases and psychiatric disorders. The book uncovers limitations of evidence-based medicine, which fosters a restrictive view of health and disease. Case studies include: the biology of migraine; limitations of biological psychiatry; conventional *versus* alternative medicine; science, religion and near-death experiences.

USA/Canada:
295 North Michigan Avenue - Suite 1B, Kenilworth, NJ 07033, USA
Call toll-free 1-800-225-3998
All other countries: Tijnmuiden 7, 1046 AK Amsterdam, The Netherlands.
Tel. ++ 31 (0)20 611 48 21, Fax ++ 31 (0)20 447 29 79
Orders@rodopi.nl **www.rodopi.nl**
Please note that the exchange rate is subject to fluctuations